APPLIED NANOTECHNOLOGY

Materials and Applications

APPLIED NANOTECHNOLOGY

Materials and Applications

Edited by

Vladimir I. Kodolov, DSc
Gennady E. Zaikov, DSc
A. K. Haghi, PhD

AAP | APPLE ACADEMIC PRESS

Apple Academic Press Inc.
3333 Mistwell Crescent
Oakville, ON L6L 0A2
Canada

Apple Academic Press Inc.
9 Spinnaker Way
Waretown, NJ 08758
USA

Printed in the United States of America on acid-free paper
International Standard Book Number-13: 978-1-77188-350-4 (Hardcover)
International Standard Book Number-13: 978-1-315-36633-3 (CRC Press/Taylor & Francis eBook)
International Standard Book Number-13: 978-1-77188-351-1 (AAP eBook)

Library and Archives Canada Cataloguing in Publication

Applied nanotechnology : materials and applications / edited by Vladimir I. Kodolov, DSc, Gennady E. Zaikov, DSc, A.K. Haghi, PhD.

Includes bibliographical references and index.
Issued in print and electronic formats.
ISBN 978-1-77188-350-4 (hardback).--ISBN 978-1-77188-351-1 (pdf)
1. Nanostructured materials. 2. Nanostructures. 3. Nanotechnology. I. Kodolov, Vladimir Ivanovich, editor I. Haghi, A. K., editor III. Zaikov, G. E. (Gennadiĭ Efremovich), 1935- editor

TA418.9.N35A66 2016 620.1'15 C2016-905465-9 C2016-905466-7

Library of Congress Cataloging-in-Publication Data

Names: Kodolov, Vladimir I. (Vladimir Ivanovich), editor. | Zaikov, G. E. (Gennadiæi Efremovich), 1935- editor. | Haghi, A. K., editor.
Title: Applied nanotechnology : materials and applications / editors, Vladimir I. Kodolov, Gennady E. Zaikov, A.K. Haghi.
Description: Toronto ; New Jersey : Apple Academic Press, 2017. | Includes bibliographical references and index.
Identifiers: LCCN 2016047916 (print) | LCCN 2016048718 (ebook) | ISBN 9781771883504 (hardcover : alk. paper) | ISBN 9781771883511 (eBook)
Subjects: | MESH: Nanostructures | Nanotechnology | Polymers Classification: LCC R856 (print) | LCC R856 (ebook) | NLM QT 36.5 | DDC 610.28/4--dc23
LC record available at https://lccn.loc.gov/2016047916

Apple Academic Press also publishes its books in a variety of electronic formats. Some content that appears in print may not be available in electronic format. For information about Apple Academic Press products, visit our website at **www.appleacademicpress.com** and the CRC Press website at **www.crcpress.com**

ABOUT THE EDITORS

Vladimir I. Kodolov, DSc

Professor and Head, Department of Chemistry and Chemical Technology, M. I. Kalashnikov Izhevsk State Technical University, Izhevsk, Russia; Chief of Basic Research, High Educational Center of Chemical Physics and Mesoscopy, Udmurt Scientific Center, Ural Division at the Russian Academy of Sciences; Scientific Head, Innovation Center at Izhevsk Electromechanical Plant, Izhevsk, Russia

Vladimir I. Kodolov, DSc, is Professor and Head of the Department of Chemistry and Chemical Technology at M. T. Kalashnikov Izhevsk State Technical University in Izhevsk, Russia, as well as Chief of Basic Research at the High Educational Center of Chemical Physics and Mesoscopy at the Udmurt Scientific Center, Ural Division at the Russian Academy of Sciences. He is also the Scientific Head of Innovation Center at the Izhevsk Electromechanical Plant in Izhevsk, Russia.

He is Vice Editor-in-Chief of the Russian journal *Chemical Physics and Mesoscopy* and also is a member of the editorial boards of several Russian journals. He is the Honorable Professor of the M. T. Kalashnikov Izhevsk State Technical University, Honored Scientist of the Udmurt Republic, Honored Scientific Worker of the Russian Federation, Honorary Worker of Russian Education, and also Honorable Academician of the International Academic Society.

Gennady E. Zaikov, DSc

Head of the Polymer Division, N. M. Emanuel Institute of Biochemical Physics, Russian Academy of Sciences, Moscow, Russia; Professor, Moscow State Academy of Fine Chemical Technology, Russia; Professor, Kazan National Research Technological University, Kazan, Russia

Gennady E. Zaikov, DSc, is Head of the Polymer Division at the N. M. Emanuel Institute of Biochemical Physics, Russian Academy of Sciences, Moscow, Russia, and Professor at Moscow State Academy of Fine Chemical Technology, Russia, as well as Professor at Kazan National Research Technological University, Kazan, Russia. He is also a prolific author, researcher, and lecturer. He has received several awards for his work, including the

Russian Federation Scholarship for Outstanding Scientists. He has been a member of many professional organizations and is on the editorial boards of many international science journals. Dr. Zaikov has recently been honored with tributes in several journals and books on the occasion of his 80th birthday for his long and distinguished career and for his mentorship to many scientists over the years.

A. K. Haghi, PhD

Associate Member of University of Ottawa, Canada; Editor-in-Chief, International Journal of Chemoinformatics and Chemical Engineering; Editor-in-Chief, Polymers Research Journal

A. K. Haghi, PhD, holds a BSc in urban and environmental engineering from the University of North Carolina (USA); a MSc in mechanical engineering from North Carolina A&T State University (USA); a DEA in applied mechanics, acoustics and materials from the Université de Technologie de Compiègne (France); and a PhD in engineering sciences from the Université de Franche-Comté (France). He is the author and editor of 165 books as well as 1000 published papers in various journals and conference proceedings. Dr. Haghi has received several grants, consulted for a number of major corporations, and is a frequent speaker to national and international audiences. Since 1983, he served as a professor at several universities. He is currently Editor-in-Chief of the *International Journal of Chemoinformatics and Chemical Engineering* and *Polymers Research Journal* and on the editorial boards of many international journals. He is a member of the Canadian Research and Development Center of Sciences and Cultures (CRDCSC), Montreal, Quebec, Canada.

CONTENTS

LIST OF CONTRIBUTORS

A. Akhalkatsi
Faculty of Exact and Natural Sciences of I. Javakhishvili Tbilisi State University, 3, I. Chavchavadze Ave. Tbilisi 0176, Georgia

K. V. Alsaraeva
Siberian State Industrial University, Novokuznetsk, Russia

D. V. Ananchenko
Ural Federal University, Institute of Physics and Technology, Mira St. 21, Yekaterinburg, Russia

E. R. Andreeva
Institute of Biomedical Problems, Russian Academy of Sciences, Moscow, Russia

J. Aneli
Institute of Machine Mechanics, 10, Mindeli str. Tbilisi, 0186, Georgia

V. Angelov
Institute of Mechanics, Bulgarian Academy of Sciences, Acad. G. Bontchev St., bl. 4, 1113 Sofia, Bulgaria

R. Anyszka
Institute of Polymer & Dye Technology, Faculty of Chemistry, Lodz University of Technology, Stefanowskiego12/16, 90-924 Lodz, Poland

I. G. Assovsky
Semenov Institute of Chemical Physics RAS, Kosygin Str. 4, Moscow 119991, Russia

E. Basarygina
Chelyabinsk State Academy of Agricultural Engineering, Faculty of Electrification and Automation of Agricultural Production, Lenin Prospect, 75, Chelyabinsk, Russian Federation

G. Basilaia
Institute of Machine Mechanics, 10, Mindeli Str. Tbilisi 0186, Georgia

Yu. V. Berestneva
L.M. Litvinenko Institute of Physical Organic and Coal Chemistry, Donetsk, Ukraine

A. A. Berlin
Semenov Institute of Chemical Physics RAS, Kosygin Str. 4, Moscow 119991, Russia

D. M. Bieliński
Institute of Polymer & Dye Technology, Faculty of Chemistry, Lodz University of Technology, Stefanowskiego 12/16, 90-924 Lodz, Poland

N. Birsa
Yuri Gagarin State Technical University of Saratov, Institute of Electronic Engineering and Mechanical Engineering, Politechnicheskaya 77, Saratov, Russia

M. Bolotashvili
Institute of Machine Mechanics, 10, Mindeli Str. Tbilisi 0186, Georgia

K. Yu. Bondarenko
Siberian State Industrial University, Novokuznetsk, Russia

O. Borges
Faculty of Pharmacy, University of Coimbra, Coimbra, Portugal; CNC, Center for Neuroscience and Cell Biology, University of Coimbra, 3004-517 Coimbra, Portugal

L. B. Buravkova
Institute of Biomedical Problems, Russian Academy of Sciences, Moscow, Russia

F. F. Chausov
Physical-Technical Institute Ural Division, RAS, Izhevsk, Russia

M. Chernova
Novosibirsk State Technical University, Faculty of Radio Engineering and Electronics, 20 K. Marksa St., Novosibirsk, Russia

G. A. Dorofeev
Physical-Technical Institute Ural Division, RAS, 132 Kirov St., Izhevsk 426000, Russia

A. B. Eresko
L.M. Litvinenko Institute of Physical Organic and Coal Chemistry, Donetsk, Ukraine

A. Y. Fedotov
Institute of Mechanics, Ural Division, RAS, Izhevsk, Russia; Kalashnikov Izhevsk State Technical University, Russia

N. Filimonova
Novosibirsk State Technical University, Faculty of Radio Engineering and Electronics, 20 K. Marksa st., Novosibirsk, Russia

V. Gavrilenko
Novosibirsk State Technical University, Faculty of Radio Engineering and Electronics, 20 K. Marksa St., Novosibirsk, Russia

A. P. Gerola
Department of Chemistry, University of Coimbra, 3004-535 Coimbra, Portugal
Grupo de Materiais Poliméricos e Compósitos, GMPC, Chemistry Department, Maringá State University, 87020-900 Maringá, Paraná, Brazil

M. A. Goldshtrakh
Semenov Institute of Chemical Physics, Russian Academy of Sciences, Moscow 119991, Kosygin str. 4 Russia.

Yu. G. Gorbunova
Frumkin Institute of Physical Chemistry and Electrochemistry, Russian Academy of Sciences, Moscow, Russia

V. N. Gorshenev
Emanuel Institute of Biochemical Physics, Russian Academy of Sciences, Moscow, Russia

V. E. Gromov
Siberian State Industrial University, Novokuznetsk, Russia

G. A. Gromova
Semenov Institute of Chemical Physics, Russian Academy of Sciences, Moscow, Russia

S. A. Gruzd'
Kalashnikov Izhevsk State Technical University, Izhevsk, Russia

N. Guarrotxena
Nanohybrid and Interactive Polymers group, Institute of Polymer Science and Technology (ICTP), Spanish National Research Council (CSIC), Juan de la Cierva 3, 28006, Madrid, Spain

K. Z. Gumargalieva
Semenov Institute of Chemical Physics, Russian Academy of Sciences, Moscow, Russia

A. K. Haghi
University of Guilan, Rasht, Iran; Canadian Research and Development Center of Sciences and Cultures

D. Il'yashenko
Yurga Institute of Technology, TPU affiliate, Yurga, Russian Federation

V. Ilyushin
Novosibirsk State Technical University, Faculty of Radio Engineering and Electronics, 20 K. Marksa St., Novosibirsk, Russia

M. Imiela
Institute of Polymer & Dye Technology, Faculty of Chemistry, Lodz University of Technology, Stefanowskiego12/16, 90-924 Lodz, Poland

A. L. Iordanskii
Semenov Institute of Chemical Physics, Russian Academy of Sciences, Moscow, Russia

Y. F. Ivanov
Institute of High-Current Electronics of Siberian Branch of Russian Academy of Sciences, Tomsk, Russia

S. Jesus
Faculty of Pharmacy, University of Coimbra, Coimbra, Portugal; CNC, Center for Neuroscience and Cell Biology, University of Coimbra, 3004-517 Coimbra, Portugal

N. Kanatnikov
State University—Education-Science-Production Complex, Branch of Karachev, 242500, Gorky Street, 1B, Karachev, Bryansk Region, Russia

E. V. Kapralov
Siberian State Industrial University, Novokuznetsk, Russia

S. G. Karpova
N.M. Emanuel Institute of Biochemical Physics, Russian Academy of Sciences, Kosygin Str. 4, Moscow 119991, Russia

A. V. Kholzakov
Physico-Technical Institute, Ural Division, Russian Academy of Sciences, Izhevsk, Russia

V. E. Khomicheva
Siberian State Industrial University, Novokuznetsk, Russia

A. V. Khvatov
Emanuel Institute of Biochemical Physics, Russian Academy of Sciences, Moscow, Russia

A. N. Kiryakov
Ural Federal University, Institute of Physics and Technology, Mira St. 21, Yekaterinburg, Russia

V. S. Klekovkin
Kalashnikov Izhevsk State Technical University, Izhevsk, Russia

I. V. Klimenko
Emanuel Institute of Biochemical Physics, Russian Academy of Sciences, Moscow

V. I. Kodolov
Kalashnikov Izhevsk State Technical University, Basic Research—High Educational Center of Chemical Physics and Mesoscopy, Udmurt Scientific Center, Ural Division, RAS, Izhevsk, Russia

V. I. Kolesnikov-Svinarev
Semenov Institute of Chemical Physics, Russian Academy of Sciences, Moscow, Russia

R. Kolmykov
Kemerovo state university, Faculty of Chemistry Kemerovo, Russia

S. V. Konovalov
Siberian State Industrial University, Novokuznetsk, Russia

G. A. Korablev
Izhevsk State Agricultural Academy, Izhevsk, Russia

M. A. Korepanov
Institute of Mechanics, Ural Division, Russian Academy of sciences, Izhevsk, Russia

V. S. Kortov
Ural Federal University, Institute of Physics and Technology, Mira St. 21, Yekaterinburg, Russia

R. Yu. Kosenko
Semenov Institute of Chemical Physics, Russian Academy of Sciences, Moscow, Russia

R. Kotsilkova
OLEM, Institute of Mechanics, Bulgarian Academy of Sciences, Acad. G. Bontchev St., bl. 4, 1113 Sofia, Bulgaria

E. V. Koverzanova
Emanuel Institute of Biochemical Physics of Russian Academy of Sciences, Moscow, Russia

G. V. Kozlov
Kh.M. Berbekov Kabardino-Balkarian State University, Nal'chik, Russian Federation

A. Krupin
Novosibirsk State Technical University, Faculty of Radio Engineering and Electronics, Novosibirsk, Russia

P. Kuzhir
Research Institute for Nuclear problems of Belarusian State University, Minsk, Belarus

G. P. Kuznetsov
Semenov Institute of Chemical Physics, Russian Academy of Sciences, Moscow, Russia

G. S Larionova
Semenov Institute of Chemical Physics, Russian Academy of Sciences, Moscow, Russia

A. M. Lipanov
Udmurt Scientific Center, Ural Division, Russian Academy of Sciences, Izhevsk, Russia

N. M. Livanova
Emanuel Institute of Biochemical Physics, Russian Academy of Sciences, Moscow, Russia

A. V. Lobanov
Semenov Institute of Chemical Physics, Russian Academy of Sciences, Moscow, Russia

S. M. Lomakin
Emanuel Institute of Biochemical Physics, Russian Academy of Sciences, Moscow, Russia

A. N. Lubnin
Physical-Technical Institute, Ural Division, RAS, Izhevsk, Russia

L. L. Lukin
Kalashnikov Izhevsk State Technical University, Izhevsk, Russia

A. M. Lukmanova
Ural Federal University, Institute of Physics and Technology, Yekaterinburg, Russia

V. Lyasnikov
Yuri Gagarin State Technical University of Saratov, Institute of Electronic Engineering and Mechanical Engineering, Saratov, Russia

Ja. Macutkevic
Vilnius University (VU), 3 Universiteto St, LT-01513 Vilnius, Lithuania

S. Makarov
Yurga Institute of Technology, TPU affiliate, Yurga, Russian Federation

S. Maksimenko
Research Institute for Nuclear problems of Belarusian State University, Minsk, Belarus

V. S. Markin
Semenov Institute of Chemical Physics, Russian Academy of Sciences, Moscow, Russia

V. Markov
State University—Education-Science-Production Complex, Branch of Karachev, Karachev, Bryansk Region, Russia

G. Martyshov
Gagarin State Technical University of Saratov, Institute of Electronic Engineering and Mechanical Engineering, Saratov, Russia

L. N. Maslov
Kalashnikov Izhevsk State Technical University, Izhevsk, Russia

M. Ya. Mel'nikov
Lomonosov Moscow State University, Russia

I. N. Menshikov
Physical-Technical Institute, Ural Division, RAS, Izhevsk, Russia

A. K. Mikitaev
Kh.M. Berbekov Kabardino-Balkarian State University, Nal'chik, Russian Federation

M. A. Mikitaev
Kh.M. Berbekov Kabardino-Balkarian State University, Nal'chik, Russian Federation

I. Y. Mittova
Department of Materials Science and Industry of Nanosystems, Voronezh State University, Faculty of Chemistry, Voronezh, Russia

K. V. Morozov
EVRAZ—Consolidated West Siberian Metallurgical Plant, Novokuznetsk, Russia

M. R. Moskalenko
Ural Federal University Named After the First President of Russia B. N. Yeltsin, Institute of Humanities and Arts, Department of History of Science and Technology, Ekaterinburg, Russia

M. V. Motyakin
Semenov Institute of Chemical Physics, Russian Academy of Sciences, Moscow, Russia

O. Muktarov
Yuri Gagarin State Technical University of Saratov, Institute of Electronic Engineering and Mechanical Engineering, Saratov, Russia

T. Muktarova
Yuri Gagarin State Technical University of Saratov, Institute of Electronic Engineering and Mechanical Engineering, Saratov, Russia

E. C. Muniz
Grupo de Materiais Poliméricos e Compósitos, GMPC, Chemistry Department, Maringá State University, 87020-900 Maringá, Paraná, Brazil

L. Nadareishvili
Institute of Cybernetics of Georgian Technical University, Tbilisi, Georgia

E. A. Naimushina
Physical-Technical Institute, Ural Division, RAS, Izhevsk, Russia

A. A. Ol'khov
Plekhanov Russian University of Economics; Semenov Institute of Chemical Physics, Russian Academy of Sciences, Moscow, Russia

A. Paddubskaya
Research Institute for Nuclear problems of Belarusian State University, Minsk, Belarus

Yu. N. Pankova
Semenov Institute of Chemical Physics, Russian Academy of Sciences, Moscow, Russia

R. Panova
Chelyabinsk State Academy of Agricultural Engineering, Faculty of Electrification and Automation of Agricultural Production, Chelyabinsk, Russian Federation

E. M. Pearce
Brooklyn Polytechnic University, Six Metrotech, Brooklin, NY, USA

Z. Pędzich
Department of Ceramics and Refractory Materials, Faculty of Materials Science & Ceramics, AGH—University of Science & Technology, Krakow, Poland

K. A. Petrovykh
Ural Federal University, Institute of Physics and Technology, Mira St. 21, Yekaterinburg, Russia

S. Pichhidze
Yuri Gagarin State Technical University of Saratov, Institute of Electronic Engineering and Mechanical Engineering, Saratov, Russia

Yu. A. Ponomareva
Siberian State Industrial University, Novokuznetsk, Russia

A. A. Popov
N.M. Emanuel Institute of Biochemical Physics, Russian Academy of Sciences, Moscow, Russia

E. S. Pushkarev
Physical-Technical Institute, Ural Division, RAS, Izhevsk, Russia

T. Putilova
Chelyabinsk State Academy of Agricultural Engineering, Faculty of Electrification and Automation of Agricultural Production, Chelyabinsk, Russian Federation

S. Rafiei
University of Guilan, Rasht, Iran

S. V. Raikov
Siberian State Industrial University, Novokuznetsk, Russia

E. V. Raksha
Donetsk National University, Donetsk; L.M. Litvinenko Institute of Physical Organic and Coal Chemistry, Donetsk

S. Z. Rogovina
Semenov Institute of Chemical Physics, Russian Academy of Sciences, Moscow, Russia

A. F. Rubira
Grupo de Materiais Poliméricos e Compósitos, GMPC, Chemistry Department, Maringá State University, 87020-900 Maringá, Paraná, Brazil

O. A. Semina
Siberian State Industrial University, Novokuznetsk, Russia

A. V. Severyukhin
Institute of Mechanics, Ural Division, Russian Academy of Sciences, Izhevsk, Russia

O. Yu. Severyukhina
Institute of Mechanics, Ural Division, Russian Academy of Sciences, Izhevsk, Russia

I. N. Shabanova
Physical Technical Institute, Ural Division, Russian Academy of Sciences, Izhevsk, Russia

O. I. Shavrin
Kalashnikov Izhevsk State Technical University, Izhevsk, Russia

I. Shihin
State University—Education-Science-Production Complex, Branch of Karachev, Karachev, Bryansk Region, Russia

M. G. Shilkina
N.N. Semenov Institute of Chemical Physics, Russian Academy of Sciences, Moscow, Russia

N. G. Shilkina
Emanuel Institute of Biochemical Physics of Russian Academy of Sciences, Moscow, Russia

A. A. Shushkov
Institute of Mechanics, Ural Division, Russian Academy of Sciences, Izhevsk, Russia

D. C. Silva
Department of Chemistry, University of Coimbra, Coimbra, Portugal

K. V. Sosnin
Siberian State Industrial University, Novokuznetsk, Russia

O. V. Staroverova
N.N. Semenov Institute of Chemical Physics, Russian Academy of Sciences, Moscow, Russia

S. N. Sykov
Institute of Mechanics, Ural Division, Russian Academy of Sciences, Izhevsk, Russia

N. S. Terebova
Physical Technical Institute, Ural Division, Russian Academy of Sciences, Izhevsk, Russia

E. V. Tomina
Department of Materials Science and Industry of Nanosystems, Voronezh State University, Faculty of Chemistry, Voronezh, Russia

V. A. Trapeznikov
Physical Technical Institute, Ural Division, Russian Academy of Sciences, Izhevsk, Russia

V. V. Trineeva
Institute of Mechanics, Ural Division, Russian Academy of Sciences, Izhevsk, Russia

A. Yu. Tsivadze
Emanuel Institute of Biochemical Physics of Russian Academy of Sciences, Moscow, Russia

N. A. Turovskij
Donetsk National University, Donetsk, Ukraine

O. O. Udartseva
Institute of Biomedical Problems, Russian Academy of Sciences, Moscow, Russia

S. V. Usachev
Emanuel Institute of Biochemical Physics of Russian Academy of Sciences, Moscow, Russia

A. V. Vakhrushev
Institute of Mechanics, Ural Division, Russian Academy of Sciences; Kalashnikov Izhevsk State Technical University, Izhevsk, Russia

L. L. Vakhrusheva
Kalashnikov Izhevsk State Technical University, Izhevsk, Russia

A. J. M. Valente
Department of Chemistry, University of Coimbra, 3004-535 Coimbra, Portugal

Yu. G. Vasiliev
Izhevsk State Agricultural Academy, Izhevsk, Russia

A. Velichko
Novosibirsk State Technical University, Faculty of Radio Engineering and Electronics, Novosibirsk, Russia

G. E. Zaikov
Emanuel Institute of Biochemical Physics, Russian Academy of Sciences, Moscow, Russia

L. S. Zelenina
Department of Materials Science and Industry of Nanosystems, Voronezh State University, Faculty of Chemistry, 394006 Voronezh, Russia

S. V. Zvonarev
Ural Federal University, Institute of Physics and Technology, Mira St. 21, Yekaterinburg, Russia

LIST OF ABBREVIATIONS

ACNFs	advantages of activated carbon nanofibers
CC	charge current
CCD	charge-coupled device
DSC	differential scanning calorimetry
FDS	fine-dispersed suspension
FEM	finite element method
GIAO	gauge-including-atomic-orbital
HD	helical drafting
HTMP	high-temperature thermomechanical processing
LMP	low melting point
MD	molecular dynamics
MEAM	modified embedded-atom method
MFI	melt flow index
MSC	mesenchymal stromal cells
NHS	*N*-hydroxysuccinimide
NPs	nanoparticles
PCL	pulsed cathodoluminescence
PS	photosensitizers
PVA	polyvinyl alcohol
SDS	sodium dodecyl sulfate
SEM	scanning electron microscopy
SEP	spatial-energy parameter
SERS	surface-enhanced Raman scattering
SPEU	segmented polyurethane
TBE	Tris–borate-EDTA
TL	thermoluminescence
TSP	thermal strain processing
XPS	X-ray photoelectron spectroscopy
XRD	X-ray analysis

PREFACE

This volume is dedicated to providing important information on new trends concerning the development of nanomaterials and nanotechnologies. The book contains the selected reviews, papers, and short communications prepared based on the lectures and abstracts at the Fifth International Conference "From Nanostructures, Nanomaterials and Nanotechnologies to Nanoindustry," which took place in Izhevsk, Udmurtia, Russia, on April 2–3, 2015. The conference scientific program covered the synthesis and production of nanostructues, nanosystems, and nanostructured materials; the methods of investigations of these products and materials; the modeling in nanoscience sphere; as well as the applications of nanoproducts and nanotechnologies in different fields including fields of nanoindustry.

The editors selected papers, including reviewed articles and short communications, on the different trends of nanochemistry and mesoscopic physics, nanomaterial science, nanomedicine, nanotechnology, modeling of nanosystems, and their obtaining processes.

The book is comprised of information on the following trends:

- Theory and synthesis of nanostructures and nanosystems for improvement of materials and objects
- Modeling in sphere on nanoscience and nanotechnology
- Investigations of structure and properties of nanostructures, nanosystems, and nanostructured materials
- Production of nanostructured materials and investigations of their properties
- Development of nanostructured materials and nanotechnologies applications.

The book provides discussion on the following topics:

1. Redox synthesis of metal/carbon nanocomposites in nanoreactors of polymeric matrices.
2. Optimized SERS-nanodumbbells on femto-sensitive nanoscale biosensing.
3. Metals combustion as method for production of ultraporous nanostructural ceramics.

4. pH-responsive modified gum arabic hydrogels for delivery of curcumin.
5. Nanosized intermolecular tuning of activity of tetrapyrrolic photosensitizers.
6. Redox properties of double-decker phthalocyanines for nanosensor development.
7. Photophysical properties of metal phthalocyanine nano-aggregates for early diagnosis.
8. Photodynamic therapy efficacy using Al-, Mg- and Zn-phthalocyanine nanoparticles.
9. Complex formation between chlorophyll and photosensitive electron acceptors.
10. The formation of graphite nanoscale structures for materials with novel functional properties.
11. A detailed review on development and application of activated carbon nanofibers in fuel cells.
12. Prospects for the use of nanomaterials in greenhouses.
13. Growth modes impact on the heterostructure's CaF_2/Si(1 0 0) electro-physical properties.
14. Nanostructure-phase states and defect substructure formation in superior-quality rails.
15. Structure and properties of Ti–Y system after electroexplosive alloying and electron beam treatment.
16. Structure and properties of wear-resistant weld deposit formed on martensitic steel using the electric-arc method.
17. The enhancement of fatigue life of Al–Si alloy by electron-beam processing.
18. X-ray diffraction characterization of mechanically alloyed high-nitrogen nanocrystalline Fe–Cr system.
19. Effect of some parameters on structure and dynamic properties of medical polymers.
20. Hexagon geometry in nanosystems.
21. Entropic and spatial energy interactions.
22. The study of nanoaerosol systems using mathematical simulation.
23. Application of unique X ray electron magnetic spectrometers for the investigation and chemical analysis of the electronic structure of the ultrathin surface layers of systems in liquid and solid state.
24. Mechanism of the metal/carbon nanocomposites super small quantities influence on structures and properties of polymeric systems and materials.

25. Polymer composites with gradient of electric and magnetic properties
26. The influence of nanofiller structure on melt viscosity of nanocomposites polymer/carbon nanotubes.
27. Nanostructured powders of solid–cobalt–nickel, their analysis by optical emission spectrometry with inductively coupled plasma and use as reference samples for the calibration of spectrometers with solid sampling
28. Polyamide 6 nanocomposites with reduced water permeability
29. Directions the perfecting of metrological ensuring a production the semiconductor materials and nanostructures.
30. Formation of composite coatings on the basis of biological hydroxyapatite plasma spraying on titanium substrate.
31. IR spectral analysis hydroxyapatite of biological origin.
32. Spectral analysis of pyrolitic carbon for heart valves.
33. X-Ray study of biological hydroxyapatite obtained by heat treatment antlers.
34. Deposition of carbon nanotubes on the implant with hydroxyapatite coating.
35. Structural features of matrixes based on ultrathin poly(3-hydroxybutyrate) fibers prepared via electrospinning.
36. Electromagnetic properties of epoxy resins filled with carbon nanotubes and golden nanoparticles.
37. Utilization of halloysite in ceramizable polymer composites.
38. Experimental and giao-calculated 1H and 13C NMR spectra of 2-(pyridin-2-yl)-1*H*-benzoimidazole.
39. Supramolecular activation of arylalkyl hydroperoxides by Et_4NBr.
40. Modeling processes of special nanostructured layers in epitaxial structures.
41. Oxidation of gallium arsenide with the modified nanoscale layer of gel V_2O_5 surface.

This book is unique and important because the new ideas and trends in nanochemistry, mesoscopic physics, nanostructured materials science and nanotechnologies as well as development perspectives on nanoindustry are discussed. The purchasers (researchers, professors, postgraduates, students, and other readers) of the book will find much interesting information and receive new knowledge on the trend of nanoscience and nanotechnology.

The book will be useful for teaching of specific courses and also for personal requalification, university/college libraries, and bookstores.

The editors and contributors will be happy to receive some comments from readers.

Vladimir I. Kodolov, DSc
Gennady E. Zaikov, DSc
A. K. Haghi, PhD

ABOUT THE CONFERENCE

From the 2nd to 3rd of April 2015, the Kalashnikov Izhevsk State Technical University hosted the Fifth International Conference "From Nanostructures, Nanomaterials and Nanotechnologies to Nanoindustry".

The conference program was comprised of the presentations from Russian and international scientists from the largest research centers and universities of 14 countries as follows: Russia, Canada, USA, Brazil, Portugal, Spain, Iran, Georgia, Bulgaria, Ukraina, Poland, Belarus, Lithuania, and Finland. The scientific program covered the synthesis and production of different nanostructures, nanosystems, and nanostructured materials; modeling in the sphere of nanomaterial science and nanotechnologies; as well as the application of nanoproducts, nanostructured materials, and nanotechnologies in different areas.

The aim of the conference was discussion of the latest achievements in the science of nanomaterials and nanotechnology and practical advances in nanoindustry development.

The Fifth International Conference "From Nanostructures, Nanomaterials and Nanotechnologies to Nanoindustry" opened at 9:00 on April 02, 2015 at Kalashnikov Izhevsk State Technical University. After the official welcoming address, the leading scientists in nanomaterial science and nanotechnologies made presentations.

Apart from plenary and oral presentations, there were computer poster presentations as well as a roundtable discussions on the results of mastering nanotechnologies, semoindustrial and industrial production of nanocomposites and nanomaterials, and practical introduction of nanocomposites and nanotechnologies into different fields, including medicine and agriculture.

The roundtable discussions combined the reports of representatives from industrial enterprises from industrial enterprises and research institutes. The exhibition of modern samples of nanoproducts, nanostructured materials as well as nanotechnological units and apparatus for the production and manufacturing of materials and objects with improved and unique properties were presented.

PART I

Nanostructures and Nanosystems for Improvement of Materials and Objects

CHAPTER 1

IMPROVED SERS-ACTIVE AG-NANOSYSTEMS FOR PRACTICAL APPLICATIONS

N. GUARROTXENA

Nanohybrid and Interactive Polymers Group, Institute of Polymer Science and Technology (ICTP), Spanish National Research Council (CSIC), Juan de la Cierva 3, 28006 Madrid, Spain

CONTENTS

ABSTRACT

The noble metal nanoparticles (NPs) exhibit optical excitations known as surface plasmons, extremely dependent on the NPs-surface, -morphology, -interaction, and -assembly which provide the basis for the molecular recognition, imaging and sensing sensitivity.

Surface-enhanced Raman scattering (SERS) is attributed primarily to the enhancement of the incident and scattered electromagnetic fields near metal surfaces through excitation of localized surface plasmons. Such field enhancement allows detection with high sensitivity (at the single-molecule level), and due to the fingerprint capabilities of SERS, also with high selectivity.

A suitable environment of high SERS activity resides at the interface between metallic nanostructures, where local fields are dramatically enhanced by plasmonic response. This condition requires the placement of SERS reporter in such "hot-spots," which can be engineered by solution methods yielding stable solutions ready to be used into a variety of sensory approaches. Despite relevant results in SERS-based metallic nanotags generation, the performances of these tags are far from the optimal. A poor control of aggregation and probe deposition in the hotspots issues still remain there.

In this sense, a rational and proper control of linker-mediated nanoassembly, ligand exchange, and surface interaction, as reported in our work, provides a suitable strategy in the development of nanostructured tools for sensory applications. These nanostructured tools enable SERS-based detection platform with high spectral specificity, sensitivity (femtoMolar detection levels), improved contrast and multiplexing capabilities. Sensitivities values situated at two orders of magnitude higher than commercially available ELISA assays. The presence of these SERS-active nanostructures can also be optimized via post-synthetic strategy, making nanodumbbell structures ideal in a wide range of tagging, sensing, and analysis applications.

1.1 INTRODUCTION

Nanotechnology is a highly promising field, where disciplines such as chemistry, engineering, biology, and medicine interact with nanostructured materials, offering potential solutions in multiple practical applications (optoelectronic, photocatalysis, sensing, and optical devices).[1–5] Diverse nanoparticles (NPs), such as quantum dots, carbon nanotubes, and magnetic NPs, functionalized with different biomolecules have been successfully used in therapy, image, and drug-delivery applications due to their versatility, ease of chemical synthesis, less toxicity, and unique properties.[2,4]

Metal NPs have recently attracted a great deal of attention and interest for potential applications in nanomedicine due to their ease of preparation, surface reactivity, biocompatibility, and optical properties.[6–8] Their unique optical properties, which are size-, shape-, and environment-dependent can also play an important role in detection and diagnosis.[6,9–12]

The particular and extremely strong adsorption and light scattering in the plasmon resonance wavelength regions enable to use metal NPs to explore optical-based detection technologies.[13] Moreover, the extraordinary optical properties of NPs due to a strong interaction between free electrons and an incident electromagnetic field allow one to develop detection formats based on surface-enhanced Raman scattering (SERS).[3,14] SERS, an ultrasensitive technique of Raman spectroscopy,[15] which takes advantage of local field enhancements for molecules near metallic nanostructures, plays an increasingly important role in this research area,[16,17] capable of molecule detection.[3]

As SERS is a vibrational spectroscopy, the internal modes of the reporter molecule can be used as specific readout signal. And, given that junctions between plasmon-coupled metal NPs produce the strongest field enhancement "hot spots",[16] the use of thiolate adsorption chemistry in SERS-active dimers generation provides a plausible pathway to design nanosystems suitable for highly sensitive assay protocols.[18–21] This condition requires the placement of SERS reporter in such "hot-spots," which can be engineered by solution methods yielding stable solutions ready to be used into a variety of sensory approaches.[18–23] Despite relevant results in SERS-based nanosystem generation, the performances of these tags are far from the optimal. A poor control of aggregation and probe deposition in the hotspots issues still remain there.

In this context, a rational and proper control of linker-mediated nanoassembly, ligand exchange and surface interaction, as studied in our work, provide a suitable strategy in the development of nanostructured tools for practical applications.[19] These nanostructured tools enable SERS-based detection nanosystems with high spectral specificity, sensitivity, improved contrast and multiplexing capabilities.[19–23] Moreover, the presence of SERS-active nanostructures can also be optimized via post-synthetic strategy,[24–28] making dimer-like structures ideal in a wide range of tagging, sensing, and analysis applications. In this work, we refer to gel electrophoresis separation to improve the content of larger SERS activity nanostructures.[24]

1.2 EXPERIMENTAL

1.2.1 MATERIALS

The following reagents were purchased from the indicated suppliers and used without further purification: biphenyl-4,4'-dithiol (DBDT, Aldrich), silver nitrate ($AgNO_3$, Aldrich), and sodium citrate ($Na_3C_3H_5O(COO)_3$, Aldrich), *N,N'*-dicyclohexylcarbodiimide (DCC, Alfa Aesar), and *N*-hydroxysuccinimide (NHS, Aldrich). Antihuman α-thrombin (Thr) mouse monoclonal antibody (MW 150 kDa) was purchased from Haematologic Technologies Inc., and bifunctional polyethylenglycol (HS-PEG-X) with X = $-OCH_3$ of MW = 2 kDa and X= –COOH of MW 3 kDa, from Rapp Polymere GmbH.

1.2.2 SYNTHESIS OF NANOPARTICLES

Citrate-capped Ag NPs were synthesized according to standard literature procedures.[29] TEM data showed a distribution of NP with diameters 35 ± 5 nm, obtained from a set of NPs greater than 100.

1.2.3 PREPARATION OF ANTITAGS

DBDT, dissolved in chloroform to a final concentration of 5 mM, was added to the NPs solution and allowed to react for 30 min at room temperature. Afterward, a solution of thiolated carboxy-functionalized polyethylene glycol (HS-$(PEG)_{65}$-COOH, MW 3 kDa) was added to the linked NPs solutions and allowed to react 30 min. The solution was then centrifuged at 4000 rpm and redissolved in water to the starting volume.

The free amino groups on the surface of antihuman α-Thr mouse monoclonal antibody was used to create amide bonds with the carboxylic moieties present on the PEG chains stabilizing the NP surface.[14,30] The amide bond formation was achieved via addition of DCC (DMF) and NHS (DMF) to the solution of PEG stabilized Ag NPs, followed by the addition of the antibody and allowed to proceed for at least 4 h in an ice bath. After centrifugation, the NPs were resuspended in PBS; and a methoxy-functionalized thiolated PEG (HS-$(PEG)_{45}$-OCH_3, 2 kDa) was previously added to Calf Thymus and Tween 20 addition to reduce unspecific events during the assay.

1.2.4 SURFACE FUNCTIONALIZATION

The assay was developed using an epoxy-functionalized glass slides. Human α-Thr mouse monoclonal antibody spots (0.7 μM in PBS buffer) were printed on the glass surface and allowed to react for 4 h in a humidity chamber. After rinsing with PBS and immersed then in a solution of blocking buffer, the slide was stored overnight between 0 and 4°C.

Human α-Thr was dissolved in PBS at concentrations of 10^2, 10^3, 10^4, 10^5, and 10^6 fM. The antigen solution (20 μL) was then printed on the microarray and allowed to react for 2 h in a humidity chamber. The slides were then rinsed in PBS containing sodium dodecyl sulfate (SDS) and in nanopure water. The same protocol was followed for the control experiments, where the solution of antigen was replaced by a solution of PBS buffer and 0.1% Tween 20.

Antitag aliquots (20 μL) were deposited on the glass slide and allowed to bind to the surface for 2 h. The slides were then rinsed with PBS containing 0.1% SDS and nanopure water.

1.2.5 GEL ELECTROPHORESIS

A volume of 45 μL PEG functionalized Ag-NPs samples were loaded into the wells of 0.75% low melting point (LMP) agarose (Promega). The gel was prepared with an immersed in 0.5× Tris–borate-EDTA (TBE) buffer and run in a horizontal electrophoresis system (Owl Easy Cast Minigel System, Fisher Scientific) at 200 V for 45 min.

1.2.6 AGAROSE GEL ELECTROPHORESIS EXTRACTION AND DIGESTION OF PEG FUNCTIONALIZED N-AG NPS

After electrophoresis, discrete bands of samples were transferred to a microcentrifuge tube where the samples were extracted by digestion of LMP–agarose gel with Agar*ACE*™ Enzyme (Promega) at 45°C in buffer for 1 h. After incubation the slightly viscous solution was purified by repeated washing (with nanopure water) and centrifuging.

1.2.7 INSTRUMENTATION

Backscattered Raman spectra were recorded on a LabRAM ARAMIS Raman Microscope system (Horiba–JobinYvon) equipped with a multichannel air-cooled charge-coupled device detector. Spectra were excited using the 633-nm line of a HeNe narrow bandwidth laser (Melles Griot). SERS maps were collected at a 49-μW laser power, with a data acquisition time of 1 s on 15 × 15 μm² areas with a 1-μm step. TEM studies were performed in a JEM 1230 transmission electron microscopy (JEOL) operated at 80 kV, and images were taken by using an AMT (Advanced Microscopy Techniques Corp.).

1.3 RESULTS AND DISCUSSION

By design, the nanosystems or multicomponent nanostructures, named "anti-tags," contain antibodies as the recognition probes and rely on SERS as the detection method (Fig. 1.1a). Basically, they consist on antibody-functionalized silver-NPs (35 nm) held together by a Raman reporter and stabilized by the presence of polyethylene glycol (PEG) chains as capping agent.

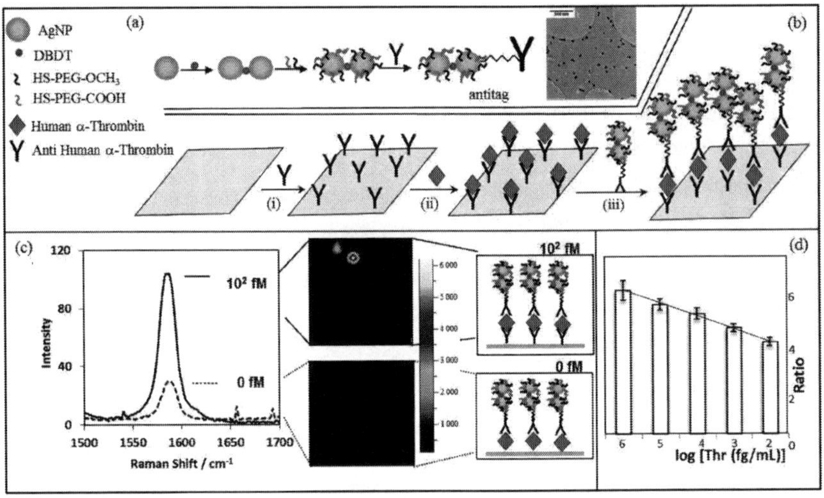

FIGURE 1.1 Schematic of preparation of an antitag with the corresponding representative TEM image (a) and SERS-based detection procedure (b). SERS maps and spectra obtained from substrates treated with a solution of thrombin targeting antitag in the presence (solid line) and absence (dashed line) of thrombin (c). SERS-based calibration curve for thrombin detection using a concentration range of (10 fM–1 nM). Ratio represents the coefficient between the signal at the most intense peak in the presence and absence of target.

In the SERS-active nanostructures generation (Fig. 1.1a), the Raman reporter molecule (DBDT) is strategically designed to chemisorb on the silver NP at the hot junction spot between closely coupled NPs, facing its NP linker role. The second step, that is surface biofunctionalization, involves first exchanging citrate at the surface of Ag NPs with a 3-kDa thiolated PEG bearing a carboxylic moiety. This process incorporates: (1) polar functionalities and steric bulk to guarantee the linkage process without aggregates formation.[31] Since on depending on the strength of NPs interactions, the linkage process may, eventually, lead to a loss of NP assembly or further agglomeration; the negative charges from the carboxylic functionalized PEG, grafted onto the Ag NPs surface through the thiol bonding, provides long-term stability of the NP organization by steric and electrostatic repulsion;[31] and (2) carboxylic functionalities to simplify bioconjugation with antibodies[14] via carbodiimide-mediated amidation.[32] The addition of a second unit of methoxy thiolated PEG after purification by centrifugation provides further stabilization.[33]

TEM image of antitags shows the efficiency on the carboxylic-PEG addition strategy, as protective coating agent, before centrifugation to yield a distribution of aggregates with smaller number of incorporated NPs (Fig. 1.1a). So, proper control of the shape and size of the aggregated NPs is critical to fabricate homogenous, nanosized probes with reproducible signals.

The capability of this detection approach was tested using an ELISA-analog sandwich to detect the presence of human α-Thr antigen (Fig. 1.1b). Recognition surfaces were developed by functionalization of the substrate with capturing antibodies (step i). Target protein was then added to the substrate and allowed to bind for 30 min (step ii). Finally, the antitags with the reporting antibody were then deposited on the target functionalized surfaces (step iii). After washing away unbound antitags, the Raman signal corresponding to surface bound Thr antigens was studied with SERS. If the antigens are present, they will bind their corresponding capturing antibodies in their respective positions. Positive detection is indicated by the presence of the corresponding SERS signal in each region (Fig. 1.1c). In this case, the SERS study requires a random screening through all the regions of the sensing substrate; however, due to the fingerprint capabilities of the SERS spectra, the vibrational modes of the Raman reporter present as linking element on SERS-active nanostructures "antitags" will serve as diagnostic signals.

The addition of the target protein (Thr) was analyzed using the diagnostic fingerprint of DBDT-Raman reporter (1589 cm^{-1}) and revealed a limit of detection of 100 fM based on a 3-fold increase in the Raman signals, compared to control experiments in the absence of the target. Figure 1.1c shows the results obtained when targeting Thr (100 fM) using our improved antitags. The SERS spectra and maps were collected on averaging the integrated intensity from at least five 15 × 15 μm substrate areas, for each test using a laser power of 633 nm. Similar results were obtained for Thr at concentrations between 1 nM and 100 fM (specifically 1 nM, 100 pM, 10 pM, 1 pM, and 100 fM; data not shown).

The statistical analysis of the antibody–antigen binding affinity results obtained using several Thr concentrations (100 fM–1 nM) validates our approach in the studied detection range and reveals the capability of the antitags in recognizing their specific target with high selectivity and efficiency. The calibration curve (Fig. 1.1d) was then fitted linearly with a resulting R squared of 0.9962 and the error bars represent the standard deviation as obtained from three different substrates.

A rational control of NP assembly is not only an important fact for effective exploitation of structure-dependent materials properties but also a key factor for generating nanostructured materials with specific activity in practical applications, such as optical (sensing) as shown in previous section and also electronic (nanodevice).

An alternative to non-ideal NPs assembly[23] would be an effective post-synthetic purification method to enrich for dimers, or any other structure, from NP-assemblies.[24–28] We used the LMP agarose gel electrophoresis technique for fractionating spherical polymer-wrapped Ag-nanoassemblies by nuclearity and shape on the basis of a rational selection of surface charge of capping agent and pore size of agarose gel matrix.[24]

Spherical AgNPs (35 nm), 40-fold concentrated by centrifugation, were brought together by virtue of metallic NP–sulfur bond formation to rend dimers and higher order aggregates. Again, the dithiolated linker molecule is also used as SERS reporter. Subsequently, a mixture of 30% of thiolated methoxy PEG (2 kDa) and 70% of thiolated carboxylic PEG (3 kDa) was absorbed onto NPs surface to impart particle stability and electrophoretic mobility. After cleaning, the sample was dissolved in TBE buffer.

The optimal surface coverage of NPs was determined by running different methoxy-PEG and carboxylic-PEG ratios on the electrophoresis setup, and monitoring their impact on the AgNPs mobility, band differentiation, and their SERS activity (results not shown).[24]

After PEGylation, the dimerized samples with an initial composition of 58% monomers, 17% dimers, 5.7% trimers, 3.8% tetramers, and 15% rods were run in a 0.75% LMP-agarose (Fig. 1.2a). After digestion at 40°C with the agarose enzyme, the electrophoretic fractions were analyzed by TEM to determine the NPs nuclearity discrimination. Fig. 1.2b and c reproduces their TEM images together with their corresponding assembly distributions.

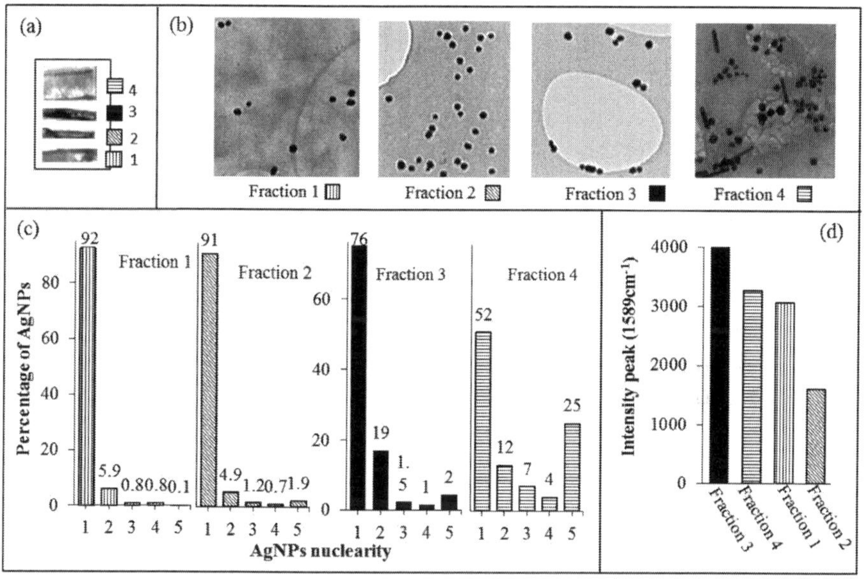

FIGURE 1.2 Electrophoretic separation of AgNPs on 0.75% LMP-agarose gel at 200 V. Sliced fractions as separated (a). TEM images of fractionated samples (b). Histograms of NPs clusters distribution (c). SERS-spectra intensity of fractions collected after electrophoretic fractionation in 0.75% LMP-agarose gel (d).

Since dimer-like NP aggregates always exhibit outstanding SERS enhancement.[16,18–21,26,34] A controlled and effective enrichment of the SERS-active structures (dimer-like) in our sample should have the highest SERS intensity (Fig. 1.2d). The correlation between SERS and TEM results showed that the highest monomer NP concentration (90%, fractions 1 and 2) gave low SERS intensity. While higher nuclearity concentration and rods (24% and 48%, in fractions 3 and 4, respectively) led to a higher Raman scattering values. Furthermore, fraction 3 with 12× higher relative content dimer/clusters than fraction 4 exhibits the strongest SERS activity. Comparison of monomer/dimer ratio between fractions 3 and 2 correlates well. Fraction 3 exhibits higher SERS intensity and lower monomer/dimer contribution by 4.5-fold. Taken together, the results give dimer and cluster structures the "hot spots" feature wherein a dimer exhibits the highest SERS activity per core.

1.4 CONCLUSION

In summary, these results report an interesting mobility-based post-synthetic route in optimizing the presence of SERS-active nanostructures, where the assembly of the metallic environment provides higher SERS enhancements (hot spots),[35] making dimer-like nanostructures ideal in a wide range of practical applications.[24]

Furthermore, the sensitivity, the robustness and the high reproducibility of the successfully engineered SERS bioassay based on these active nanoassemblies provide a suitable platform for developing multiplexed sensitive detection of proteins and other target molecules; not possible with the colorimetric and fluorometric ELISA analogs.[19–21] This sensing approach incorporates important improvements in SERS-tags sensitivity (femtoMolar detection level) by properly managing the interaction between Ag-NPs within nanoassemblies.

ACKNOWLEDGMENTS

N.G. thanks the support of the Ministry of Education of Spain (MEC) through Programa Nacional de Movilidad de Recursos Humanos de Investigación del Plan Nacional I+D+I 2008–2011 and Ministry of Economy and Competitiveness (MAT2011-22861). Author also thanks Professor G.C. Bazan (Institute of Polymer and Organic Solids, University of California, Santa Barbara, USA) for his support and fruitful contribution in this work.

KEYWORDS

- **surface-enhanced Raman spectroscopy (SERS)**
- **SERS-active dimers or "hotspots"**
- **antitags; biosensor**
- **AgNPs-nuclearities fractionation**

REFERENCES

1. Raghavendra, R.; Arunachalam, K.; Annamalai, S.; Arunachalam, A. Diagnostics and Therapeutic Application of Gold Nanoparticles. *Int. J. Pharm. Pharm. Sci.* **2014,** *6*(2), 74–87.
2. Paull, J. Nanotechnology: The Next Challenge for Organics. *J. Org. Syst.* **2008,** *3*(1), 3–22.
3. Kneipp, K.; Kneipp, H.; Itzkan, I.; Dasari, D. R.; Feld, M. S. Ultrasensitive Chemical Analysis by Raman Spectroscopy. *Chem. Rev.* **1999,** *99*, 2957–2975.
4. Mirkin, C. A.; Letsinger, R. L.; Mucic, R. C.; Storhoff, J. J. A DNA based Method for Rationally Assembling Nanoparticles into Macroscopic Materials. *Nature* **1996,** *382*, 607–609.
5. Cai, W.; Gao, T.; Hong, H.; Sun, J. Applications of Gold Nanoparticles in Cancer Nanotechnology. *Nanotechnol. Sci. Appl.* **2008,** *1*, 17–32.
6. Khan, A. K.; Rashid, R.; Murtaza, G.; Zahra, A. Gold Nanoparticles: Synthesis and Applications in Drug Delivery. *Trop. J. Pharm. Res.* **2014,** *13*(7), 1169–1177.
7. Abdelhalim, M. A. K.; Mady, M. M. Liver Uptake of Gold Nanoparticles after Intraperitoneal Administration In Vivo: A Fluorescence Study. *Lipids Health Dis.* **2011,** *10*, 195–202.
8. Popovtzer, R.; Agrawal, A.; Kotov, A. N.; Popovtzer, A.; Balter, J. Targeted Gold Nanoparticles Enable Molecular CT Imaging of Cancer. *Int. J. Nanomed.* **2008,** *8*(9120), 4593–4596.
9. Misra, R.; Acharya, S.; Sahoo, S. K. Cancer Nanotechnology: Application of Nanotechnology in Cancer Therapy. *Drug Deliv. Today* **2010,** *15*(19/20), 842–850.
10. El-Sayed, M. A. Some Interesting Properties of Metals Confined in Time and Nanometer Space of Different Shapes. *Acc. Chem. Res.* **2001,** *34*, 257–264.
11. Sun, Y.; Xia, Y. Increased Sensitivity of Surface Plasmon Resonance of Gold Nanoshells Compared to that of Gold Solid Colloids in Response to Environmental Changes. *Anal. Chem.* **2002,** *74*, 5297–5305.
12. Liz-Marzán, L. M.; Giersig, M.; Mulvaney, P. Synthesis of Nanosized Gold–Silica Core–Shell Particles. *Langmuir* **1996,** *12*, 4329–4335.
13. Xiang, M. H.; Xu, X.; Liu, F.; Li, K. A. Gold Nanoparticle based Plasmon Resonance Light-scattering Method as a New Approach for Glycogen-biomacromolecule Interactions. *J. Phys. Chem. B* **2009,** *113*, 2734–2738.
14. Hirsh, L. R.; Jackson, J. B.; Lee, A.; Halas, N. J.; West, J. L. A Whole Blood Immunoassay using Gold Nanoshells. *Ann. Chem.* **2003,** *75,* 2377–2381.
15. Moskovits, M. J. Surface Enhanced Raman Spectroscopy. *Raman Spectrosc.* **2005,** *36,* 485–496.
16. Aroca, R. Surface-enhanced Vibrational Spectroscopy, Wiley: New York, 2006.
17. Willets, K. A.; Van Duyne, R. P. Supplemental Material *Annu. Rev. Phys. Chem.* **2007,** *58*, 267–297.
18. Fabris, L.; Schierhorn, M.; Moskovits, M.; Bazan, G. C. Aptatag-based Multiplexed Assay for Protein Detection by Surface Enhanced Raman Spectroscopy. *Small* **2010,** *6*, 1550–1557.
19. Guarrotxena, N.; Bazan, G. C. Antibody-functionalized SERS Tags with Improved Sensitivity. *Chem. Commun.* **2011,** *47*, 8784–8786.

20. Guarrotxena, N.; Liu, B.; Fabris, L.; Bazan, G. C. Antitags: Nanostructured Tools for Developing SERS-based ELISA Analogs. *Adv. Mater.* **2010,** *22*, 4954–4958.
21. Guarrotxena, N.; Bazan, G. C. Antitags: SERS-encoded Nanoparticle Assemblies that Enable Single-spot Multiplex Protein Detection. *Adv. Mater.* **2014,** *26*, 1941–1946.
22. Pallaoro, A.; Braun, G. B.; Reich, N. O.; Moskovits, M. Mapping Local pH in Live Cells Using Encapsulated Fluorescent SERS Nanotags. *Small* **2010,** *6*, 618–622.
23. Braun, G. B.; Lee, S. J.; Laurence, T.; Fera, N.; Fabris, L.; Bazan, G. C.; Moskovits, M.; Reich, N. O. Generalized Approach to SERS-active Nanomaterials via Controlled Nanoparticle Linking, Polymer Encapsulation, and Small Molecule Infusion. *J. Phys. Chem. C* **2009,** *113*, 13622–13629.
24. Guarrotxena, N.; Braun, G. Ag-nanoparticle Fractionation by Low Melting Point Agarose Gel Electrophoresis. *J. Nanopart. Res.* **2012,** *14*, 1199–1207.
25. Chen, G.; Wang, Y.; Tan, L. H.; Yang, M.; Tan, L. S.; Chen, Y.; Che, H. High-purity Separation of Gold Nanoparticle Dimers and Trimers. *J. Am. Chem. Soc.* **2009,** *131*, 4218–4219.
26. Chen, G.; Wang, Y.; Yang, M.; Xu, J.; Goh, S. J.; Pan, M.; Chen, H. Measuring Ensemble-averaged Surface-enhanced Raman Scattering in the Hotspots of Colloidal Nanoparticle Dimers and Trimers. *J. Am. Chem. Soc.* **2010,** *132*, 3644–3645.
27. Sun, X.; Tabakman, S. M.; Seo, W. S.; Zhang, L.; Zhang, G.; Sherlock, S.; Bai, L.; Dai, H. Separation of Nanoparticles in a Density Gradient: feCo@C and Gold Nanocrystals. *Angew. Chem.* **2009,** *48*, 939–942.
28. Steinigeweg, D.; Schutz, M.; Salehi, M.; Schlucker, S. Gold Nanoparticles: Fast and Cost-effective Purification of Gold Nanoparticles in the 20–250 nm Size Range by Continuous Density Gradient Centrifugation. *Small* **2011,** *7*, 2443–2448.
29. Lee, P. C.; Meisel, D. J. Adsorption and Surface-enhanced Raman of Dyes on Silver and Gold Sols. *J. Phys. Chem.* **1982,** *86*, 3391–3395.
30. Wang, S.; Mamedova, N.; Kotov, N. A.; Chen, W.; Studer, J. Antigen/Antibody Immunocomplex from CdTe Nanoparticle Bioconjugates, *Nano Lett.* **2002,** *2*, 817–822.
31. Parak, W. J.; Gerion, D.; Zanchet, D.; Woerz, A. S.; Pellegrino, T.; Michael, C.; Williams, S. C.; Seitz, M.; Bruehl, R. E.; Bryant, Z.; Bustamante, C.; Bertozzi, C. R.; Alivisatos, A. P. Conjugation of DNA to Silanized Colloidal Semiconductor Nanocrystalline Quantum Dots. *Chem. Mater.* **2002,** *14*, 2113–2119.
32. a) Wang, J. Nanoparticle-based Electrochemical DNA Detection. *Anal. Chim. Acta* **2003,** *500*, 247–257. (b) Sperling, R. A.; Pellegrino, T.; Li, J. K.; Cang, W. H.; Parak, W. J. Electrophoretic Separation of Nanoparticles with a Discrete Number of Functional Groups. *Adv. Funct. Mater.* **2006,** *16*, 943–948.
33. Qian, X. M.; Nie, S. M. Single-molecule and Single-nanoparticle SERS: From Fundamental Mechanisms to Biomedical Applications. *Chem. Soc. Rev.* **2008,** *37*, 912–920.
34. Moskovits, M. Surface-enhanced Spectroscopy. *Rev. Mod. Phys.* **1985,** *57*, 783–826.
35. Guarrotxena, N.; Ren, Y.; Mikhailovsky, A. Raman Response of Dithiolated Nanoparticle Linkers. *Langmuir* **2011,** *27*(1), 347–351.

CHAPTER 2

REDOX SYNTHESIS OF METAL/ CARBON NANOCOMPOSITES IN NANOREACTORS OF POLYMERIC MATRICES

V. V. TRINEEVA[1] and V. I. KODOLOV[2*]

[1]*Institute of Mechanics, Ural Division, Russian Academy of Sciences, Izhevsk, Russia*

[2]*Kalashnikov Izhevsk State Technical University, Basic Research—High Educational Center of Chemical Physics and Mesoscopy, Udmurt Scientific Center, Ural Division, RAS, Izhevsk, Russia*

**Corresponding author. E-mail: vkodol.av@mail.ru*

CONTENTS

ABSTRACT

This paper dedicates to the explanation on the base of mesoscopic physics principles for metal/carbon nanocomposites obtaining in the nanoreactors of polymeric matrices. The conditions of mesoscopic physics principles applications are proved. These conditions include the following phenomena as the interference, spectrum, and charge quantization, which take place at the action of mesoscopic (nano)particles having sizes from 1000 to 0.1 nm on different media.

In addition to that the correspondent particles can only vibrate and conduct electrons. It's shown that these particularities of clusters or nanoparticles are possible during the redox synthesis of metal/carbon nanocomposites in nanoreactors of polymeric matrices.

2.1 INTRODUCTION

At present, some hypothesis and the investigations results are explained by means of mesoscopic physics principles[1,2] and also by the fractal theory equations.[3–5] Self organization in system and self similarity (image) are the ground notions in mesoscopic physics as well as in the fractal theory and nanotechnology.

Therefore, it is interesting in this paper to consider the application of these disciplines for the synthesis of nanostructures, including metal/carbon nanocomposites. Mesoscopic Physics deals with mesoscopic particles which have linear size from 1000 to 0.1 nm, because the length of coherence wave equals to 1000 nm, and the electron wave length in different materials changes from 100 to 0.1 nm. For these particles the following particularities are known as the interference, spectrum quantization and charge quantization.[1] The application Mesoscopic Physics principles is possible, if the mesoscopic particle will be established within small volume space, for instance, nanoreactor, in which this nanoparticle is coordinated on active groups of nanoreactor walls in correspondence with realized conditions. In these cases the correspondent particle can only vibrate and conduct the electrons across oneself. Therefore, the nanoreactor can be presented as the nanosized hollow or the limited free volume in polymeric matrix. Then chemical particles are directly self organized in this hollow for the creation of transitional (mesoscopic) state before the nanoproduct formation. Also the nanoreactor definition can be proposed as the specific porous nanostructure, in which the distance between walls changes from 1 to 100 nm.

2.2 THEORETICAL AND EXPERIMENTAL INVESTIGATIONS OF SYNTHESIS WITHIN NANOREACTORS OF POLYMERIC MATRICES

To our mind the formation of metal/carbon nanocomposites within nanoreactors of polymeric matrices may be similar to the photography processes. At the beginning, when metal containing phase is mixed up with polymeric phase, the nanoreactor within polymeric matrix is filled by metal clusters of metal containing phase. The process is accompanied with cluster coordination on active centers of nanoreactor walls. It's photography process. Then during process of xerogels formation the development flows. And then at the heating of obtained product the fixation occurs.

Thus, at the metal/carbon nanocomposites obtaining there are following processes: the coordination reactions, the redox reactions, and certainly the self organization processes.

Let us discuss this picture of the nanostructures formation. Now the active chemical metal containing particle is taken root in the free nanosized hollow of polymeric matrices. It coordinates with functional groups of nanoreactor walls and loses the ability to diffusion and rotation. This muddled particle can be characterized by vibration motion and also electron motion including the transport electrons across it. However, if the electrotransport across our particle (cluster) will be carried out, the redox reactions will be stimulated and the metal reduction will be occurred. This process is direct and lead to self organization which may be observed at AFM investigations. For the further stimulation of metal/carbon nanocomposites formation process, it is necessary the regular regime of unfinished product heating.

Consequently, the redox synthesis of metal/carbon nanocomposites can be explained by basic principles of Mesoscopic Physics.

The investigation of redox synthesis of Metal/Carbon nanocomposites in nanoreactors of polymeric matrices is realized[9,10] in three stages (Fig. 2.1).

Before the experimental investigations the computational designing of nanoreactors filled by metal containing phase as well as quantum chemical modeling of processes within nanoreactors are carried out.

Polymeric Matrices, for example, Polyvinyl Alcohol → *Nanoreactors of Polymeric Matrices for the filling by Metal Compounds Phase* ← *Metal Compounds, for instance, Clusters of Copper or Nickel Oxides*

Coordination Reactions and, Partially, Redox Processes ↓

The obtaining of xerogels containing the nanoreactors of different forms in dependence on metal nature/
The estimation of energetic and geometric characteristics of Nanoreactors

↓

Thermo chemical Stage (Redox Synthesis of Metal/Carbon Nanocomposites)
DTA-TG investigations for the determination of temperature conditions and also the heating duration (the middle temperature for the Copper/Carbon Synthesis films equals to 200°C, at the middle duration equals to 3 hour).

FIGURE 2.1 The scheme of metal/carbon nanocomposites synthesis within nanoreactors of polymeric matrices.

Then the experimental designing and nanorectors filling by metal containing phase are realized with using two methods:

1. The mixing of salt solution with the solution of functional polymer, for example, polyvinyl alcohol (PVA) solution.
2. The common degenaration of polymeric phase with metal containing phase (metal oxides) in active medium (water).

The following stage includes the obtaining of xerogels filled by metal containing phase. The third stage is the properly redox synthesis of metal/carbon nanocomposites in nanoreactors of polymeric matrices (obtained xerogels) at narrow temperature intervals.

The first and second stages concern to preperatiory stages. On the second stage the functional groups in nanoreactor walls participate in coordination reactions with metal ions (first method) or with the clusters of metal containing phase (second method).

The computational experiment was carried out with software products Gamess and HyperChem with visualization. The definite result of the computational experiment is obtained stage by stage. Any transformations at the initial stages are taken into account at further stages. The prognostic possibilities of the computational experiment consist in defining the probability of the formation of nanostructures of definite shapes. The optimal dimensions and shape of internal cavity of nanoreactors, optimal correlation between metal containing and polymeric components for obtaining the necessary nanoproducts are found with the help of quantum-chemical modeling.[9]

If the method 1 is used, for instance, the interaction of PVA and metal chloride solutions where metals are iron, cobalt, nickel, and copper, than the colored xerogels is formed as follows: iron—brown-red, nickel—pale-green, cobalt—blue, and copper—yellow-green. Consequently, PVA interacts with metal ions with the formation of complex compounds.

For the corresponding correlations "polymer–metal containing phase," the dimensions, shape, and energetic characteristics of nanoreactors are found with the help of coordination, the structure, and relief of xerogel surface change. The comparison of phase contrast pictures on the corresponding films indicates greater concentration of the extended polar structures in the films containing copper, in comparison with the films containing nickel and cobalt (Fig. 2.2).

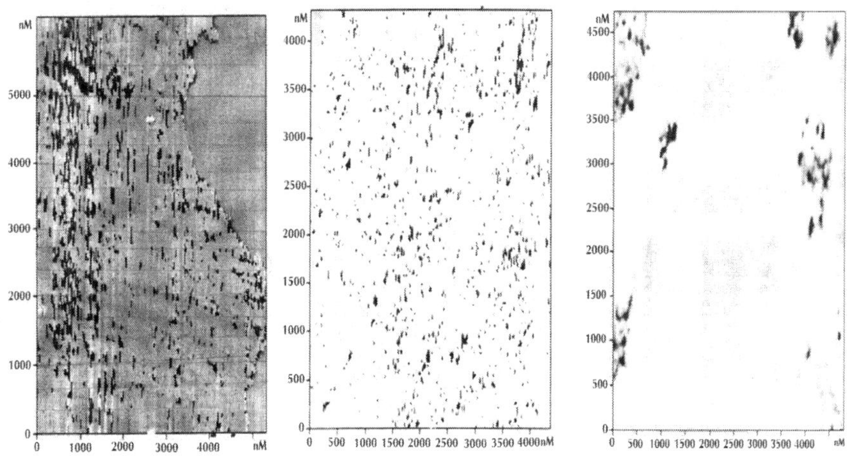

FIGURE 2.2 Pictures of phase contrast of PVA surfaces containing copper (a), nickel (b) and cobalt (c).

The processing of the pictures of phase contrast to reveal the regions of energy interaction of cantilever with the surface in comparison with the background produces practically similar result with optical transmission microscopy.[11,12]

Corresponding to data of AFM the sizes of nanoreactors obtained from solutions of metal chlorides and the mixture of PVA with polyethylene polyamine (PEPA) are determined (Table 2.1).

TABLE 2.1 Sizes of Nanoreactors Found with the Help of Atomic Force Microscopy.[11]

Composition	Sizes of AFM formations				
	Length	Width	Height	Area	Density
PVA:PEPA:$CoCl_2$ = 2:1:1	400–800	150–400	30–40	60–350	5.5
PVA:PEPA:$NiCl_2$ = 2:1:1	80–100	80–100	25–35	6–12	120
PVA:PEPA:$CuCl_2$ = 2:1:1	80–100	80–100	20–30	6–20	20
PVA:PEPA:$CoCl_2$ = 2:2:1	600–900	300–600	100–120	180–500	3.0
PVA:PEPA:$NiCl_2$ = 2:2:1	40–80	40–60	10–30	2–4	350

The results of AFM investigations of xerogels films obtained from metal oxides and PVA[13,14] is distinguished (Fig. 2.3) in comparison with previous data, that testify to difference in reactivity of metal chlorides and metal oxides.

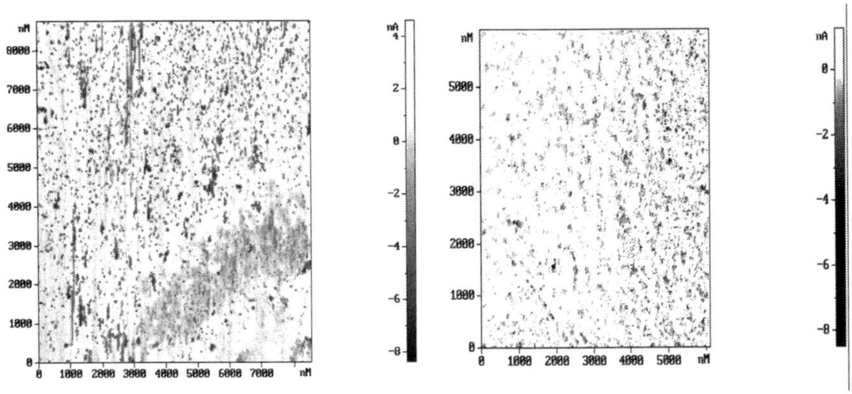

a *b*

FIGURE 2.3 Phase contrast pictures of xerogels films PVA—Ni (a) and PVA—Cu (b).

According to AFM results investigation the addition of Ni/C nanocomposite in PVA leads to more strong coordination in comparison with analogous addition Cu/C nanocomposite.

The mechanism of formation of nanoreactors filled with metals was found with the help of IR spectroscopy.

Depending on metal coordinating ability and conditions for nanostructure obtaining (in liquid or solid medium with minimal content of liquid), we obtain "embryos" of future nanostructures of different shapes, dimensions, and composition. It is advisable to model coordination processes and further redox processes with the help of quantum chemistry apparatus.

The availability of d metal in polymeric matrix results, in accordance with modeling results, in its regular distribution in the matrix and self-organization of the matrix.

At the same time, the metal orientation proceeds in interface regions and nanopores of polymeric phase which stipulates further direction of the process to the formation of metal/carbon nanocomposites. In other words, the birth and growth of nanosized structures occur during the process in the same way as known from the macromolecule physics and fractal theory.[16]

The mechanism of formation of nanoreactors filled with metals was found with the help of IR spectroscopy.

Thus, at the second stage the coordination of metal containing phase and corresponding orientation in nanoreactor take place.

To investigate the processes at the second stage of obtaining metal/carbon nanocomposites X-ray photoelectron spectroscopy, transmission electron microscopy, and IR spectroscopy are applied. The sample for IR spectroscopy was prepared when mixing metal/carbon nanocomposite powder with one drop of vaseline oil in agate mortar to obtain a homogeneous paste with further investigation of the paste obtained on the appropriate instrument. As the vaseline oil was applied when the spectra were taken, we can expect strong bands in the range 2750–2950 cm^{-1}.

At the third stage, it is required to give the corresponding energy impulse to transfer the "transition state" formed into carbon/metal nanocomposite of definite size and shape. To define the temperature ranges in which the structuring takes place, DTA-TG investigation is applied.

To define the temperature ranges in which the structuring takes place, DTA-TG investigation is applied. It is known that small changes of weight loss (TG curve) at invariable exothermal effect (DTA curve) testify to the self organization (structural formation) in system.

According to data of DTA-TG investigation optimal temperature field for film nanostructure obtaining is 230–270°C, and for spatial nanostructure obtaining—325–410°C.

It is found that in the temperature range under 200°C nanofilms, from carbon fibers associated with metal phase as well, are formed on metal or metal oxide clusters. When the temperature elevates up to 400°C, 3D nanostructures are formed with different shapes depending on coordinating ability of the metal.

Assuming that the nanocomposites obtained can be considered as oscillators transferring their oscillations onto the medium molecules, we can determine to what extent the IR spectrum of liquid medium will change. It is proposed to consider the obtaining of metal/carbon nanocomposites in nanoreactors of polymeric matrixes as self-organization process similar to the formation of ordered phases.

The perspectives of this investigation are looked through in an opportunity of thin regulation of processes and the entering of corrective amendments during processes.

This process can be considered as the recrystallization and can be described by the fractal theory with the application of Avrami–Kolmogorov equations. However, these equations were adopted to the conditions of metal/carbon nanocomposites redox synthesis. At the beginning, let us discuss the possibilities of Avrami equations for the determination of the heating regime conditions at the redox synthesis of metal/carbon nanocomposites.

The difference of potentials between the interacting particles and object walls stimulating these interactions is the driving force of self-organization processes (formation of nanoparticles with definite shapes). The potential jump at the boundary "nanoreactor wall-reacting particles" is defined by the wall surface charge and reacting layer size. If we consider the redox process as the main process preceding the nanostructure formation, the work for charge transport corresponds to the energy of nanoparticle formation process in the reacting layer. Then the equation of energy conservation for nanoreactor during the formation of nanoparticle mol will be as follows:

$$nF\Delta\varphi = RT\ln\left(\frac{N_p}{N_r}\right), \tag{2.1}$$

where n is the number reflecting the charge of chemical particles moving inside "the nanoreactor"; F is the Faraday number; $\Delta\varphi$ is the difference of potentials between the nanoreactor walls and flow of chemical particles; R is the gas constant; T is the process temperature; N_p is the mol share of nanoparticles obtained; N_r is the mol share of initial reagents from which the nanoparticles are obtained.

Using the above equation, we can determine the values of equilibrium constants when reaching the certain output of nanoparticles, sizes of nanoparticles and shapes of nanoparticles formed with the appropriate equation modification. The internal cavity sizes or nanoreactor reaction zone and its geometry significantly influence the sizes and shapes of nanostructures.

The sequence of the processes is conditioned by the composition and parameters (energy and geometry) of nanoreactors. To accomplish such processes it is advisable to preliminarily select the polymeric matrix containing the nanoreactors in the form of nanopores or crazes as process appropriate. Such selection can be realized with the help of computer chemistry. Further the computational experiment is carried out with the reagents placed in the nanoreactor with the corresponding geometry and energy parameters. Examples of such computations were given in [3]. The experimental confirmation of polymer matrix and nanoreactor selection to obtain carbon/metal containing nanostructures was given in [4,5]. Kolmogorov–Avrami equations are widely used for such processes which usually reflect the share of a new phase produced.

When the metal ion moves inside the nanoreactor with redox interaction of ion (mol) with nanoreactor walls, the balance setting in the pair "metal containing-polymeric phase" can apparently be described with the following equation:

$$zF\Delta\phi = RT\ln K = RT\ln\left(\frac{N_p}{N_r}\right) = RT\ln(1-W), \qquad (2.2)$$

where z is the number of electrons participating in the process; $\Delta\varphi$ is the difference of potentials at the boundary "nanoreactor wall-reactive mixture"; F is the Faraday number; R is the universal gas constant; T is the process temperature; K is the process balance constant; N_p is the number of moles of the product produced in nanoreactor; N_r is the number of moles of reagents or atoms (ions) participating in the process which filled the nanoreactor; W is the share of nanoproduct obtained in nanoreactor.

In turn, the share of the transformed components participating in phase interaction can be expressed with the equation which can be considered as a modified Avrami equation

$$W = 1-\exp\left[-\tau^{n}\exp\left(\frac{zF\Delta\varphi}{RT}\right)\right], \qquad (2.3)$$

where τ is the duration of the process in nanoreactor; n is the number of degrees of freedom changing from 1 to 6. When n equals 1, one-dimensional nanostructures are obtained (linear nanostructures, nanofibers). If n equals 2 or changes from 1 to 2, flat nanostructures are formed (nanofilms, circles, petals, wide nanobands). If n changes from 2 to 3 and more, spatial nanostructures are formed as n also indicates the number of degrees of freedom. The selection of the corresponding equation recording form depends on the nanoreactor (nanostructure) shape and sizes and defines the nanostructure growth in the nanoreactor.

During the redox process connected with the coordination process, the character of chemical bonds changes. Therefore, correlations of wave numbers of the changing chemical bonds can be applied as the characteristic of the nanostructure formation process in nanoreactor

$$W = 1 - \exp\left\{ -\tau^n \cdot \frac{v_{is}}{v_{fs}} \right\}, \tag{2.4}$$

where v_{in} corresponds to wave numbers of initial state of chemical bonds, and v_f wave numbers of chemical bonds changing during the process. By the analogy with the above calculations the parameters a in Equation (5) should be considered as a value that reflects the transition from the initial to final state of the system and represents the ratios of activities of system states.

The experimental modeling of obtaining nanofilms after the alignment of copper compounds with PVA at 200°C revealed that optimal duration when the share of nanofilms approaches 100% equals 2.5 h. This corresponds to the calculated value based on the aforesaid Avrami equation. The calculations are made supposing the formation of copper nanocrystals on the nanofilms. It is pointed out that copper ions are predominantly reduced to metal. Therefore, it was accepted for the calculations that n equals 2 (two-dimensional growth), potential of redox process during the ion reduction to metal ($\Delta\varphi$) equals 0.34 V, temperature (T) equals 473 K, Faraday number (F) corresponds to 26.81 (A h/mol), gas constant R equals 2.31 (W h/mol·degree). The analysis of the dimensionality shows the zero dimension of the ratio—$zF/\Delta\phi RT$. The calculations are made when changing the process duration with a half-hour increment. Results of calculations on the equation 10 are brought in table.

If nanofillms are scrolled together with copper nanowires, β is taken as equaled to 3, the temperature increases up to 400°C, the optimal time when the transformation degree reaches 99.97%, corresponds to the duration of

2 h, thus also coinciding with the experiment. According to the calculation results if following the definite conditions of the system exposure, the duration of the exposure has the greatest influence on the value of nanostructure share.

The selection of the corresponding equation form depends upon the shape and sizes of nanoreactor (nanostructure) and defines the nanostructure growth in nanoreactor or the influence distribution of the nanostructure on the structurally changing medium. At the same time v_{is} corresponds to the frequency of skeleton oscillations of C–C bond at 1100 cm^{-1}, v_{fs}—symmetrical skeleton oscillations of C=C bond at 1050 cm^{-1}. In this case the equation looks as Equation (11).

For the example discussed the content of nanofilms in % will be changing together with the changes in the duration as follows in table.

Under the aforesaid conditions the linear sizes of copper (from ion radius to atom radius) and carbon–carbon bond (from C–C to C=C) are changing during the process. Apparently, the structure of copper ion and electron interacts with electrons of the corresponding bonds forming the layer with linear sizes $r_i + l_{\text{C–C}}$ in the initial condition and the layer with the size $r_a + l_{\text{C=C}}$ in the final condition. Then the equation for the content of nanofilms can be written down as follows:

$$W = 1 - \exp\left\{ -\tau^n \times \frac{r_\alpha + l_{\text{C=C}}}{r_i + l_{\text{C-C}}} \right\} \tag{2.5}$$

At the same time r_i for Cu^{2+} equals 0.082 nm, r_a for four-coordinated copper atom corresponds to 0.113 nm, bond energy C–C equals 0.154 nm, and C=C bond—0.142 nm. Representing the ratio of activities as the ratio of corresponding linear sizes and taking the value n as equaled to 2, at the same time changing τ in the same intervals as before, we get the following change in the transformation degree based on the process duration (Table 2.2).

TABLE 2.2 The Changes of Nanofilms Content Calculated Dependent on the Duration of Redox Synthesis.

Duration (h)		**0.5**	**1.0**	**1.5**	**2.0**	**2.5**
Content of nano-films, (%)	Eq. (10)	22.5	63.8	89.4	98.3	99.8
	Eq. (12)	23.0	64.9	90.5	98.5	99.9
	Eq. (13)	23.7	66.0	91.2	98.7	99.9

Modified Kolmogorov–Avrami equations were tested to prognosticate the duration of the processes of obtaining metal/carbon nanofilms in the system "Cu–PVA" at 200°C.[2] The calculated time (2.5 h) correspond to the experimental duration of obtaining carbon nanofilms on copper clusters.

Thus, with the help of Kolmogorov–Avrami equations or their modified analogs, we can determine the optimal duration of the process to obtain the required result. It opens up the possibility of defining other parameters of the process and characteristics of nanostructures obtained (by shape and sizes).

The methods of optical spectroscopy and X-ray photoelectron spectroscopy allow to determine the energy of the interaction of the chemical particles in the nanoreactors with the active centers of the nanoreactor walls, which stimulate reduction–oxidation processes.

Depending on the nature of the metal salt and the electrochemical potential of the metal, different metal reduction nanoproducts in the carbon shells differing in shape are formed. Based on this result, we may speak about a new scientific branch—nanometallurgy. The stages of nanostructures synthesis may be represented by the following scheme (Fig. 2.4):

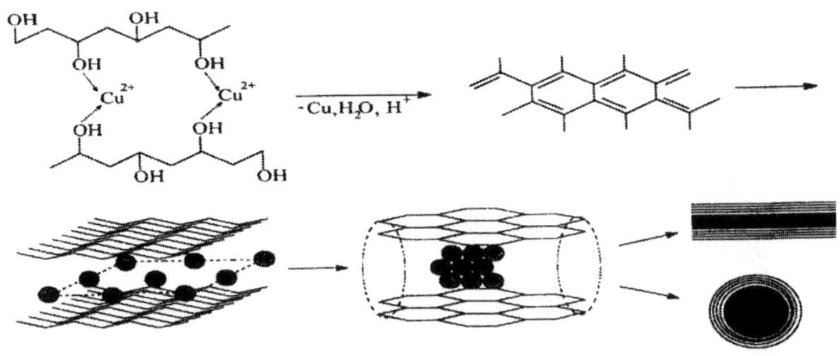

FIGURE 2.4 Scheme of Copper/carbon nanostructures obtaining from copper ions and polyvinyl alcohol.

Otherwise, as follows from this scheme, the First step (stage) of metal/carbon nanocomposite formation is the growth carbon fibers associated with metal clusters. The second stage is the formation of carbon film which covers metal clusters. And then this film transforms in the 3D nanostructures with metal containing clusters inside.

The possible ways for obtaining metallic nanostructures in carbon shells have been determined. The investigation results allow to speak about the possibility of the isolation of metallic and metal-containing nanoparticles in the carbon shells differing in shape and structure. However, there are still problems related to the calculation and experiment because using the existing investigation methods it is difficult unambiguously to estimate the geometry and energy parameters of nanoreactors under the condition of 'erosion' of their walls during the formation of metallic nanostructures in them.

It is interesting that the scheme (Fig. 2.5) and the succession of stages on fractal dimension correspond to experimental morphology of nanocomposites obtained after corresponding heating stages (Fig. 2.6).

It is found that in the temperature range under 200°C nanofilms, from carbon fibers associated with metal phase as well (Fig. 2.5), are formed on metal or metal oxide clusters. When the temperature elevates up to 400°C, 3D nanostructures are formed with different shapes depending on coordinating ability of the metal.

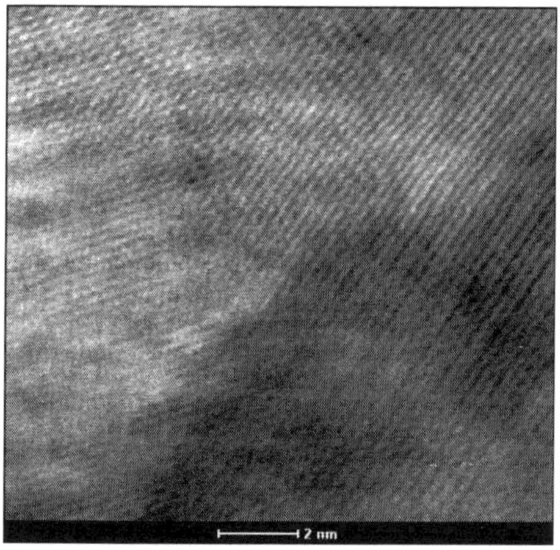

FIGURE 2.5 Microphotographs obtained with the help of transmission electron microscopy. Cu/C nanocomposite.

Two types of nanocomposites rather widely applied during the modification of various polymeric materials were investigated. These were copper/carbon nanocomposite and nickel/carbon nanocomposite specified below. In turn, the nanopowders obtained were tested with the help of high-resolution transmission electron microscopy, electron microdiffraction, laser analyzer, X-ray photoelectron spectroscopy and IR spectroscopy.

The method of metal/carbon nanocomposite synthesis applied has the following advantages:

1. Originality of stage-by-stage obtaining of Metal/ Carbon nanocomposites with intermediary evaluation of the influence of initial mixture composition on their properties.
2. Wide application of independent modern experimental and theoretical analysis methods to control the technological process.
3. Technology developed allows synthesizing a wide range of metal/ carbon nanocomposites depending on the process conditions.
4. Process does not require the use of inert or reduction atmospheres and specially prepared catalysts.
5. Method of obtaining metal/carbon nanocomposites allows applying secondary raw materials.

So, the nanoreactor walls, containing the functional groups, have different potentials. When the mesoscopic particles (in our case, ions or metal oxide clusters) are coordinated with functional groups of nanoreactor walls, their potentials change too. Now, according to mesoscopic physics principles, the conditions is created for electrons transport across mesoscopic particles (clusters) in which there are reducers (positively charged metal or metal ions). The redox processes are began and the d metal electron structure changes with the growth of unpaired electron number. d Metal becomes paramagnetic. At the same time the hydrocarbon shells, corresponding to nanoreactor walls, are oxidized with the aromatic cycles or hexagon formation in which π electrons are interacted with metal unpaired electrons. Otherwise there is possibility of unpaired electrons formation on the carbon shell of metal/carbon nanocomposite obtained. Let us discuss the properties and especially energetic characteristics of metal/carbon nanocomposites.

2.3 METAL/CARBON NANOCOMPOSITES PROPERTIES, ESPECIALLY ENERGETIC CHARACTERISTICS

Metal/carbon nanocomposites will be more active than carbon or silicon nanostructures because their masses are bigger at identical sizes and shapes. Therefore, the vibration energy transmitted to the medium is also high.

Metal/carbon nanocomposite (Me/C) represents metal nanoparticles stabilized in carbon nanofilm structures.[12–14] In turn, nanofilm structures are formed with carbon amorphous nanofibers associated with metal containing phase. As a result of stabilization and association of metal nanoparticles with carbon phase, the metal chemically active particles are stable in the air and during heating as the strong complex of metal nanoparticles with carbon material matrix is formed.

For the corresponding correlations "polymer–metal containing phase" the dimensions, shape and energy characteristics of nanoreactors are found with the help of AFM.[9,15] Depending on a metal participating in coordination, the structure and relief of xerogel surface change.

To investigate the processes at the second stage of obtaining metal/carbon nanocomposites, X-ray photoelectron spectroscopy, transmission electron microscopy, and IR spectroscopy are applied.

In turn, the nanopowders obtained were tested with the help of high-resolution transmission electron microscopy, electron microdiffraction, laser analyzer, X-ray photoelectron spectroscopy, and IR spectroscopy.

The test results of nanocomposites obtained are given in Table 2.3.

TABLE 2.3 Characteristics of Metal/carbon Nanocomposites (Met/C NC)

Type of Met/C NC	Cu/C	Ni/C	Co/C	Fe/C
Composition, metal/carbon (%)	50/50	60/40	65/35	70/30
Density (g/cm^3)	1.71	2.17	1.61	2.1
Average dimension (nm)	20(25)	11	15	17
Specific surface (m^2/g)	160 (average)	251	209	168
Metal nanoparticle shape	Close to spherical, there are dodecahedrons	There are spheres and rods	Nanocrystals	Close to spherical
Caron phase shape (shell)	Nanofibers associated with metal phase forming nanocoatings	Nanofilms scrolled in nanotubes	Nanofilms associated with nanocrystals of metal containing phase	Nanofilms forming nanobeads with metal containing phase
Atomic magnetic moment (metal)[8] (μB)	0.0	0.6	~1.0	2.1
Atomic magnetic moment (nanocomposite) (μB)	0.6	1.6	1.7	2.3

The distinctive feature of the considered technique for producing metal/carbon nanocomposites is a wide application of independent, modern, experimental and theoretical analysis methods to substantiate the proposed technique and investigation of the composites obtained (quantum-chemical calculations, methods of transmission electron microscopy and electron diffraction, method of X-ray photoelectron spectroscopy, X-ray phase analysis, etc.). The technique developed allows synthesizing a wide range of metal/carbon nanocomposites by composition, size and morphology depending on the process conditions. In its application, it is possible to use secondary metallurgical and polymer raw materials. Thus, we can adjust the nanocomposite structure to extend the function of its application without pre-functionalization. Controlling the sizes and shapes of nanostructures by changing the metal-containing phase, we can, to some extent, apply completely new, practicable properties to the materials which sufficiently differ from conventional materials.

The essence of the method[11] consists in coordination interaction of functional groups of polymer and compounds of 3d-metals as a result of grinding of metal-containing and polymer phases. Further, the composition obtained undergoes thermolysis following the temperature mode set with the help of thermogravimetric and differential thermal analyses. At the same time, we observe the polymer carbonization, partial or complete reduction of metal compounds and structuring of carbon material in the form of nanostructures with different shapes and sizes.

The conditions of metal/carbon nanocomposites redox synthesis such as nature of metal containing phase and also polymeric phase (e.g., molecular mass and acetate group's number for PVA) influence on the composition and structure of metal/carbon nanocomposites.

Below the examples for the phase compositions changes (Fig. 2.6) and the morphology of copper/carbon nanocomposites obtained in different conditions (Table 2.4) are presented.

Metal/carbon nanocomposite (Me/C) represents metal nanoparticles stabilized in carbon nanofilm structures.[12,13] In turn, nanofilm structures are formed with carbon amorphous nanofibers associated with metal containing phase. As a result of stabilization and association of metal nanoparticles with carbon phase, the metal chemically active particles are stable in the air and during heating as the strong complex of metal nanoparticles with carbon material matrix is formed.

Figure 2.7 demonstrates the microphotographs of transmission electron microscopy specific for different types of metal/carbon nanocomposites.

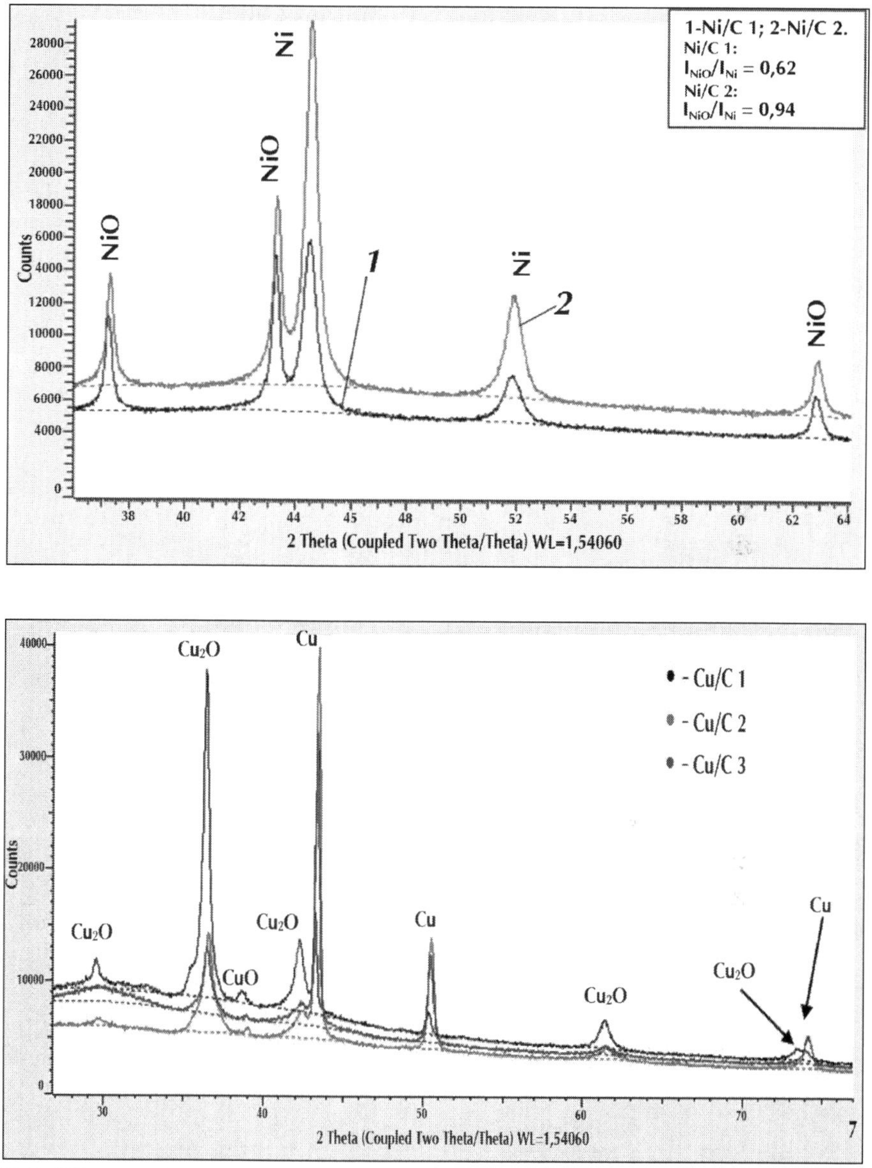

FIGURE 2.6 The roentgenograms of copper/carbon and nickel/carbon nanocomposites obtained in different conditions.

TABLE 2.4 The Example of Copper/Carbon Nanocomposites Morphology Changes Stipulated by Different Conditions.

Nanocomposite code	Reagents of synthesis	Temperature regime of synthesis (°C)	Morphology of nanocomposites	Particles sizes of nanocomposites
Cu/C1	Copper oxide + PVA	420	Metal crystal phase within carbon nanofilms	Min—5 nm Max—35 nm Individual nanoparticles near to 100 nm
Cu/C2	Copper oxide + PVA	390	Metal spherical nanoparticles within carbon nanofilms	Min—2 nm Max—40 nm Individual nanoparticles near to 70–100 nm

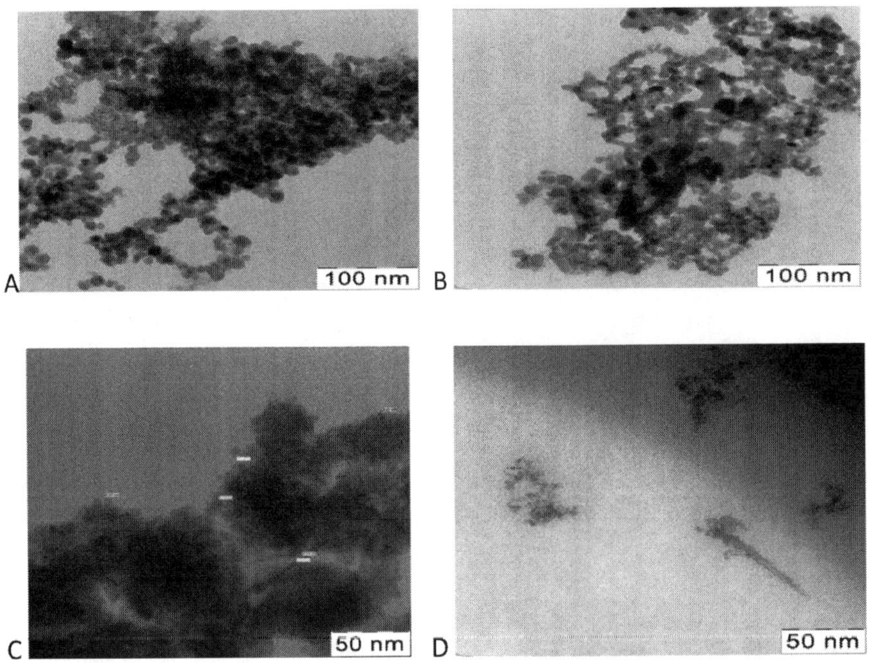

FIGURE 2.7 Microphotographs of metal/carbon nanocomposites: (A) Cu/C, (B) Ni/C, (C) Co/C, and (D) Fe/C.

If the metal/carbon nanocomposites are considered as super molecules, their surface energies analogously energy of usual molecules consists portions of energy which correspond to progressive, rotation, and vibration motions and also electronic motion:

$$\varepsilon_s^{NC} = \varepsilon_{prog} + \varepsilon_{rot} + \varepsilon_{vib} + \varepsilon_{elm}, \quad (2.6)$$

where ε_s^{NC} is the surface energy of nanocomposite; ε_{prog} is the nanocomposite surface energy portion which corresponds to the progressive motion; ε_{rot} is the nanocomposite surface energy portion which corresponds to the rotation motion; ε^{vib} is the nanocomposite surface energy portion which corresponds to the vibration motion, and ε^{elm} is the nanocomposite surface energy portion which corresponds to electron motion for the interactions of nanocomposites with surroundings molecules.

When the nanocomposite progressive motion portion increases, the nanoparticles diffusion processes importance grows in media that leads to their coagulation with the decreasing of the surface energy aggregate obtained. The coagulation of nanoparticles increases at the increasing of their quantities in medium which has little viscosity.

From the metal/carbon nanocomposites Raman and IR spectra analysis, it follows that the skeleton vibration of them on the vibrations frequencies corresponds to ultrasonic vibrations. The nanocomposite vibrations energy values are determined by the corresponding nanoparticles sizes and masses. The mass of nanoparticle (metal/carbon nanocomposite) depends on their content and types of metal containing phase clusters.

Usually metal/carbon nanocomposites have the great dipole moment. Therefore, it is possible the proposition that nanocomposite is vibrator which radiates electromagnetic waves. The nanocomposite vibration emission in medium is determined by their dielectric characteristics and the corresponding functional groups presence in medium.

At the metal/carbon nanocomposites obtaining the interaction of polymeric matrix with metal containing phase leads to formation of metal clusters covered by carbon shells accompanied by metal electron structure changes. In some cases, the medium characteristics influence on nanocomposites throw into the increasing of nanocomposite surface energy portion, which concerns with changes of their electron structure and equally with electron structure of medium.

In these cases, the growth of metal atomic magnetic moment is observed (Table 2.5) that corresponds to the unpaired electron number increasing. The considerable changes of metal atomic magnetic moments in nanocomposites proceed when the phosphorus atoms include to carbon shells of nanocomposites.

TABLE 2.5 The Comparison of Atomic Magnetic Moments of Some 3d Metals in Metal/Carbon Nanocomposites and their Functionalized Analogs.[6,13]

Metal in nanocomposite	Cu	Ni	Co	Fe
Metal atomic magnetic moment for nanocomposite	0.6	1.6	1.7	2.3
Metal atomic magnetic moment for phosphorus containing nanocomposite	2.0	3.0	2.5	2.5

The appearance or increasing of paramagnetic properties is linked with the possibility of chemical bonds formation, and also the growth of metal/carbon nanocomposites reactivity into polar media.

Vibration portion of surface energy of metal/carbon nanocomposites can be presented as

$$\varepsilon_{vib} = \frac{mu_{vib}^2}{2} \tag{2.7}$$

where m is the mass of metal/carbon nanocomposite, which includes summary mass of carbon or carbon, polymeric shell and cluster of metal containing phase, u_{vib} is the vibration velocity correspondent to product of vibration amplitude on vibration frequency.

Usually the fine dispersed suspensions or sols of nanostructures are used for the modification of polymeric compositions (matrixes) to distribute nanoparticles into polymeric matrix. The liquids, which are used for polymeric materials production, are represented as dispersion medium. For the obtaining of fine dispersed suspensions small and super small quantities of metal/carbon nanocomposites are introduced into dispersed media at the mechanical and further the ultrasonic mixing.

The nanocomposite vibration amplitude is the product of wave number on velocity of light. Therefore, Equation (1) can be written as

$$\varepsilon_{vib} = \frac{(m_{cl} + m_{sh})dvc}{2} \tag{2.8}$$

where m_{cl} is the mass of metal containing phase; m_{sh} is the mass of carbon or carbon polymeric shell; d – the average linear size of metal/carbon nanocomposite nanoparticles; v is the wave number of nanocomposite skeleton vibration; and c is the velocity of light.

Under the influence by electromagnetic radiation arising at the ultrasonic vibration of nanocomposite, the self-organization of polymeric compositions macromolecules is possible. According to Equation (2), the nanocomposite vibration energy is determined by the average mass and size of nanocomposite (Table 2.6).

TABLE 2.6 Energetic Characteristics of Metal/Carbon Nanocomposites.[11–13]

Metal in metal/carbon nanocomposites	Cu	Ni	Co	Fe
Summary mass of metal/carbon nanocomposite (au)	36.75	40.01	42.50	42.69
Specific surface of metal/carbon nanocomposite, S (m^2/g)	160	251	209	163
Skeleton vibration frequency of metal/carbon nanocomposite, ν_{vib} (s^{-1})	4×10^{11}	4.2×10^{11}	4.1×10^{11}	4×10^{11}
Average vibration velocity of metal/carbon nanocomposites, u_{vib} (cm/s)	8.5×10^5	4.6×10^5	6.1×10^5	6.9×10^5
Average vibration energy of metal/carbon nanocomposites, ε_{vib} (erg)	1.6×10^{13}	4.3×10^{12}	7.6×10^{12}	9.9×10^{12}

The vibration motion portion energy essentially determines by the vibration amplitude because the nanocomposite skeleton vibration frequencies practically equal (the proximity of metal nature).

If the vibration amplitudes and the nanocomposite sizes are correlated, the nanocomposites sizes show the great influence on the vibration energy and also on the vibration velocity. Therefore, according to the increasing of vibration energy as well as the increasing of metal/carbon nanocomposites linear sizes, the correspondent nanocomposites (M/C nc) are displaced in the following order:

$$\varepsilon_{vib}\,(\text{Cu/C nc}) > \varepsilon_{vib}\,(\text{Fe/C nc}) > \varepsilon_{vib}\,(\text{Co/C nc}) > \varepsilon_{vib}\,(\text{Ni/C nc})$$

This disposition of nanocomposites on the vibration energy values is possible, when the vibration motion portion is greatly bigger on the comparison with other portions of surface energy. However, the common surface energy of nanocomposites obtained differs from previous row and corresponds to values of specific surface

$$S_{sp}\,(\text{Ni/C nc}) > S_{sp}\,(\text{Co/C nc}) > S_{sp}\,(\text{Fe/C nc}) > S_{sp}\,(\text{Cu/C nc})$$

Certainly, in dependence on methods and conditions of the metal/carbon nanocomposites synthesis the changes of forms and contents for nanocomposites are possible. At the same time, the fundamentals of the metal/carbon nanocomposite activity open the new perspectives for their interaction prognosis in the different media.

The nanocomposites described above were investigated with the help of IR spectroscopy by the technique indicated above. In this paper the IR spectra of Cu/C and Ni/C nanocomposites are discussed (Figs. 2.8 and 2.9), which find a wider application as the material modifiers.

On IR spectra of two nanocomposites the common regions of IR radiation absorption are registered. Further, the bands appearing in the spectra and having the largest relative area were evaluated. We can see the difference in the intensity and number of absorption bands in the range 1300–1460 cm^{-1}, which confirms the different structures of composites.

In the range 600–800 cm^{-1}, the bands with a very weak intensity are seen, which can be referred to the oscillations of double bonds (π-electrons) coordinated with metals.

In the range 600–800 cm^{-1}, the bands with a very weak intensity are seen, which can be referred to the oscillations of double bonds (π-electrons) coordinated with metals.

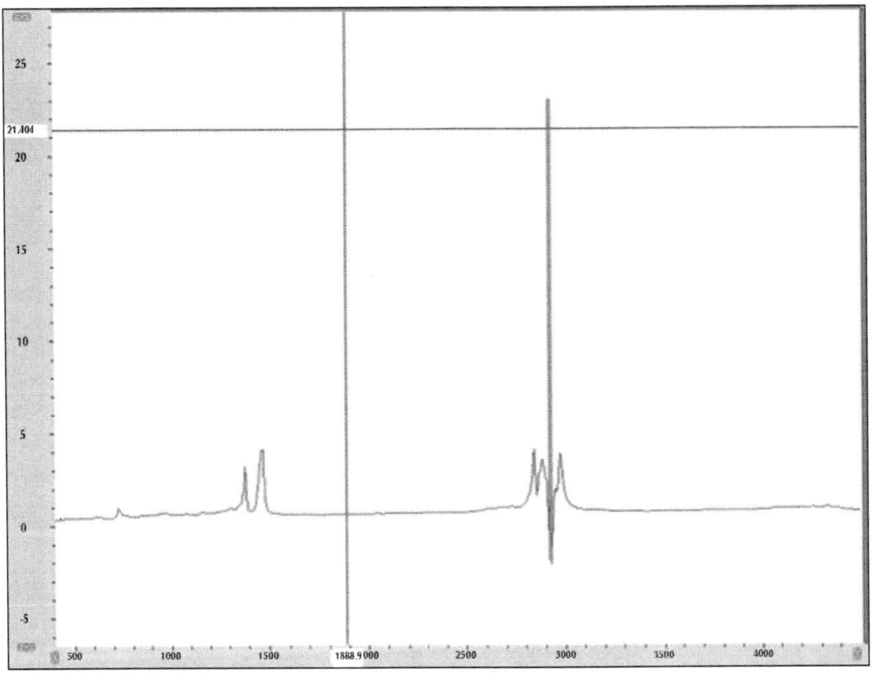

FIGURE 2.8 IR spectra of copper/carbon nanocomposite powder.

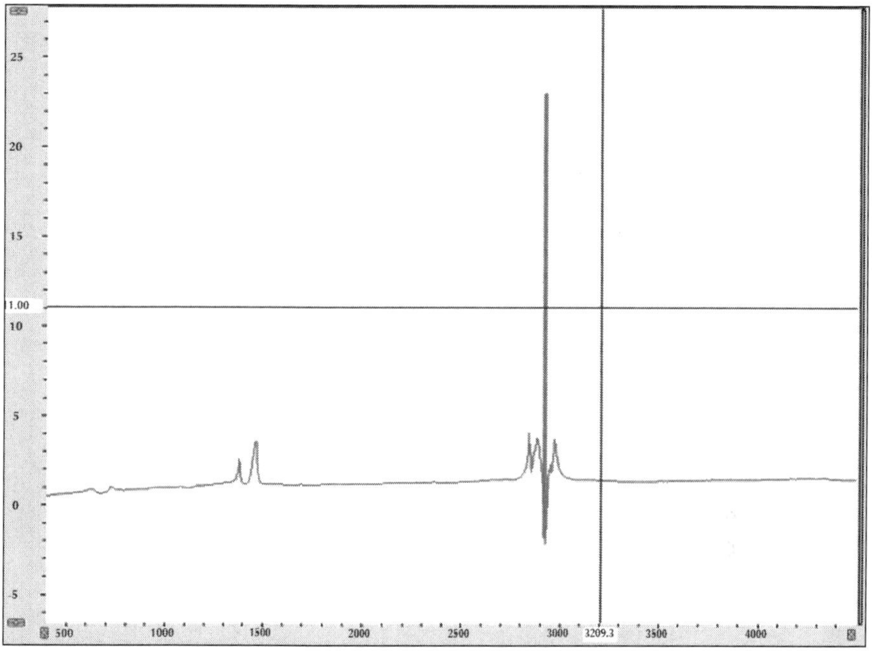

FIGURE 2.9 IR spectra of nickel/carbon nanocomposite powder.

In case of Cu/C nanocomposite a weak absorption is found at 720 cm^{-1}, and in case of Ni/C nanocomposite, the absorption at 620 cm^{-1} is also observed.

In IR spectrum of copper/carbon nanocomposite two bands with a high relative area are found:

at 1323 cm^{-1} (relative area—9.28)

at 1406 cm^{-1} (relative area—25.18).

These bands can be referred to skeleton oscillations of polyarylene rings.

In IR spectrum of nickel/carbon nanocomposite the band mostly appears at 1406 cm^{-1} (relative area—14.47).

The investigations of carbon films in metal/carbon nanocomposites and their peculiarities are carried out by Raman spectroscopy with the using of Laser Spectrometer Horiba LabRam HR 800. Below the Raman spectra and PEM microphotograph of copper/carbon (Cu/C) nanocomposite are represented (Fig. 2.10).

Wave numbers and intensity relation testify to presence of nanoparticles containing the Copper atoms coordinated. At the same time, the comparison of IR and Raman spectra shows their closeness on wave numbers and intensities relation.

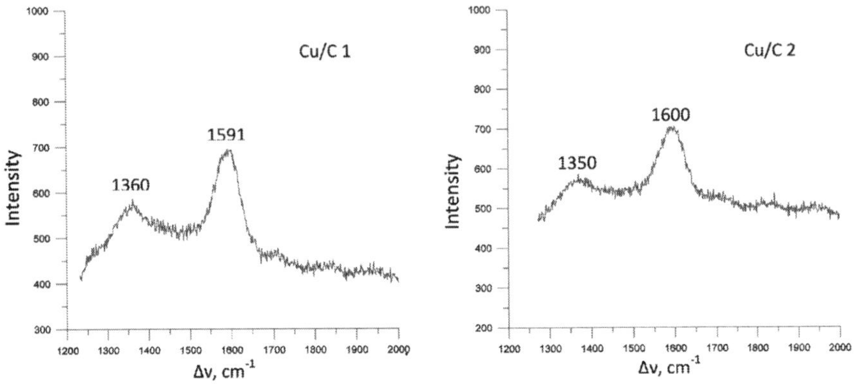

FIGURE 2.10 Raman spectra of Cu/C nanocomposites: Cu/C 1 (28 nm) and Cu/C 2 (25 nm).

According to the investigations with transmission electron microscopy, the formation of carbon nanofilm structures consisting of carbon threads is characteristic for copper/carbon nanocomposite. In contrast, carbon fiber structures, including nanotubes, are formed in nickel/carbon nanocomposite. There are several absorption bands in the range 2800–3050 cm^{-1}, which are attributed to valence oscillations of C–H bonds in aromatic and aliphatic compounds. These absorption bonds are connected with the presence of vaseline oil in the sample. It is difficult to find the presence of metal in the composite as the metal is stabilized in carbon nanostructure. At the same time, it should be pointed out that apparently nanocomposites influence the structure of vaseline oil in different ways. The intensities and number of bands for Cu/C and Ni/C nanocomposites are different:

1. For copper/carbon nanocomposite in the indicated range—5 bands, and total intensity corresponds by the relative area—64.63.
2. For nickel/carbon nanocomposite in the same range—4 bands with total intensity (relative area)—85.6.

The distribution of nanoparticles in water, alcohol and water–alcohol suspensions prepared based on the above technique are determined with the help of laser analyzer. In Figs. 2.10 and 2.11, we can see distributions of copper/carbon nanocomposite in the media different polarity and dielectric penetration.

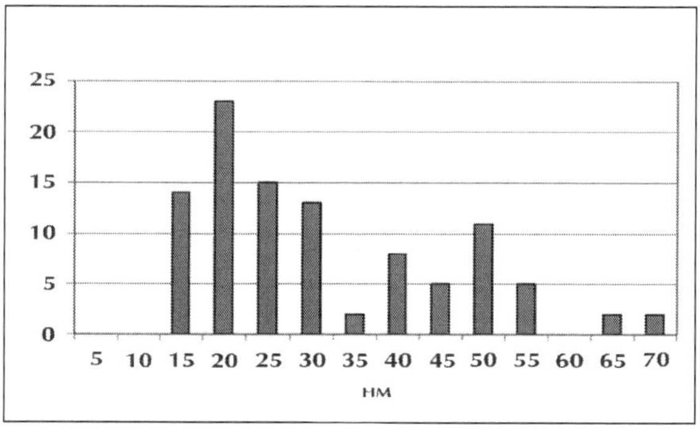

FIGURE 2.11 Distribution of copper/carbon nanocomposites in alcohol.

When comparing the figures, we can see that ultrasound dispergation of one and the same nanocomposite in media different by polarity results in the changes of distribution of its particles. In water solution, the average size of Cu/C nanocomposite equals 20 nm, and in alcohol medium—greater by 5 nm.

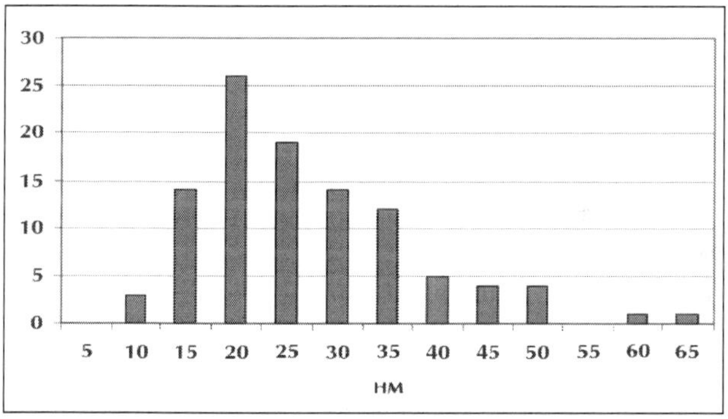

FIGURE 2.12 Distribution of copper/carbon nanocomposites in water.

Assuming that the nanocomposites obtained can be considered as oscillators transferring their oscillations onto the medium molecules, we can determine to what extent the IR spectrum of liquid medium will change, for example, PEPA applied as a hardener in some polymeric compositions, when we introduce small and super small quantities of nanocomposite into it.

The introduction of a modifier based on metal/carbon nanocomposites into the composition results in medium structuring, decrease in the number of defects, thus improving the material physical and mechanical characteristics.[5,6] The availability of metal compounds in nanocomposites can provide the final material with additional characteristics, such as magnetic susceptibility and electric conductivity. Data of EPR spectra investigations for copper/carbon and nickel/carbon nanocomposites are given below.

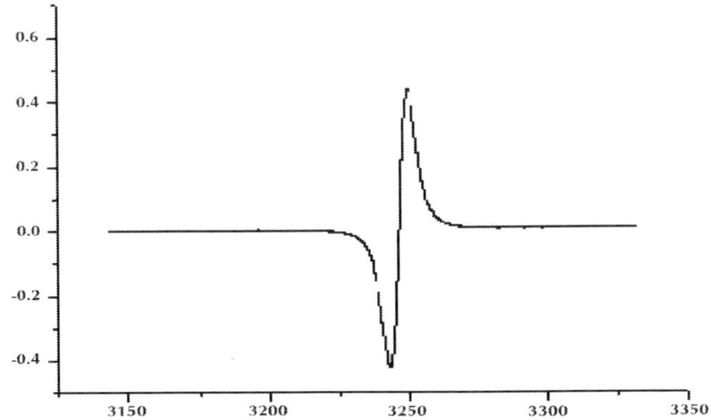

FIGURE 2.13 EPR spectrum of copper/carbon nanocomposite.

EPR spectrum of copper/carbon nanocomposite (Fig. 2.13) is presented as a singlet spectrum in which the distance between points of incline maximum ΔH_{pp} = 6.8 Hz, *g*-factor equals to *g*-factor of diphenyl picril hydrazyl (*g* = 2.0036), and the unpaired electrons number corresponds to value 1.2×10^{17} spin/*g* (Fig. 2.12). In the comparison with this spectrum, the EPR spectrum for nickel/carbon nanocomposite has ΔH_{pp} corresponding to 2400 Hz, *g*—2.46 and the unpaired electrons number—10^{22} spin/*g* (Table 2.7).

TABLE 2.7 Data of EPR Spectra for Copper/Carbon and Nickel/Carbon Nanocomposites.

Type of metal/carbon nanocomposite	*g*-Factor	Number of unpaired electrons (spin/*g*)
Copper/Carbon nanocomposite	2.0036	1.2×10^{17} spin/*g*
Nickel/Carbon nanocomposite	2.46	10^{22} spin/*g*

It is possible; the spectra difference explains the carbon shape difference for these nanocomposites.

Thus, the metal/carbon nanocomposites are stable radicals with the migration of unpaired electrons from metal to carbon shell and back. These nanoparticles can stimulate the transport of electrons within media.

2.4 CONCLUSION

In this paper, the possibilities of developing new ideas on base of Mesoscopic Physics principles for self-organization processes during Redox synthesis within nanoreactors of polymeric matrices are discussed on the examples of metal/carbon nanocomposites. It is proposed to consider the obtaining of metal/carbon nanocomposites in nanoreactors of matrices as the self-organization process similar to the formation ordered phases analogous to crystallization. The perspectives of investigations are looked through in an opportunity of thin regulation of processes and the entering of corrective amendments during processes stages.

According to the analysis of metal/carbon nanocomposites characteristics, which are determined by their sizes and content, their activities are stipulated the correspondent dipole moments and vibration energies.

The introduction of phosphorus in nanocomposites by the method similar to redox synthesis in nanoreactors leads to the growth of dipole moments and metal magnetic moments for corresponded nanocomposites. It's shown that the nanocomposites vibration energies depend on their average masses.

It's noted that the specific surface of metal/carbon nanocomposites particles changes in dependence on the nature of nanocomposite in other order than the correspondent order of the vibration energies. Therefore, the energetic characteristics of nanocomposites are more important for the activity determination in comparison with their size characteristics.

KEYWORDS

- **Kolmogorov–Avrami equations**
- **nanoreactors**
- **metal–carbon**
- **nanocomposite**

REFERENCES

1. IMRI. *Introduction in Mesoscopic Physics*; M.: Physmatlit, 2009; 304 p.
2. Moskalets, M. V. *Fundamentals of Mesoscopic Physics*. Khar'kov: NTU KhPI, 2010.

3. Morozov, A. D. *Introduction in Fractal Theory*; ICT: M.-Izhevsk, 2002; 160p.
4. Kolmogorov, A. N.; Fomin, S. V. *Introductory Real Analysis*; Prentice Hall: Portland, 2009; p 403.
5. Wunderlikh, B. *Physics of Macromolecules*. Mir: M., 1979; Vol 2, 422 p.
6. Brown, J. F., Jr.; White, D. M. *J. Am. Chem. Soc.* **1960,** *82*, 5671.
7. Morgan, P. W. Condensation Polymers by Interfacial and Solution Methods (translated) L.; *Chemistry* **1970,** 448.
8. Buchachenko, A. L.; Nanochemistry—Direct way to High Technologies. *Uspechi Chimii* **2003,** *72*(5), 419–437.
9. Kodolov, V. I.; Khokhriakov, N. V. *Chemical Physics of Formation and Transformation Processes of Nanostructures and Nanosystems*; publ. IzhSACA: Izhevsk, 2009; Vol 1, 361; Vol 2, 415p.
10. Shabanova, I. N.; Kodolov, V. I.; Terebova, N. S.; Trineeva, V. V. *X-ray Photoelectron Spectroscopy Investigations of Metal/Carbon Nanosystems and Nanostructured Materials*; Publ. "Udmurt University": M., Izhevsk, 2012; 252 p.
11. Kodolov, V. I.; Khokhriakov, N. V.; Trineeva, V. V.; Blagodatskikh, I. I Nanostructure Activity and its Display in Nanoreactors of Polymeric Matrices and in Active Media. *Chem. Phys. Mesosc.* **2008,** *10*(4), 448–460.
12. Kodolov, V. I. The Addition to Previous Paper. *Chem. Phys. Mesosc.* **2009,** *11*(1), 134–136.
13. Trineeva, V. V.; Vakhrushina, M. A.; Bulatov, D. I.; Kodolov, V. I. The Obtaining of Metal/Carbon Nanocomposites and Investigation of their Structure Phenomena. *Nanotechnics* **2012,** *4*, 50–55.
14. Trineeva, V. V.; Lyakkhovich, A. M.; Kodolov, V. I. Forecasting of the Formation Processes of Carbon Metal Containing Nanostructures using the Method of Atomic Force Microscopy. *Nanotechnics* **2009,** *4*(20), 87–90.
15. Kodolov, V. I.; Blagodatskikh, I. I.; Lyakhovich, A. M.; et al. Investigation of the Formation Processes of Metal Containing Carbon Nanostructures in Nanoreactors of Polyvinyl Alcohol at Early Stages. *Chem. Phys. Mesosc.* **2007,** *9*(4), 422–429.
16. Kodolov, V. I.; Khokhriakov, N. V.; Kuznetsov, A. P. To the Issue of the Mechanism of the Influence of Nanostructures on Structurally Changing Media at the Formation of "intellectual" Composites. *Nanotechnics* **2006,** *3*(7), 27–35.
17. Kodolov, V. I.; Khokhriakov, N. V.; Trineeva, V. V.; Blagodatskikh, I. I. Problems of Nanostructure Activity Estimation, Nanostructures Directed Production and Application. In *Nanomaterials Yearbook—2009. From Nanostructures, Nanomaterials and Nanotechnologies to Nanoindustry*; Nova Science Publishers, Inc.: New York, 2010; pp 1–18.
18. Fedorov, V. B.; Khakimova, D. K.; Shipkov, N. N.; Avdeenko, M. A. To Thermodynamics of Carbon Materials. *Doklady AS USSR* **1974,** *219*(3), 596–599.
19. Fedorov, V. B.; Khakimova, D. K.; Shorshorov, M. H.; et al. To Kinetics of Graphitation. *Doklady AS USSR* **1975,** *222*(2), 399–402.
20. Khokhriakov, N. V.; Kodolov, V. I. Quantum-chemical Modeling of Nanostructure Formation. *Nanotechnics* **2005,** *2*, 108–112.
21. Lipanov, A. M.; Kodolov, V. I.; Khokhriakov, N. V.; et al, Challenges in Creating Nanoreactors for the Synthesis of Metal Nanoparticles in Carbon Shells. *Alternat. Energy Ecol.* **2005,** *2*(22), 58–63.
22. Kodolov, V. I.; Didik, A. A.; Volkov, A. Yu.; Volkova, E. G. Low-temperature Synthesis of Copper Nanoparticles in Carbon Shells. HEIs' news. *Chem. Chem. Eng.* **2004,** *47*(1), 27–30.

23. Serkov, A. T., Ed. *Theory of Chemical Fiber Formation*; M.: Himiya, 1975; 548 p.
24. Palm, V. A. *Basics of Quantitative Theory of Organic Reactions*; L: Himiya, 1967; 356 p.
25. Kodolov, V. I. *On Modeling Possibility in Organic Chemistry. Organic Reactivity*. Tartu: TSU, 1965; Vol 2, 4; pp 11–18.
26. Chashkin, M. A.; Kodolov, V. I.; Zakharov, A. I.; et al. Metal/Carbon Nanocomposites for Epoxy Compositions: Quantum-chemical Investigations and Experimental Modeling. *Polym. Res. J.* **2011,** *5*(1), 5–19.
27. Kodolov, V. I.; Trineeva, V. V. Fundamental Definitions for Domain of Nanostructures and Metal/Carbon Nanocomposites. In *Nanostructure, Nanosystems and Nanostructured Materials: Theory, Production and Development*; Apple Academic Press: Toronto–New Jersey, 2013; pp 1–42.
28. Kodolov, V. I.; Trineeva, V. V.; Blagodatskikh, I. I.; Vasil'chenko, Yu. M.; Vakhrushina, M. A.; Bondar, A. Yu. The Nanostructures Obtaining and the Synthesis of Metal/carbon Nanocomposites in Nanoreactors. In *Nanostructure, Nanosystems and Nanostructured Materials: Theory, Production and Development*; Apple Academic Press: Toronto–New Jersey, 2013; pp 101–145.
29. Shabanova, I. N.; Terebova, N. S. Dependence of the Value of the Atomic Magnetic Moment of d Metals on the Chemical Structure of Nanoforms. In *The problems of Nanochemistry for the Creation of New Materials*; IEPMD: Torun, Poland, 2012; pp 123–131.
30. Kodolov, V. I.; Trineeva, V. V. Perspectives of Idea Development about Nanosystems Self-organization in Polymeric Matrixes. In *The Problems of Nanochemistry for the Creation of New Materials*; IEPMD: Torun, Poland, 2012; pp 75–100.

CHAPTER 3

GROWTH MODES IMPACT ON THE HETEROSTRUCTURE'S CaF_2/Si(1 0 0) ELECTRO-PHYSICAL PROPERTIES

A. VELICHKO, V. GAVRILENKO, V. ILYUSHIN, A. KRUPIN, N. FILIMONOVA, and M. CHERNOVA*

Novosibirsk State Technical University, Faculty of Radio Engineering and Electronics, 20 K. Marksa st., Novosibirsk, Russia

**Corresponding author. E-mail: vkodol.av@mail.ru*

CONTENTS

ABSTRACT

The analysis of dielectric parameters of the calcium fluoride films grown on Si(1 0 0) via solid-phase epitaxy (group 2) and without it (group 1) is carried out. The CaF_2/Si(1 0 0) structures were grown in two-chamber molecular-beam epitaxy installation "Katun'-100." Experiments showed that the deposition mode of CaF_2, directly after the growth of buffer Si layer at substrate temperature (T_s) 530°C, is not suitable to obtain dielectric layers with sufficient quality. Reflected high-energy electron diffraction (RHEED) of the first-group films are electron diffraction patterns of the mixed type. It is alignment of point reflection's grid with Debye half-rings that indicate polycrystalline and mosaic spots on the surface. Such diffraction patterns are reproduced by rotating the substrate to 90°—fold angle; therefore, crystallographic orientation of the calcium fluoride film is the same as orientation of substrate (1 0 0).

Use of solid-phase epitaxy (SPE) method at the nucleation stage of the CaF_2 layers allows to obtain single-crystal uniform-thickness films with high dielectric characteristics. At all stages of CaF_2 films growth after SPE, the diffraction patterns are presented strips, which means the smoother surface of the second*group samples compared with the first group. Unfortunately, strips are paired and remain during the whole film growth. This can be interpreted as a superposition of two diffraction patterns from areas that are deployed with respect to each other by 90°. Such RHEED images are repeated after rotation to 30°-fold angle. This means that by using SPE, the CaF_2 films grown on Si(1 0 0) have an orientation (1 1 1).

3.1 INTRODUCTION

One of the ways to create the "silicon-on-insulator" structure is to use the heteroepitaxial system Si/CaF_2 where the calcium fluoride layer is used as dielectric. It has face-centered cubic lattice and its lattice constant (a_{CaF_2}=0.546nm)almost coincide with silicon lattice constant (a = 0.543 nm). Good agreement of the structural properties and parameters of the calcium fluoride and silicon crystal lattice make possible the epitaxial growth of one material layer on the surface of another material.

Orientation [1 1 1] is preferable for CaF_2 growth orientation (as for all alkali-earth metals) because the surface energy for CaF_2(1 1 1) is two times smaller than for CaF_2(1 0 0). Consequently, during the epitaxy of the CaF_2

on the Si(1 1 1) surface, two-dimensional growth mechanism in a wide temperature range is observed.

But in the integrated-circuits production, the silicon substrates with orientation (1 0 0) are used. This is due to the fact that surface-state density in structures with substrate orientation (1 0 0) is about an order of magnitude smaller than in structures with orientation (1 1 1).[1]

However, because of large value of the free surface energy CaF_2(1 0 0) that is determined by dipole moment on the interface CaF_2/Si(1 0 0), the growth on the Si(1 0 0) substrates is carried out by a three-dimensional mechanism with formation the faceted surface. It has the form of growth pyramid with height about 20–50 nm.[2]

Such growth leads to heterogeneity in thickness of the film that in turns leads to degradation of its dielectric values and makes it more difficult to obtain next layers of the Si on its surface. Thus, it is very important to develop the technology of growth CaF_2 layers on the Si(1 0 0) surface with atomically smooth surface.

Model representations of the growth processes show that decreasing the substrate temperature lead to slowing of the surface migration processes and, as result, to increasing of the nucleation centers number. This suggests that the most homogeneous films are to be obtained by deposition at low temperatures. However, the films are amorphous and polycrystalline. If the amorphous film is formed on a monocrystalline substrate and its thickness is not too large, then it can be crystallized under the influence of the substrate-orienting effect at high temperature. This process is called solid-phase epitaxy (SPE).

In the works [3–6], we already had investigated the growth processes of the CaF_2 on Si(1 0 0). Therefore, the aim of this work is to analyze the dielectric parameters of the calcium fluoride films grown on Si(1 0 0) substrate by using with SPE technique and without it.

3.2 EXPERIMENTAL TECHNIQUE

The CaF_2/Si(1 0 0) structures have been grown at the double-chamber molecular-beam epitaxy system "Katun-100". For the growth had been used the substrates with following parameters: orientation is (1 0 0), diameter is 100 mm, an average microroughness height is 1.5–2 nm. After standard chemical processing the substrates with formed on clean silicon surface the passivation oxide has been placed at the growth unit of the epitaxial system. Protective oxide (field oxide) has been removed by annealing at the substrate temperature 800°C till clear diffraction pattern Si(1 0 0)-(2 × 1) would appear. Then Si buffer layer with thickness ~20 nm has been grown. The control of the annealing and growth processes had been held by reflected high-energy electron diffraction (RHEED). RHEED patterns have been videotaped. All stages of the technological process for every sample were carried out at the same conditions.

The silicon substrates with buffer layer prepared by this method were used for investigation of the CaF_2 heteroepitaxy on Si. Given by CaF_2 molecular source temperature, the growth rate was same for all samples 0.27 A/s. Growth duration was 2 h and thicknesses of obtained films ranged from 160 to 170 nm. To minimize the impact of fast electrons beam on CaF_2 surface morphology, the diffraction have been observed only during first 10 min of the growth process.[7,8] Then diffractometer has been switched off and substrate rotating mechanism has been switched on. This resulted in more homogeneous thickness of the film in the entire substrate area.

Two groups of wafers have been grown in this set of experiments. The samples from the first group have been grown by described above technique at substrate temperature T_s = 530°C. An outstanding feature of the second-group samples was the first stage of the calcium fluoride growth that was carried out by SPE technique. After the buffer-layer growth process, the Si wafers were cooled down to room temperature within 3 h. Then the CaF_2 deposition with thickness of 4 nm were carried out, followed by crystallization of 5 min and at temperature T_s = 630°C. After that the SPE film's thickness was adjusted by the first-group growth mode.

Whereat on the substrates with dielectric films, the aluminum contacts 0.7 × 0.7 mm have been deposited through the mask. The deposition was carried out at the magnetron sputtering set "Oratoria-5." Then the measurements of the passing through the dielectric in Al/CaF_2/Si structure at 1-V current and of the dielectric break-down voltage were carried out. The break-down voltage measurements were obtained at the semiconductor devices characteristics measure L2–56; current measurements were carried out at the Shch 300. On every wafer, 36 current measurements and 10 break-down voltage measurements have been carried out. The break-down field was calculated from an average break-down voltage.

3.3 RESULTS AND DISCUSSION

Typical RHEED pattern from calcium fluoride surface, after 2-h growth of the first-group sample, is presented in Figure 3.1. This RHEED pattern is mixed type electron-diffraction pattern. It represents the set of the point reflection grid and Debye half-rings which show that the surface has both polycrystalline and mosaic areas.[9] Such diffraction diagram can be reproduced at the substrate rotation about a 90°-multiple angle. Consequently, in this case, the calcium fluoride films grow with an orientation of the substrate, that is, (1 0 0) orientation.

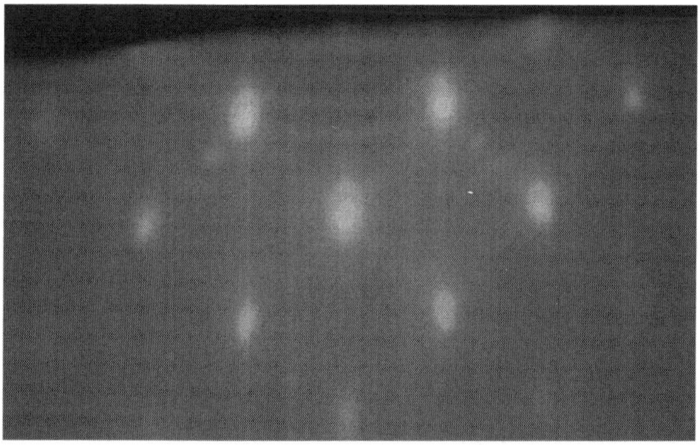

FIGURE 3.1 Typical RHEED pattern from calcium fluoride surface after 2-h growth of the first-group sample.

Statistical distribution of the microroughness on the first-group sample surface is presented in Figure 3.2.

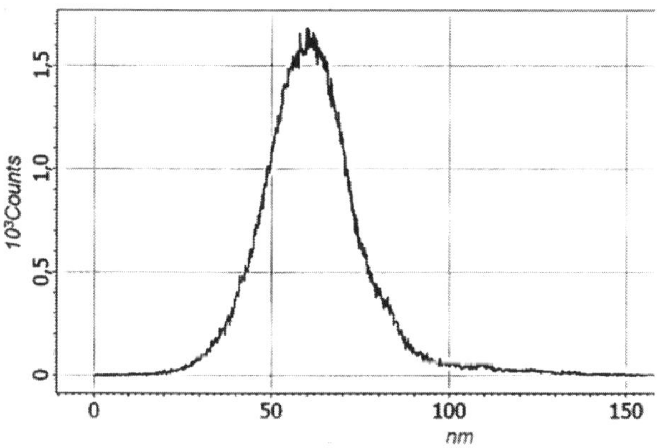

FIGURE 3.2 Statistical distribution of the microroughness on the first group film surface at the 1.5 × 1.5 scanned area.

An average microroughness on the first-group films' surface at the 1.5 × 1.5 scanned area is 60 nm.

Structures of the second group are presented by sample C3. RHEED pattern from calcium fluoride film after SPE is presented in Figure 3.3. It shows not only film's single-crystal properties but also significant improving of surface morphology. An average microroughness height of this film is 20 nm. The following growth at 530°C doesn't change the diffraction pattern significantly.

FIGURE 3.3 RHEED pattern from calcium fluoride surface of the C3 sample.

It is essential that at every stages of CaF_2 films growth after SPE the diffraction patterns presented like strands which show that surface of the second group samples are significantly smoother than the first group samples' surface. The strands are paired and remain during whole film growth process. It can be interpret as an overlaying of two diffraction patterns from areas that rotated one in relation to another at 90°.[10] Such RHEED patterns are repeats after rotation to the 30°-multiple angle. This points to the fact that using the SPE, the CaF_2 films grows on the Si(1 0 0) with orientation (1 1 1).

An average microroughness of the second-group sample surface is presented in Figure 3.4.

In our experiment, when the SPE method for calcium fluoride deposition on Si(1 0 0) is used, the surface microroughness is lowered from 60 nm to 23 nm at the scanning area 1.5×1.5 μm.

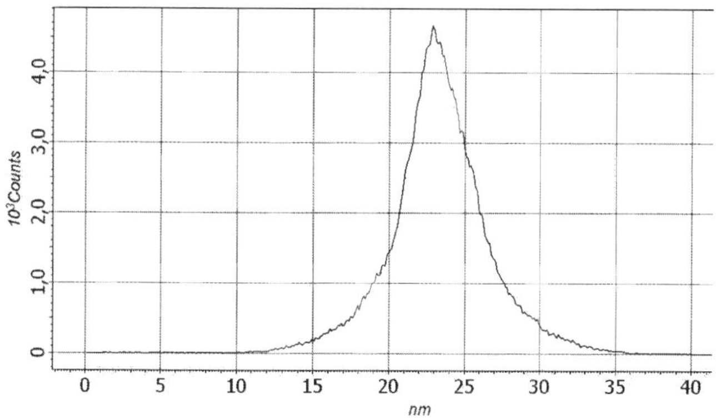

FIGURE 3.4 Statistical distribution of the microroughness on the second-group film surface at the 1.5 × 1.5 scanned area.

Table 3.1 lists growth parameters, changes of the films thickness, and electrical parameters.

TABLE 3.1 Parameters of the CaF_2/Si(1 0 0) Structures.

Wafer no.	Growth parameters	CaF_2 thickness (nm)	Breakdown field E_{cp} (V/cm)	Break-down voltage U_{cp} (V)	Leakage currents (at $U = 1$ V), I_{cp} (nA/cm²)
C1	2 h at $T_s = 530°C$	164	4.4×10^6	47	1200
C4	SPE + 2 h at T_s = 530°C	160	1.51×10^7	210	34.7

3.4 CONCLUSION

The investigation allows to make the following conclusions:

1. Calcium fluoride films, grown at $T_s = 530°C$, have rough surface morphology with microroughness height about 60 nm and have big leakage currents. Such growth mode is not optimal to obtain high-quality dielectric layers.
2. Calcium fluoride films grown in two stages (the first stage is SPE deposition of CaF_2 to increase the number of nucleation centers and the second stage is the growth at $T_s = 530°C$) with the same film thickness have a smoother surface. An average microroughness height is

about 20 nm. Leakage currents of such films for two orders of magnitude smaller than leakage currents of the first group samples.

KEYWORDS

- **molecular-beam epitaxy**
- **calcium fluoride**
- **silicon**
- **solid-phase epitaxy**
- **breakdown voltage**
- **leakage current**
- **diffraction pattern**
- **monocrystal**

REFERENCES

1. Filimonova, N. I. *Molekulyarno-luchevaya epitaksiya dielektricheskikh sloev BaF$_2$/CaF$_2$/Si(1 0 0) dlya struktur "poluprovodnik na dielektrike".* Diss. kand. tekh. nauk [Molecular-beam Epitaxy of the Dielectric Layers BaF_2/CaF_2/Si(1 0 0) for "Semiconductor on Dielectric" Structures, Ph.D. Diss.] Tomsk, 2011, pp 6–7.
2. Fathauer, R. W. Surface Morphology of Epitaxial CaF_2 films on Si Substrates. *Appl. Phys. Lett.* **1984,** *45*(5), 519–521.
3. Ilyushin, V. A.; Velichko, A. A.; Filimonova, N. I. Vliyanie temperaturnykh rezhimov rosta na morfologiyu poverkhnosti plenok CaF_2/Si(1 0 0) poluchennykh MLE [An Influence of the Growth Temperature Modes on MBE Obtained CaF_2/Si(1 0 0) Film Surface Morphology]. *Sci. Bull. NSTU* **2007,** *3*(28), 197–202.
4. Velichko, A. A.; Ilyushin, V. A.; Filimonova, N. I. Vliyanie mikronerovnostey poverkhnosti podlozhki Si(1 0 0) na morfologiyu poverkhnosti epitaksial'nykh sloev CaF_2 v nizkotemperaturnom rezhime rosta [An Influence of the Si(1 0 0) Substrate Surface Microroughness on Surface Morphology of the CaF_2 Epitaxial Layers Grown in Low Temperature Mode]. *Sci. Bull. NSTU* **2010,** *3*(28), 111–118.
5. Velichko, A. A.; Ilyushin, V. A.; Filimonova, N. I.; Katsyuba, A. V. Vliyanie mikronerovnostey poverkhnosti podlozhki Si(1 0 0) na morfologiyu epitaksial'nykh sloev CaF_2 v vysokotemperaturnom rezhime rosta [An Influence of the Si(1 0 0) Substrate Surface Microroughness on Surface Morphology of the CaF_2 Epitaxial Layers Grown in High Temperature Mode]. *Poverkhnost'* **2013,** *5*, 95–100.
6. Velichko, A. A.; Ilyushin, V. A.; Filimonova, N. I. Influence of Substrate Surface Morphology on CaF_2 on Si(1 0 0) in Low and High Temperature Modes, 12th

International Conference on Actual Problems of Electronic Instrument Engineering. NSTU: Novosibirsk, 2014, Vol 1, pp 22–28.
7. Katsyuba, A. V.; Krupin, A. Yu. [Measuring of the CaF_2 Film Morphology by the Electron Beam During the Molecular Beam Process] *3 Vserossijskaja molodezhnaja konferencija s jelementami nauchnoj shkoly "Funkcional'nye nanomaterialy i vysokochistye veshhestva"* [III International Conference and Scientific School for Young Scientists, "Functional Nanomaterials and High-purity Substances], Mendeleev UCTR: Moscow, 2012, pp 57–58 (in Russian).
8. Velichko, A. A.; Ilyushin, V. A.; Ostertak, D. I.; Peysakhovich, Yu. G.; Filimonova, N. I. Vliyanie elektronnogo puchka difraktometra bystrykh elektronov na morfologiyu poverkhnosti geterostruktur CaF_2/Si(1 0 0) [Influence of Diffractometer Electron Beam of Fast Electrons on CaF_2/Si(1 0 0) Heterostructures Surface Morphology]. *J. Surf. Invest. X-ray, Synchrotron Neutron Tech.* **2007,** *8*, 50–59.
9. Zhukova, L. A.; Gurevich, M. A. *Elektronografiya poverkhnostnykh sloev i plenok poluprovodnikovykh materialov* [*Electron Diffraction of the Surface Layers and Semiconductor Material Films*]; Metallurgiya Publ.: Moscow, 1971; pp 68–83.
10. Belenchuk, A.; Fedorov, A.; Huhtinen, H.; Kantser, V.; Laiho, R.; Shapoval, O.; Zakhvalinskii, V. Growth of (1 1 1)-Oriented PbTe Films on Si(0 0 1) Using a BaF_2 Buffer. *Thin Solid Films* **2000,** *358*, 277–282.

CHAPTER 4

SHORT COMMUNICATIONS: RESEARCH NOTE ON NANOSTRUCTURES AND NANOSYSTEMS

A. B. ERESKO[1], A.V. LOBANOV[2], A.YU.TSIVADZE[3], E. R. ANDREEVA[4], E. V. RAKSHA[1], G.A. GROMOVA[1], G. E. ZAIKOV[3], G.S. LARIONOVA[2], I.V. KLIMENKO[3], L. B. BURAVKOVA[4], M. YA. MEL'NIKOV[5], N.A. TUROVSKIJ[6], O. O. UDARTSEVA[4], V.N. GORSHENEV[3], YU. V. BERESTNEVA[1], and YU.G. GORBUNOVA[7]

[1]L. M. Litvinenko Institute of Physical Organic and Coal Chemistry, Donetsk National University, Donetsk, Ukraine

[2]Semenov Institute of Chemical Physics, Russian Academy of Sciences, Moscow, Russia

[3]Emanuel Institute of Biochemical Physics, Russian Academy of Sciences, Moscow, Russia

[4]Institute of Biomedical Problems, Russian Academy of Sciences, Moscow, Russia

[5]Lomonosov Moscow State University, Moscow, Russia

[6]Donetsk National University, Donetsk, Ukraine

[7]Frumkin Institute of Physical Chemistry and Electrochemistry, Russian Academy of Sciences, Moscow, Russia

CONTENTS

4.1 NANOSIZED INTERMOLECULAR TUNING OF ACTIVITY OF TETRAPYRROLIC PHOTOSENSITIZERS

Tetrapyrrolic photosensitizers (PSs) based on complexes of phthalocyanines and chlorins with d^0 and d^{10} elements II, III, and IV (e.g., Mg, Zn, Al, Ga, Si) have rare long-lived (0.1–2 ms) triplet excited states and absorption in the transparency region of biological tissues. Molecular design of PS macroheterocycles leads to a long-wavelength shift of their absorption band only. At the same time, PS association leading to the formation of nanosized intermolecular H- or J-aggregates significantly effects on manifestation of photosensitizing (energy transfer), photocatalytic (electron transfer) or fluorescent activity of PS. Association of PS can be controlled by including them in various supramolecular and nanosized systems.

In complexes with biocompatible carriers H-aggregates of PS exhibit selective photochemical activity in electron transfer to O_2 with the formation of ROS but the triplet–triplet energy transfer and fluorescence are impossible. Monomers and J-aggregates of PS have fluorescence and participate in the triplet–triplet energy transfer forming $^1O_2$. A competing electron transfer is suppressed in this case.

The spectral properties of PS J-aggregates on (λ_{abs}(PS) = 740–830 nm, $\tau(^T PS)$ = 0.7 ms) excellently correspond to the requirements of the agents for photodynamic therapy. The carriers quenching of triplet states of J-aggregates of phthalocyanines were found. The test of these PS for fluorescent diagnostics showed that they do not have dark cytotoxic effects (apoptosis, necrosis, post-apoptotic necrosis), do not change the granularity of lymphocytes and possess intense fluorescence.

This work was supported by RFBR (project no. 15-03-03591).

4.2 REDOX PROPERTIES OF DOUBLE-DECKER PHTHALOCYANINES FOR NANOSENSOR DEVELOPMENT

Double-decker phthalocyanine (Pc) lanthanide complexes have a high stability, intense absorption, and can be used in various technological and biomedical applications. Therefore, it is necessary to study the nature of the bands in the electronic absorption spectra, as well as the possibility of changing the spectral properties depending on the composition of multicomponent systems.

In this report, a correlation between the ionic radius of the metal in double-decker phthalocyanines and positions of Q-band maxima in the electronic absorption spectra was determined in dimethylformamide and chloroform. The increasing of ionic radius from holmium to lutetium resulted in a regular change in the position of the maxima of Q-bands in the absorption spectra. In DMF Pc, complexes exist in the negative (blue) form $[Pc^{2-}–M^{III}–Pc^{2-}]^{-}$. Complexes in solutions of $CHCl_3$ unlike solutions of DMF exist in the of neutral (green) form $[Pc^{2-}–M^{III}–Pc^{-}]^{0}$. This form is characterized by a band at $\lambda \sim 470$ nm, which corresponds to the transition to e^{-} on π-orbital with the HOMO level.

In the case of lutetium and ytterbium diphthalocyanine, partial oxidation in albumin solutions results in the appearance of the band at 660 nm. The intensity of this band increases proportional to the concentration of albumin. Such specific behavior of double-decker phthalocyanine complexes allows to consider them as promising materials for the development of sensor devices to determine albumin. Highly sensitive detection of albumin is of interest for the diagnosis of inflammatory processes, diabetes, and liver diseases.

Silica nanoparticles can be toxic to humans. The reduction process of the neutral form of phthalocyanine is observed in the presence of silica. This process can also be considered as a response to the silica nanoparticles. The observed spectral changes occur on the mechanisms of redox transitions between two forms (blue and green). The studied system can be regarded as a prototype sensor to albumin and silica nanoparticles to create sensors.

This work was supported by RFBR (project no. 15-03-03591).

4.3 PHOTOPHYSICAL PROPERTIES OF METAL PHTHALOCYANINE NANOAGGREGATES FOR EARLY DIAGNOSIS

Double-decker phthalocyanine (Pc) lanthanide complexes have a high stability, intense absorption, and can be used in various technological and biomedical applications. Therefore, it is necessary to study the nature of the bands in the electronic absorption spectra, as well as the possibility of changing the spectral properties depending on the composition of multicomponent systems.

In this report, a correlation between the ionic radius of the metal in double-decker phthalocyanines and positions of Q-band maxima in the electronic absorption spectra was determined in dimethylformamide and chloroform. The increasing of ionic radius from holmium to lutetium resulted in a regular change in the position of the maxima of Q-bands in the absorption spectra. In DMF Pc, complexes exist in the negative (blue) form $[Pc^{2-}–M^{III}–Pc^{2-}]^{-}$. Complexes in solutions of $CHCl_3$ unlike solutions of DMF exist in the of neutral (green) form $[Pc^{2-}–M^{III}–Pc^{-}]^0$. This form is characterized by a band at $\lambda \sim 470$ nm, which corresponds to the transition to e^- on π-orbital with the HOMO level.

In the case of lutetium and ytterbium diphthalocyanine, partial oxidation in albumin solutions results in the appearance of the band at 660 nm. The intensity of this band increases proportional to the concentration of albumin. Such specific behavior of double-decker phthalocyanine complexes allows to consider them as promising materials for the development of sensor devices to determine albumin. Highly sensitive detection of albumin is of interest for the diagnosis of inflammatory processes, diabetes, and liver diseases.

Silica nanoparticles can be toxic to humans. The reduction process of the neutral form of phthalocyanine is observed in the presence of silica. This process can also be considered as a response to the silica nanoparticles. The observed spectral changes occur on the mechanisms of redox transitions between two forms (blue and green). The studied system can be regarded as a prototype sensor to albumin and silica nanoparticles to create sensors.

This work was supported by RFBR (project no. 15-03-03591).

4.4 PHOTODYNAMIC THERAPY EFFICACY USING AL-, MG-, AND ZN-PHTHALOCYANINE NANOPARTICLES

Photodynamic treatment (PDT) represents a noninvasive method of target-cell elimination that includes using of nontoxic dyes or photosensitizers in combination with harmless visible light. Originally developed as a tumor therapy, now this approach may become a useful tool for treatment of different diseases like cardiovascular or infectious diseases. In present time, phthalocyanines (Pc) and Pc nanoparticles are proposed to be the most promising dyes (photosensitizers) owing to their strong phototoxicity effect, easy availability, and low cost. The present study compares PDT efficacy using Al-, Mg-, Zn-phthalocyanine-loaded (Al-, Mg-, Zn-Pc) nanoparticles and water-soluble Al-phthalocyanine (Photosens®) on lymphocytes, macrophages, endothelial, and mesenchymal stromal cells.

Human macrophages, human endothelial (HUVEC), and mesenchymal stromal cells (MSC) were obtained and cultured as described elsewhere. Lymphocytes (Lymph) were isolated from whole peripheral blood by Ficoll–Hystopaque gradient centrifugation. Pc-loaded nanoparticles were obtained from metal phthalocyanine complexes (Acros Organics, USA) using poly-*N*-vinylpyrrolidone (PVP), dimethylformamide and silica particle solution (60 nm) in 70% ethanol. Fabricated nanoparticles were characterized and used in 0.1–3 μg/ml concentrations. To provide PDT, Pc-nanoparticles or Photosens® were added to culture medium 24 h before irradiation. Then cells were washed carefully and illuminated with 675 nm light. Dark phototoxicity and PDT efficacy were evaluated by MTT-test. Pc-PVP nanoparticles in 0.3–3 μg/ml concentrations and Al-, Mg-Pc-silica nanoparticles didn't affect cell viability while essential dark phototoxicity of Zn-Pc-silica nanoparticles (in 100-ng/ml concentration and over) were demonstrated for HUVEC and MSC. Accumulation of Mg-Pc-PVP nanoparticles was accompanied by reduction of mitochondrial potential in Lymph. Cell susceptibility to PDT with different Pc-nanoparticles varied. However, PDT with Zn-Pc-PVP and Mg-Pc-silica was noneffective for all examined cell types, while irradiation of Al-Pc-loaded cells with 675-nm light significantly decreased cell viability.

Thus novel nanoparticles with different properties were fabricated. Zn-Pc-PVP and Mg-Pc-silica nanoparticles may become a useful tool for diagnostics of different diseases, while Al-Pc-silica nanoparticles may significantly increase PDT efficacy.

This work was supported by RFBR (project no. 15-03-03591).

4.5 COMPLEX FORMATION BETWEEN CHLOROPHYLL AND PHOTOSENSITIVE ELECTRON ACCEPTORS

Photodynamic treatment (PDT) represents a noninvasive method of target-cell elimination that includes using of nontoxic dyes or photosensitizers in combination with harmless visible light. Originally developed as a tumor therapy, now this approach may become a useful tool for treatment of different diseases like cardiovascular or infectious diseases. In present time, phthalocyanines (Pc) and Pc nanoparticles are proposed to be the most promising dyes (photosensitizers) owing to their strong phototoxicity effect, easy availability, and low cost. The present study compares PDT efficacy using Al-, Mg-, Zn-phthalocyanine-loaded (Al-, Mg-, Zn-Pc) nanoparticles and water-soluble Al-phthalocyanine (Photosens®) on lymphocytes, macrophages, endothelial, and mesenchymal stromal cells.

Human macrophages, human endothelial (HUVEC), and mesenchymal stromal cells (MSC) were obtained and cultured as described elsewhere. Lymphocytes (Lymph) were isolated from whole peripheral blood by Ficoll–Hystopaque gradient centrifugation. Pc-loaded nanoparticles were obtained from metal phthalocyanine complexes (Acros Organics, USA) using poly-*N*-vinylpyrrolidone (PVP), dimethylformamide and silica particle solution (60 nm) in 70% ethanol. Fabricated nanoparticles were characterized and used in 0.1–3 μg/ml concentrations. To provide PDT, Pc-nanoparticles or Photosens® were added to culture medium 24 h before irradiation. Then cells were washed carefully and illuminated with 675 nm light. Dark phototoxicity and PDT efficacy were evaluated by MTT-test. Pc-PVP nanoparticles in 0.3–3 μg/ml concentrations and Al-, Mg-Pc-silica nanoparticles didn't affect cell viability while essential dark phototoxicity of Zn-Pc-silica nanoparticles (in 100-ng/ml concentration and over) were demonstrated for HUVEC and MSC. Accumulation of Mg-Pc-PVP nanoparticles was accompanied by reduction of mitochondrial potential in Lymph. Cell susceptibility to PDT with different Pc-nanoparticles varied. However, PDT with Zn-Pc-PVP and Mg-Pc-silica was noneffective for all examined cell types, while irradiation of Al-Pc-loaded cells with 675-nm light significantly decreased cell viability.

Thus novel nanoparticles with different properties were fabricated. Zn-Pc-PVP and Mg-Pc-silica nanoparticles may become a useful tool for diagnostics of different diseases, while Al-Pc-silica nanoparticles may significantly increase PDT efficacy.

This work was supported by RFBR (project no. 15-03-03591).

4.6 THE FORMATION OF GRAPHITE NANOSCALE STRUCTURES FOR MATERIALS WITH NOVEL FUNCTIONAL PROPERTIES

The increased interest to nanoobjects is related with unusual physical and chemical properties due to the occurrence of quantum size effects. In accordance with the conventional division of nanomaterials into compact structures nanoelements of which are in contact with each other, and the nanodispersion with nanoscale particles are distributed in a dispersion medium, the results of the preparation of such materials are reported. Biocomposite samples of collagen fibers containing hydroxyapatite nanoparticles distributed on the surface were synthesized and their properties were investigated. Magnetic graphite suspensions were prepared by strong oxidation of vibration-milled graphite, modification of metal salts, thermal, and microwave expansion.

Modification of samples of porous materials (paper, plastic foam, porous ceramics) by chemical reagents with precipitation of the reaction products on dielectric surface allowed to obtain materials with novel functional properties. Colloidal graphite suspensions and magnetic fluids are also used to modify the surface of the dielectric materials. Electrical properties of the obtained samples of materials and the magnetic susceptibility of graphite modified by metal salts were determined. Methods of photochemical preparation of silver nanoparticles were used to modify the particles hydroxyapatite and comparative study of the antibacterial activity of phthalocyanines in combination with silver nanoparticles was carried out.

Expanded vibro-milled graphites (like graphene structure with specific surface of 1000 m^2/g by BET theory) were used for supercapacitor device with capacitance of 200 F/g. Graphite particles and carbon fabrics modified by solutions of phthalocyanines showed high specific surface and conductive properties of graphite material.

This work was supported by RFBR (project no. 15-03-03591).

4.7 EXPERIMENTAL AND GIAO-CALCULATED ^{1}H AND ^{13}C NMR SPECTRA OF 2-(PYRIDIN-2-YL)-1*H*-BENZOIMIDAZOLE

2-Pyridin-2-yl-1*H*-benzimidazole (PBI) is a very versatile multidonor ligand displaying three potential donor atoms, one sp^3- and two sp^2-hybridized N-donors and is particularly interesting in view of its photochemical and photophysical properties. This provides PBI as well as PBI derivatives using as a model compound for the determination of water content and proton transfer processes investigations in membrane fuel cells. The efficiency and selectivity of these systems will depend on the conformational properties of the pyridinyl as well as benzimidazole fragments. The aim of this work is a comprehensive study of the PBI intramolecular dynamics as well as its NMR ^{1}H and ^{13}C spectra by DFT and MP2 methods.

The potential energy pathway for internal rotation in PBI molecule was estimated by optimizing the molecular geometries with different dihedral angle between the two aromatic planes (see Fig. 4.1). The barriers to internal rotation were calculated within the framework of B3LYP/6-31G, B3LYP/6-311G(d,p), and MP2/6-31G methods. The analysis of vibrational frequency was also performed at the same level of theory, and the calculation results revealed that the PBI conformers have no imaginary frequency.

Isotropic ^{1}H and ^{13}C nuclear magnetic shielding constants of PBI molecule conformers have been calculated by the gauge-including-atomic-orbital method at the same DFT and MP2 levels. The obtained magnetic shielding tensors were converted into chemical shift (δ, ppm) by setting the absolute shielding value for TMS as standard at exactly the same calculation level.

Comparison of experimental and calculated ^{1}H and ^{13}C NMR chemical shifts was performed (see Fig. 4.2). Calculated ^{1}H and ^{13}C NMR chemical shifts for both conformers were considered. The best correspondence between calculated and experimental values was observed for the Conformer 1 of the PBI molecule.

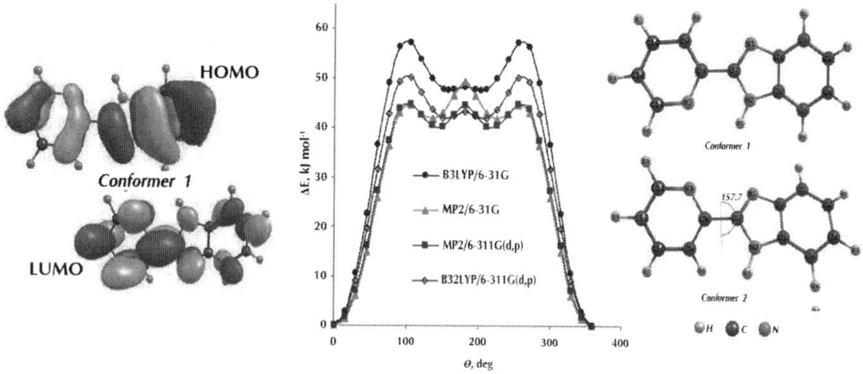

FIGURE 4.1 The potential energy pathway for internal rotation in PBI molecule.

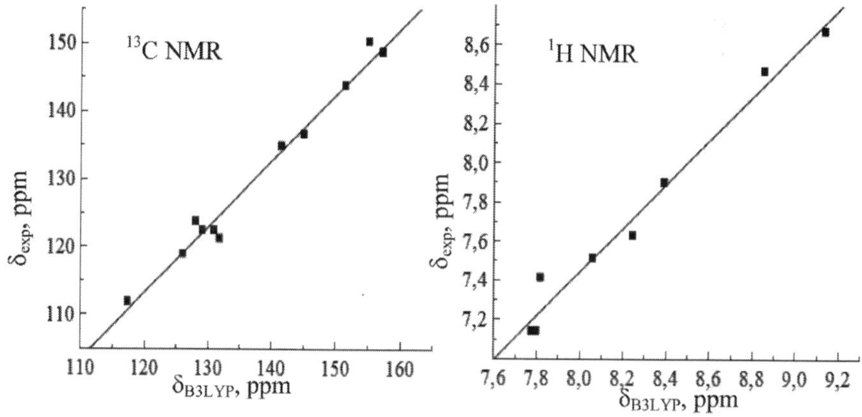

FIGURE 4.2 Comparison of experimental and calculated 1H and 13C NMR chemical shifts.

4.8 SUPRAMOLECULAR ACTIVATION OF ARYLALKYL HYDROPEROXIDES BY ET_4NBR

Arylalkyl hydroperoxides are useful starting reagents in the synthesis of surface-active peroxide initiators for the preparation of polymeric colloidal systems as well as nanocomposites with improved stability. Hydroperoxide compounds are also widely used as chemical source of the active oxygen species in the processes of surface modification and chemical functionalization of carbon nanomaterials. Chemical activation by quaternary ammonium salts is the perspective method to regulate the reactivity of commercially available organic hydroperoxides.

The present paper reports on new results of kinetic investigations of arylalkyl hydroperoxides (ROOH) catalytic decomposition in the presence of tetraethylammonium bromide (Et_4NBr), NMR spectroscopic studies of the hydroperoxides interactions with Et_4NBr as well as the results of molecular modeling of ROOH-catalyst interactions. Arylalkyl hydroperoxides under consideration are dimethylbenzylmethyl hydroperoxide, 1,1-dimethyl-3-phenylpropyl hydroperoxide, 1,1-dimethyl-3-phenylbutyl hydroperoxide, 1,1,3-trimethyl-3-(*p*-methylphenyl) butyl hydroperoxide.

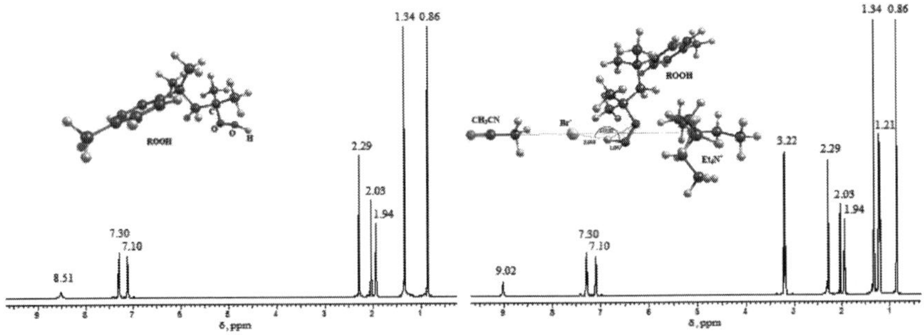

Reaction of the arylalkyl hydroperoxides supramolecular catalytic decomposition in the presence of Et_4NBr includes the stage of ion-molecular reagents recognition. Two-step model for reagent separated ion pairs formation due to ROOH and solvent separated ion pairs interaction is supposed on the base of spectroscopy, kinetics, and the molecular modeling data.[1–3]

The ROOR interaction with $Q^+LCH_3CNLX^-$ is the first stage of the process. The complex formed further reorganozation leads to the reagent separated ion pairs formation. The peroxide bond chemical activation depends on the salt anion and cation nature as well as on solvent molecule.

KEYWORDS

- **Nanosized Intermolecular Tuning**
- **Nanosensor Development**
- **Metal Phthalocyanine Nanoaggregates**
- **Zn-Phthalocyanine Nanoparticles**
- **Graphite Nanoscale Structures**
- **Supramolecular Activation**

REFERENCES

1. Turovskij, N. A.; Raksha, E. V.; Berestneva, Yu. V.; et al. In *Polymer Products and Chemical Processes. Techniques, Analysis, and Applications*; Pethrick, R. A., Pearce, E. M., Zaikov, G. E., Eds.; Apple Academic Press: Toronto, New Jersey, 2013; Vol 323, pp 269–284.
2. Turovskij, N. A.; Berestneva, Yu. V.; Raksha, E. V.; et al. *Monatsh. Chem.* **2014,** *145*(8), 1443–1448.
3. Turovskij, N. A.; Berestneva, Yu. V.; Raksha, E.V.; et al. *Polym. Res. J.* **2014,** *8*(2), 85–90.

PART II

Modeling in Sphere of Nanoscience and Nanotechnology

CHAPTER 5

HEXAGONAL GEOMETRY IN NANOSYSTEMS

G. A. KORABLEV[1], YU. G. VASILIEV[2], V. I. KODOLOV[3], and G. E. ZAIKOV[4*]

[1]*Department of Physics, Izhevsk State Agricultural Academy, Izhevsk, Russia, E-mail: korablevga@mail.ru*

[2]*Department of Physiology and Animal Sanitation, Izhevsk State Agricultural Academy, Izhevsk, Russia, E-mail: devugen@mail.ru*

[3]*Department of Chemistry and Chemical Engineering, Kalashnikov Izhevsk State Technical University, Basic Research—High Educational Center of Chemical Physics and Mesoscopy, Udmurt Scientific Center, Ural Division, RAS, Izhevsk, Russia, Tel.: +7 3412 582438, E-mail: kodol@istu.ru*

[4]*Emanuel Institute of Biochemical Physics, Russian Academy of Sciences, Moscow, Russia, E-mail: chembio@sky.chph.ras.ru*

[*]*Corresponding author. E-mail: GEZaikov@Yahoo.com*

CONTENTS

ABSTRACT

Some principles of forming carbon cluster nanosystems are analyzed based on spatial-energy ideas. The dependence nomogram of the degree of structural interactions on coefficient α is given, the latter is considered as an analog of entropic characteristic. The attempt is made to explain the specifics of forming hexagonal cell clusters in biosystems.

5.1 INTRODUCTION

Main components of organic compounds constituting 98% of cell elemental composition are carbon, oxygen, hydrogen, and nitrogen. The polypeptide bond formed by COOH and NH_2 groups of amino acid CONH acts as the binding base of cell protein biopolymers.

Thus, carbon is the main conformation center of different structural ensembles, including the formation of cluster compounds. By analyzing the mechanism of formation of carbon clusters can be understood by analogy and the geometry of hexagonal cell structures.

In the Nobel lecture in physiology, Edvard Moser[1] pointed out such analogy and presented some trial data, which, probably need to have additional theoretical confirmation. For further discussion of these problems, the idea of spatial-energy parameter (*P*-parameter) is introduced in this paper.

5.2 INITIAL CRITERIA

The idea of spatial-energy parameter (*P*-parameter) which is the complex characteristic of the most important atomic values responsible for interatomic interactions and having the direct bond with the atom electron density is introduced based on the modified Lagrangian equation for the relative motion of two interacting material points.[2]

The value of the relative difference of *P*-parameters of interacting atoms components—the structural interaction coefficient α is used as the main numerical characteristic of structural interactions in condensed media:

$$\alpha = \frac{P_1 - P_2}{(P_1 + P_2)/2} 100\% \quad (5.1)$$

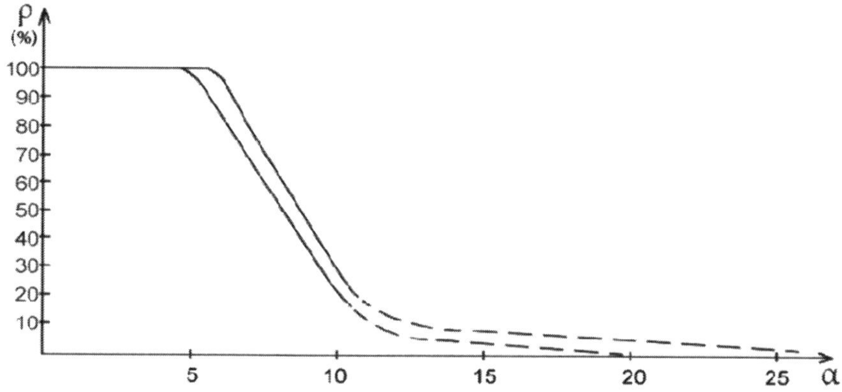

FIGURE 5.1 Nomogram of structural interaction degree dependence (ρ) on coefficient α.

Applying the reliable experimental data, we obtain the nomogram of structural interaction degree dependence (ρ) on coefficient α, the same for a wide range of structures (Fig. 5.1). This approach gives the possibility to evaluate the degree and direction of the structural interactions of phase formation, isomorphism, and solubility processes in multiple systems, including molecular ones.

Such nomogram can be demonstrated[2] as a logarithmic dependence:

$$\alpha = \beta (\ln \rho)^{-1}, \tag{5.2}$$

where coefficient β is the constant value for the given class of structures. β can structurally change mainly within ±5% from the average value. Thus coefficient α is reversely proportional to the logarithm of the degree of structural interactions and therefore, by analogy with Boltzmann equation, can be characterized as the entropy of spatial-energy interactions of atomic–molecular structures.[3]

Actually the more is ρ, the more probable is the formation of stable ordered structures (e.g., the formation of solid solutions), that is, the less is the process entropy. But also the less is coefficient α.

Equation (5.2) does not have the complete analogy with entropic Boltzmann's equation as in this case not absolute but only relative values of the corresponding characteristics of the interacting structures are compared, which can be expressed in percent. This refers not only to coefficient α but also to the comparative evaluation of structural interaction degree (ρ), for example—the percent of atom content of the given element in the solid solution relatively to the total number of atoms.

Conclusion: The relative difference of spatial-energy parameters of the interacting structures can be a quantitative characteristic of the interaction entropy: $\alpha \equiv S$.

5.3 FORMATION OF CARBON NANOSTRUCTURES

After different allotropic modifications of carbon nanostructures (fullerenes, tubules) have been discovered, a lot of papers dedicated to the investigations of such materials, for instance were published, determined by the perspectives of their vast application in different fields of material science.

The main conditions of stability of these structures formulated based on modeling the compositions of over 30 carbon clusters are given[4]:

1. Stable carbon clusters look like polyhedrons where each carbon atom is three-coordinated.
2. More stable carbopolyhedrons containing only 5- and 6-term cycles.
3. 5-Term cycles in polyhedrons—isolated.
4. Carbopolyhedron shape is similar to spherical.

Let us demonstrate some possible explanations of such experimental data based on the application of spatial-energy concepts. The approximate equality of effective energies of interacting subsystems is the main condition for the formation of stable structure in this model based on the following equation:

$$\left(\frac{P_0}{KR}\right)_1 \approx \left(\frac{P_0}{KR}\right)_2; \quad P_1 \approx P_2 \tag{5.3}$$

where K—coordination number and R—bond dimensional characteristic.

At the same time, the phase-formation stability criterion (coefficient α) is the relative difference of parameters P_1 and P_2 that is calculated following Equation (5.1) and is $\alpha ST < (20–25)\%$ (according to the nomogram).

During the interactions of similar orbitals of homogeneous atoms $P_0' = P_0''$ we have

$$K_1 R_1 \approx K_2 R_2 \tag{5.3}$$

Let us consider these initial notions as applicable to certain allotropic carbon modifications:

1. *Diamond*. Modification of structure where $K_1 = 4, K_2 = 4;\ P_0' = P_0'', R_1 = R_2, P_1 = P_2$ and $\alpha = 0$. This is absolute bond stability.
2. Non-diamond carbon modification for which $P_0' = P_0'', K_1 = 1; R_1 = 0.77$ Å; $K_2 = 4;\ R_2^{4+} = 0.2$ Å, $\alpha = 3.82\%$. Absolute stability due to ionic-covalent bond.
3. *Graphite*. $P_0' = P_0'', K_1 = K_2 = 3, R_1 = R_2, \alpha = 0$—absolute bond stability.
4. Chains of hydrocarbon atoms consisting of the series of homogeneous fragments with similar values of *P*-parameters.
5. Cyclic organic compounds as a basic variant of carbon nanostructures. Apparently, not only inner-atom hybridization of valence orbitals of carbon atom takes place in cyclic structures, but also total hybridization of all cycle atoms.

But not only the distance between the nearest similar atoms by bond length (d) is the basic dimensional characteristic, but also the distance to geometric center of cycle interacting atoms (D) as the geometric center of total electron density of all hybridized cycle atoms.

Then the basic stabilization equation for each cycle atom will take into account the average energy of hybridized cycle atoms:

$$\left(\frac{\sum P_0}{Kd}\right)_i' \approx \left(\frac{\sum P_0}{KD}\right)_i''; \quad P' \approx P'' \tag{5.4}$$

where $\Sigma P_0 = P_0 N$; N—number of homogeneous atoms, P_0—parameter of one cycle atom, K—coordination number relatively to geometric center of cycle atoms. Since in these cases $K' = K''$ and $N' = N''$, $K = N$, the following simple correlation for paired bond appears:

$$\frac{P_0'}{d} \approx \frac{P_0''}{D}; \quad P_E' \approx P_E'' \tag{5.5}$$

During the interactions of similar orbitals of homogeneous atoms $P_0' \approx P_0''$, and then:

$$d \approx D \tag{5.6}$$

Equation (5) reflects a simple regularity of stabilization of cyclic structures:

In cyclic structures, the main condition of their stability is an approximate equality of effective interaction energies of atoms along all bond directions.

The corresponding geometric comparison of cyclic structures consisting of 3, 4, 5, and 6 atoms results in the conclusion that only in 6-term cycle (hexagon) the bond length (d) equals the length to geometric center of atoms (D): $d = D$.

Such calculation of α following the equation analogous to (1), gives for hexagon $\alpha = 0$ and absolute bond stability. And for pentagon $d \approx 1.17D$ and the value of $\alpha = 16\%$, that is this is the relative stability of the structure being formed. For the other cases $\alpha > 25\%$—structures are not stable. Therefore, hexagons play the main role in nanostructure formation and pentagons are additional substructures, spatially limited with hexagons. Based on stabilization, equation hexagons can be arranged into symmetrically located conglomerates consisting of several hexahedrons.

It is assumed that defectless carbon nanotubes (NT) are formed as a result of rolling the bands of flat atomic graphite net. The graphite has a lamellar structure, each layer of which is composed of hexagonal cells. Under the center of hexagon of one layer, there is an apex of hexagon of the next layer.

The process of rolling flat carbon systems into NT is, apparently, determined by polarizing effects of cation–anion interactions resulting in statistic polarization of bonds in a molecule and shifting of electron density of orbitals in the direction of more electronegative atoms.

Thus, the aforesaid spatial-energy notions allow characterizing in general the directedness of the process of carbon nanosystem formation.[5]

5.4 HEXOGONAL STRUCTURES IN BIOSYSTEMS

In the full-on report by Edvard Moser[1], the following problem results can be pointed out:

1. Cluster structures of cells form geometrically symmetrical hexagonal systems.
2. Cells themselves statistically concentrate along coordinate axes of symmetry with deviations not exceeding 7.5% (Fig. 5.2).

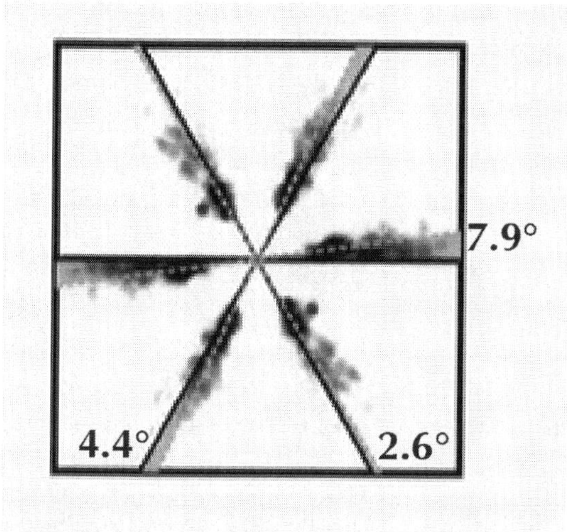

FIGURE 5.2 Statistic distribution of cells along coordinate axes.[1]

3. For independent cluster systems in different excitation, activity phase four modules which differ scale-wisely on coefficients can be pointed out: 1.4–1.421.

Cluster C_{60} containing 60 three-coordinated atoms and 180 effective bonds is the smallest stable carbon cluster. The similar structure is most probable in biosystems, even with the availability of 3-coordinated bonds of nitrogen atoms due to its $2P^3$-orbitals. This is cluster K_{60}. The second module of clusters with the coefficient 1.4 has 252 bonds, which means 72 additional bonds and corresponds to 12 new hexagons or 24 new atoms of the system forming cluster K_{84}. By the way, such cluster as C_{84} is among stable ones in carbon systems. If the rectified coefficient 1.421 is used in calculations, four more additional bonds are formed, which, apparently, play a binding role between the cluster subsystems.

The carbon cluster C_{60} contains 12 pentagons separated with 20 hexagons. The pentagons can be considered as a defect of graphite plane but structurally stabilizing the whole system. Still it is unknown if there are similar formations in biosystems.

It can be assumed that entropic statistics of distribution of activity degree of structural interactions given in Section 5.2 based on the nomogram (Fig. 5.1) is also fulfilled in biosystems. Thus, according to the nomogram if $\alpha < 7\%$, the maximum of structural interactions is observed and their sharp decrease—if $> 7\%$.

Therefore, the maximal deviation angle of cell statistic distribution from coordinate axes equal to 7.5° can be considered the demonstration of entropic regularity.

Apart from the aforesaid, a number of facts of hexagonal formation of biological systems can be also given as examples. For instance, the collocation of thin and thick myofilaments in skeletal muscle fibers and cardiomyocytes. At the same time, six thin myofilaments are revealed around each thick one. This system of functionally linked macromolecular complexes is constituted of calcium-dependent transient bonds between myosins and actins.

Also mechanotoropic interactions in surface layers of multilayer flat epitheliums conjugated with ample desmosomal contacts creating force fields on the background of the available hydrostatic pressure in epithelial cells are naturally followed by the formation of ordered epidermal columns with flat cells in the surface mainly of hexagonal shape and more rarely—of pentagonal one. There are also other options of revealing the aforesaid regularities.

5.5 GENERAL CONCLUSION

The comparisons and calculations carried out based on spatial-energy ideas allow explaining some features of forming hexagonal structures in biosystems.

KEYWORDS

- **spatial-energy parameter**
- **hexagonal clusters**
- **cell systems**
- **entropy**

REFERENCES

1. Nobel lecture by E. Moser in physiology: 11.03.2015, TV channel "Science".
2. Korablev, G. A. *Spatial-Energy Principles of Complex Structures Formation*; Brill Academic Publishers and VSP: Netherlands, 2005; 426 pp.
3. Korablev, G. A.; Petrova, N. G.; Korablev, R. G.; Osipov, A. K.; Zaikov, G. E. On Diversified Demonstration of Entropy. *Polym. Res. J.* **2014,** *8*(3), 145–153.
4. Sokolov, V. I.; Stankevich, I. F. Fullerenes—New Allotropic Forms of Carbon: Structure, Electron Composition and Chemical Properties. *Success. Chem.* **1993,** *62*(5), 455–473.
5. Korablev, G. A.; Zaikov, G. E. Formation of Carbon Nanostructures and Spatial-energy Criterion of Stabilization. *Mech. Composite Mater. Struct.* **2009,** *15*(1), 106–118.

CHAPTER 6

ENTROPIC AND SPATIAL-ENERGY INTERACTIONS

G. A. KORABLEV[1], V. I. KODOLOV[2], and G. E. ZAIKOV[3*]

[1]*Department of Physics, Izhevsk State Agricultural Academy, Izhevsk, Russia, E-mail: korablevga@mail.ru*

[2]*Kalashnikov Izhevsk State Technical University, Basic Research—High Educational Center of Chemical Physics and Mesoscopy, Udmurt Scientific Center, Ural Division, Studencheskaya St. 7, RAS, 426000 Izhevsk, Russia, Tel.: +7 3412 582438; E-mail: kodol@istu.ru*

[3]*Department of Chemistry, Emanuel Institute of Biochemical Physics, Russian Academy of Sciences, Russia, E-mail: chembio@sky.chph.ras.ru*

[*]*Corresponding author. E-mail: GEZaikov@Yahoo.com*

CONTENTS

ABSTRACT

The concept of the entropy of spatial-energy interactions is used similarly to the ideas of thermodynamics on the static entropy.

The idea of entropy appeared based on the second law of thermodynamics and ideas of the adduced quantity of heat.

These rules are general assertions independent of microscopic models. Therefore, their application and consideration can result in a large number of consequences which are most fruitfully used in statistic thermodynamics.

In this research, we are trying to apply the concept of entropy to assess the degree of spatial-energy interactions using their graphic dependence, and in other fields. The nomogram to assess the entropy of different processes is obtained. The variability of entropy demonstrations is discussed, in biochemical processes, economics, and engineering systems, as well.

6.1 INTRODUCTION

In statistic thermodynamics the entropy of the closed and equilibrious system equals the logarithm of the probability of its definite macrostate:

$$S = k \ln W, \tag{6.1}$$

where W is the number of available states of the system or degree of the degradation of microstates; k is the Boltzmann's constant.

Or,

$$W = e^{S/k} \tag{6.2}$$

These correlations are general assertions of macroscopic character; they do not contain any references to the structure elements of the systems considered and they are completely independent of microscopic models.[1]

Therefore, the application and consideration of these laws can result in a large number of consequences.

At the same time, the main characteristic of the process is the thermodynamic probability W. In actual processes in the isolated system, the entropy growth is inevitable—disorder and chaos increase in the system, the quality of internal energy goes down.

The thermodynamic probability equals the number of microstates corresponding to the given macrostate.

Since the system degradation degree is not connected with the physical features of the systems, the entropy statistic concept can also have other applications and demonstrations (apart from statistic thermodynamics).

"It is clear that out of the two systems completely different by their physical content, the entropy can be the same if their number of possible microstates corresponding to one macroparameter (whatever parameter it is) coincide. Therefore the idea of entropy can be used in various fields. The increasing self-organization of human society ... leads to the increase in entropy and disorder in the environment that is demonstrated, in particular, by a large number of disposal sites all over the earth".[2]

In this research, we are trying to apply the concept of entropy to assess the degree of spatial-energy interactions using their graphic dependence, and in other fields.

6.2 ON TWO PRINCIPLES OF ADDING ENERGY CHARACTERISTICS OF INTERACTIONS

The analysis of kinetics of various physical and chemical processes shows that in many cases the reciprocals of velocities, kinetic, or energy characteristics of the corresponding interactions are added.

Some examples are ambipolar diffusion, resulting velocity of topochemical reaction, change in the light velocity during the transition from vacuum into the given medium, effective permeability of biomembranes.

In particular, such supposition is confirmed by the formula of electron transport possibility (W_∞) due to the overlapping of wave functions 1 and 2 (in steady state) during electron-conformation interactions:

$$W_\infty = \frac{1}{2}\frac{W_1 W_2}{W_1 + W_2} \tag{6.3}$$

Equation (6.3) is used when evaluating the characteristics of diffusion processes followed by nonradiating transport of electrons in proteins.[3]

And also: "From classical mechanics it is known that the relative motion of two particles with the interaction energy $U(r)$ takes place as the motion of material point with the reduced mass μ:

$$\frac{1}{\mu} = \frac{1}{m_1} + \frac{1}{m_2} \tag{6.4}$$

in the field of central force $U(r)$, and general translational motion—as a free motion of material point with the mass:

$$m = m_1 + m_2 \tag{6.5}$$

Such things take place in quantum mechanics as well".[4]

The task of 2-particle interactions taking place along the bond line was solved in the times of Newton and Lagrange:

$$E = \frac{m_1 v_1^2}{2} + \frac{m_2 v_2^2}{2} + U(\overline{r_2} - \overline{r_1}) \tag{6.6}$$

where E is the total energy of the system, first and second elements—kinetic energies of the particles, third element—potential energy between particles 1 and 2, vectors $\overline{r_2}$ and $\overline{r_1}$ and—characterize the distance between the particles in final and initial states.

For moving thermodynamic systems the first commencement of thermodynamics is as follows[5]:

$$\delta E = d\left(U + \frac{mv^2}{2}\right) \pm \delta A, \tag{6.7}$$

where δE is the amount of energy transferred to the system; element $d(U + mv^2/2)$—characterize the changes in internal and kinetic energies of the system; $+\delta A$—work performed by the system; $-\delta A$—work performed with the system.

As the work value numerically equals the change in the potential energy, then:

$$+\delta A = -\Delta U \tag{6.8}$$

$$-\delta A = +\Delta U \tag{6.9}$$

It is probable that not only in thermodynamic but in many other processes in the dynamics of moving particles interaction not only the value of potential energy is critical, but its change as well. Therefore, similar to Equation (6.4), the following should be fulfilled for 2-particle interactions:

$$\delta E = d\left(\frac{m_1 v_1^2}{2} + \frac{m_2 v_2^2}{2}\right) \pm \Delta U \tag{6.10}$$

Here,

$$\Delta U = U_2 - U_1, \tag{6.11}$$

where U_2 and U_1 is the potential energies of the system in final and initial states.

At the same time, the total energy (E) and kinetic energy ($mv^2/2$) can be calculated from their zero value, then only the last element is modified in Equation (6.6).

The character of the change in the potential energy value (ΔU) was analyzed by its sign for various potential fields and the results are given in Table 6.1.

From the table it is seen that the values— ΔU and accordingly $+\delta A$ (positive work) correspond to the interactions taking place along the potential gradient, and ΔU and $-\delta A$ (negative work) occur during the interactions against the potential gradient.

The solution of 2-particle task of the interaction of two material points with masses m_1 and m_2 obtained under the condition of the absence of external forces, corresponds to the interactions flowing along the gradient, the positive work is performed by the system (similar to the attraction process in the gravitation field).

TABLE 6.1 Directedness of the Interaction Processes.

No	Systems	Type of potential field	Process	U	r_2/r_1 (x_2/x_1)	U_2/U_1	Sign ΔU	Sign δA	Process directedness in potential field
1	Opposite electrical charges	Electrostatic	Attraction	$-k\frac{q_1q_2}{r}$	$r_2 < r_1$	$U_2 > U_1$	–	+	Along the gradient
			Repulsion	$-k\frac{q_1q_2}{r}$	$r_2 > r_1$	$U_2 < U_1$	+	–	Against the gradient
2	Similar electrical charges	Electrostatic	Attraction	$k\frac{q_1q_2}{r}$	$r_2 < r_1$	$U_2 > U_1$	+	–	Against the gradient
			Repulsion	$k\frac{q_1q_2}{r}$	$r_2 > r_1$	$U_2 < U_1$	–	+	Along the gradient

TABLE 6.1 *(Continued)*

No	Systems	Type of potential field	Process	U	r_2/r_1 (x_2/x_1)	U_2/U_1	Sign ΔU	Sign δA	Process directedness in potential field
3	Elementary masses m_1 and m_2	Gravitational	Attraction	$-\gamma\frac{m_1 m_2}{r}$	$r_2 < r_1$	$U_2 > U_1$	–	+	Along the gradient
			Repulsion	$-\gamma\frac{m_1 m_2}{r}$	$U_2 < U_1$	$U_2 < U_1$	+	–	Against the gradient
4	Spring deformation	Field of elastic forces	Compression	$k\frac{\Delta x^2}{2}$	$x_2 < x_1$	$U_2 > U_1$	+	–	Against the gradient
			Extension	$k\frac{\Delta x^2}{2}$	$x_2 > x_1$	$U_2 > U_1$	+	–	Against the gradient
5	Photoeffect	Electrostatic	Repulsion	$k\frac{q_1 q_2}{r}$	$r_2 > r_1$	$U_2 < U_1$	–	+	Along the gradient

The solution of this equation via the reduced mass (μ)[6] is the Lagrange equation for the relative motion of the isolated system of two interacting material points with masses m_1 and m_2, which in coordinate x is as follows:

$$\mu \times x'' = -\frac{\partial U}{\partial x}; \quad \frac{1}{\mu} = \frac{1}{m_1} + \frac{1}{m_2}.$$

Here U is the mutual potential energy of material points; μ is the reduced mass. At the same time, $x'' = a$ (feature of the system acceleration). For elementary portions of the interactions, Δx can be taken as follows:

$$\frac{\partial U}{\partial x} \approx \frac{\Delta U}{\Delta x} \qquad \text{That is,} \qquad \mu\, a\Delta x = \Delta U.$$

Then,

$$\frac{1}{1/(a\Delta x)}\frac{1}{1/m_1 + 1/m_2} \approx -\Delta U; \qquad \frac{1}{1/(m_1 a\Delta x) + 1/(m_2 a\Delta x)} \approx -\Delta U$$

Or,

$$\frac{1}{\Delta U} \approx \frac{1}{\Delta U_1} + \frac{1}{\Delta U_2} \tag{6.12}$$

where ΔU_1 and ΔU_2 are the potential energies of material points on the elementary portion of interactions, ΔU is the resulting (mutual) potential energy of these interactions.

Thus,

1. In the systems in which the interactions proceed along the potential gradient (positive performance), the resulting potential energy is found based on the principle of adding reciprocals of the corresponding energies of subsystems.[7] Similarly, the reduced mass for the relative motion of 2-particle system is calculated.
2. In the systems in which the interactions proceed against the potential gradient (negative performance), the algebraic addition of their masses as well as the corresponding energies of subsystems is performed (by the analogy with Hamiltonian).

6.3 SPATIAL-ENERGY PARAMETER (*P*-PARAMETER)

From Equation (6.12), it is seen that the resulting energy characteristic of the system of two material points interaction is found based on the principle of adding reciprocals of initial energies of interacting subsystems.

"Electron with the mass m moving near the proton with the mass M is equivalent to the particle with the mass: $\mu = \frac{mM}{m+M}$".[8]

Therefore, when modifying Equation (6.12), we can assume that the energy of atom valence orbitals (responsible for interatomic interactions) can be calculated[7] by the principle of adding reciprocals of some initial energy components based on the following equations:

$$\frac{1}{q^2/r_i} + \frac{1}{W_i n_i} = \frac{1}{P_E} \tag{6.13}$$

or

$$\frac{1}{P_0} = \frac{1}{q^2} + \frac{1}{(Wrn)_i}; \tag{6.14}$$

$$P_E = P_0 / r_i \tag{6.15}$$

Here W_i is the electron orbital energy;[9] r_i is the orbital radius of i-orbital;[10] $q = Z^*/n^*$,[11,12] n_i is the number of electrons of the given orbital, Z^* and n^* are the nucleus effective charge and effective main quantum number, r is the bond dimensional characteristics.

P_0 was called a spatial-energy parameter (SEP), and P_E is the effective P-parameter (effective SEP). Effective SEP has a physical sense of some averaged energy of valence electrons in the atom and is measured in energy units, for example, electron-volts (eV).

The values of P_0-parameter are tabulated constants for the electrons of the given atom orbital.

For dimensionality SEP can be written down as follows:

$$[P_0] = [q^2] = [E] \times [r] = [h] \times [v] = \mathrm{kg\,m^3/s^2} = \mathrm{J\,m},$$

where $[E]$, $[h]$, and $[v]$ are the dimensions of energy, Planck constant, and velocity. Thus P-parameter corresponds to the processes going along the potential gradient.

The introduction of P-parameter should be considered as further development of quasi-classical notions using quantum-mechanical data on atom structure to obtain the criteria of energy conditions of phase formation. At the same time, for the systems of similarly charged (e.g., orbitals in the given atom), homogeneous systems the principle of algebraic addition of such parameters is preserved:

$$\sum P_E = \sum (P_0 / r_i) \tag{6.16}$$

$$\sum P_E = \frac{\sum P_0}{r} \tag{6.17}$$

or,

$$\sum P_0 = P_0' + P_0'' + P_0''' + \cdots \tag{6.18}$$

$$r \sum P_E = \sum P_0 \tag{6.19}$$

Here P-parameters are summed on all atom valence orbitals.

To calculate the values of P_E-parameter at the given distance from the nucleus depending on the bond type either atom radius (R) or ion radius (r_I) can be used instead of r.

Briefly about the reliability of such approach, the calculations demonstrated that the values of P_E-parameters numerically equal (within 2%) total energy of valence electrons (U) by the atom statistic model. Using the known correlation between the electron density (β) and interatomic potential by the atom statistic model,[13] we can obtain the direct dependence of P_E-parameter upon the electron density at the distance r_i from the nucleus.

The rationality of such approach is confirmed by the calculation of electron density using wave functions of Clementi[14] and its comparison with the value of electron density calculated via the value of P_E-parameter.

6.4 WAVE EQUATION OF *P*-PARAMETER

To characterize atom spatial-energy properties two types of P-parameters are introduced. The bond between them is a simple one:

$$P_E = \frac{P_0}{R}$$

where R is the atom dimensional characteristic. Taking into account additional quantum characteristics of sublevels in the atom, this equation can be written down in coordinate x as follows:

$$\Delta P_E \approx \frac{\Delta P_0}{\Delta x} \quad \text{or} \quad \partial P_E = \frac{\partial P_0}{\partial x}$$

where the value ΔP equals the difference between P_0-parameter of i orbital and P_{CD}-countdown parameter (parameter of main state at the given set of quantum numbers).

According to the established[7] rule of adding P-parameters of similarly charged or homogeneous systems for two orbitals in the given atom with different quantum characteristics and according to the energy conservation rule we have

$$\Delta P''_E - \Delta P'_E = P_{E,\lambda}$$

where $P_{E,\lambda}$ is the SEP of quantum transition.

Taking for the dimensional characteristic of the interaction $\Delta\lambda = \Delta x$, we have

$$\frac{\Delta P_0''}{\Delta\lambda} - \frac{\Delta P_0'}{\Delta\lambda} = \frac{P_0}{\Delta\lambda} \quad \text{or} \quad \frac{\Delta P_0'}{\Delta\lambda} - \frac{\Delta P_0''}{\Delta\lambda} = -\frac{P_0\lambda}{\Delta\lambda}$$

Let us again divide by Δλ term by term:

$$\frac{\left(\Delta P_0'/\Delta\lambda - \Delta P_0''/\Delta\lambda\right)}{\Delta\lambda} = -\frac{P_0}{\Delta\lambda^2}, \quad \text{where:}$$

$$\frac{\left(\Delta P_0'/\Delta\lambda - \Delta P_0''/\Delta\lambda\right)}{\text{Ä}\lambda} \sim \frac{\mathrm{d}^2 P_0}{\mathrm{d}\lambda^2} \quad \text{i.e.,} \quad \frac{\mathrm{d}^2 P_0}{\mathrm{d}\lambda^2} + \frac{P_0}{\Delta\lambda^2} \approx 0$$

Taking into account only those interactions when $2\pi\Delta x = \Delta\lambda$ (closed oscillator), we have the following equation:

$$\frac{\mathrm{d}^2 P_0}{4\pi^2 \mathrm{d}x^2} + \frac{P_0}{\Delta\lambda^2} = 0 \quad \text{or} \quad \frac{\mathrm{d}^2 P_0}{\mathrm{d}x^2} + 4\pi^2 \frac{P_0}{\Delta\lambda^2} \approx 0$$

Since $\Delta\lambda = \frac{h}{mv}$, then:

$$\frac{\mathrm{d}^2 P_0}{\mathrm{d}x^2} + 4\pi^2 \frac{P_0}{h^2} m^2 v^2 \approx 0$$

Or,

$$\frac{\mathrm{d}^2 P_0}{\mathrm{d}x^2} + \frac{8\pi^2 m}{h^2} P_0 E_k = 0 \tag{6.20}$$

where $E_k = mV^2/2$ is the electron kinetic energy.

Schrodinger equation for the stationery state in coordinate x:

$$\frac{d^2\psi}{dx^2} + \frac{8\pi^2 m}{h^2}\psi E_k = 0$$

When comparing these two equations, we see that P_0-parameter numerically correlates with the value of Ψ-function: $P_0 \approx \Psi$, and is generally proportional to it: $P_0 \sim \Psi$. Taking into account the broad practical opportunities of applying the P-parameter methodology, we can consider this criterion as the materialized analog of Ψ-function.[15,16]

Since P_0-parameters like Ψ-function have wave properties, the superposition principles should be fulfilled for them, defining the linear character of the equations of adding and changing P-parameter.

6.5 STRUCTURAL EXCHANGE SPATIAL-ENERGY INTERACTIONS

In the process of solid solution formation and other structural equilibrium-exchange interactions, the single electron density should be set in the points of atom-component contact. This process is accompanied by the redistribution of electron density between the valence areas of both particles and transition of the part of electrons from some external spheres into the neighboring ones. Apparently, frame atom electrons do not take part in such exchange.

Obviously, when electron densities in free atom-components are similar, the transfer processes between boundary atoms of particles are minimal; this will be favorable for the formation of a new structure. Thus, the evaluation of the degree of structural interactions in many cases means the comparative assessment of the electron density of valence electrons in free atoms (on averaged orbitals) participating in the process, which can be correlated with the help of P-parameter model.

The less the difference ($P'_0/r'_i - P''_0/r''_i$), the more favorable is the formation of a new structure or solid solution from the energy point.

In this regard, the maximum total solubility, evaluated via the coefficient of structural interaction α, is determined by the condition of minimum value α, which represents the relative difference of effective energies of external orbitals of interacting subsystems:

$$\alpha = \frac{P'_0/r'_i - P'_0/r''_i}{\left(P'_0/r'_i + P'_0/r''_i\right)/2} 100\% \qquad \alpha = \frac{P'_C - P''_C}{P'_C + P''_C} 200\%,$$

where P_S—structural parameter is found by the equation:

$$\frac{1}{P_C} = \frac{1}{N_1 P'_E} + \frac{1}{N_2 P''_E} + \cdots,$$

here N_1 and N_2 are the number of homogeneous atoms in subsystems.

The isomorphism degree and mutual solubility are evaluated in many (over 1000) simple and complex systems (including nanosystems). The calculation results are in compliance with theoretical and experimental data.

The nomogram of dependence of structural interaction degree (ρ) upon coefficient α, the same for a wide range of structures, was constructed based on all the data obtained.

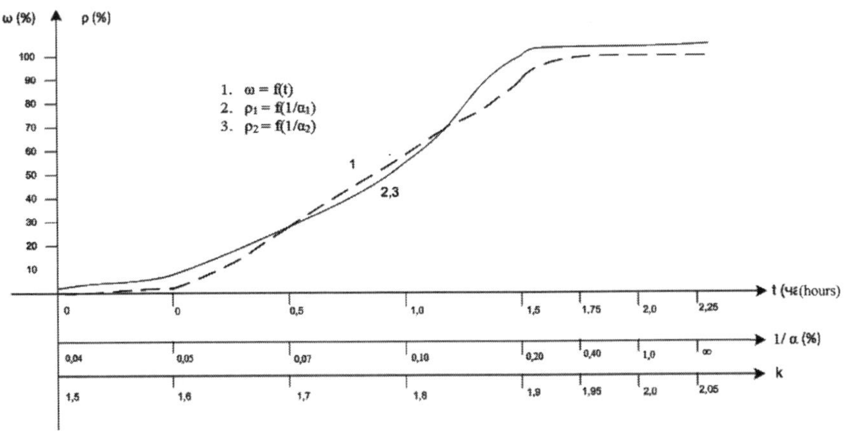

FIGURE 6.1 Dependence of the carbonization rate on the coefficient α.

This approach gives the possibility to evaluate the degree and direction of the structural interactions of phase formation, isomorphism and solubility processes in multiple systems, including molecular ones.

Such nomogram can be demonstrated[7] as a logarithmic dependence:

$$\alpha = \beta \, (\ln \rho)^{-1}, \tag{6.21}$$

where coefficient β is the constant value for the given class of structures. β can structurally change mainly within ±5% from the average value. Thus coefficient α is reversely proportional to the logarithm of the degree of structural interactions and therefore can be characterized as the entropy of spatial-energy interactions of atomic-molecular structures.

Actually the more is ρ, the more probable is the formation of stable ordered structures (e.g., the formation of solid solutions), that is, the less is the process entropy. But also, the less is coefficient α.

Equation (6.21) does not have the complete analogy with Boltzmann's equation (6.1) as in this case not absolute but only relative values of the corresponding characteristics of the interacting structures are compared which can be expressed in percent. This refers not only to coefficient α but also to the comparative evaluation of structural interaction degree (ρ), for example—the percent of atom content of the given element in the solid solution relatively to the total number of atoms. Therefore, in equation (6.21), coefficient $k = 1$.

Thus, the relative difference of spatial-energy parameters of the interacting structures can be a quantitative characteristic of the interaction entropy: $\alpha \equiv S$.

6.6 ENTROPIC NOMOGRAM OF SURFACE-DIFFUSIVE PROCESSES

As an example, let us consider the process of carbonization and formation of nanostructures during the interactions in polyvinyl alcohol gels and metal phase in the form of copper oxides or chlorides. At the first stage, small clusters of inorganic phase are formed surrounded by carbon containing phase. In this period, the main character of atomic-molecular interactions needs to be assessed via the relative difference of *P*-parameters calculated through the radii of copper ions and covalent radii of carbon atoms.

In the next main carbonization period, the metal phase is being formed on the surface of the polymeric structures.

From this point, the binary matrix of the nanosystem C → Cu is being formed.

The values of the degree of structural interactions from coefficient α are calculated, that is $\rho_2 = f(1/\alpha_2)$—curve 2 given in Figure 6.1. Here, the graphical dependence of the degree of nanofilm formation (ω) on the process time is presented by the data from[7]—curve 1 and previously obtained nomogram in the form $\rho_1 = f(1/\alpha_1)$—curve 3.

The analysis of all the graphical dependencies obtained demonstrates the practically complete graphical coincidence of all three graphs: $\omega = f(t)$, $\rho_1 = f(1/\alpha_1)$, $\rho_2 = f(1/\alpha_2)$ with slight deviations in the beginning and end of the process. Thus, the carbonization rate, as well as the functions of many other physical–chemical structural interactions, can be assessed via the values of the calculated coefficient α and entropic nomogram.

6.7 NOMOGRAMS OF BIOPHYSICAL PROCESSES

6.7.1 ON THE KINETICS OF FERMENTATIVE PROCESSES

"The formation of ferment-substrate complex is the necessary stage of fermentative catalysis. At the same time, *n* substrate molecules can join the ferment molecule."[3] (p. 58).

For ferments with stoichiometric coefficient n not equal one, the type of graphical dependence of the reaction product performance rate (μ) depending on the substrate concentration (c) has[3] a sigmoid character with the specific bending point (Fig. 6.2).

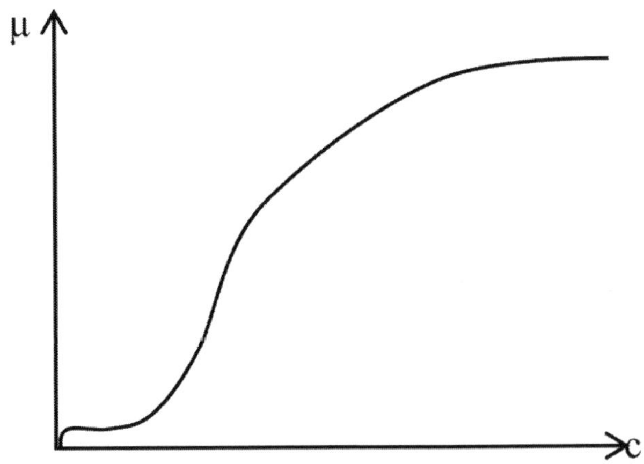

FIGURE 6.2 Dependence of the fermentative reaction rate (μ) on the substrate concentration (c).

In Figure 6.2, it is seen that this curve generally repeats the character of the entropic nomogram in Figure 6.1.

The graph of the dependence of electron transport rate in biostructures on the diffusion time period of ions is similar (p. 278).[3]

In the procedure of assessing fermentative interactions (similarly to the previously applied in par. 5 for surface-diffusive processes), the effective number of interacting molecules over 1 is applied.

In the methodology of *P*-parameter, a ferment has a limited isomorphic similarity with substrate molecules and does not form a stable compound with them, but, at the same time, such limited reconstruction of chemical bonds which "is tuned" to obtain the final product is possible.

6.7.2 DEPENDENCE OF BIOPHYSICAL CRITERIA ON THEIR FREQUENCY CHARACTERISTICS

1. The passing of alternating current through live tissues is characterized by the dispersive curve of electrical conductivity—this is the graphical dependence of the tissue total resistance (z-impedance) on the alternating current frequency logarithm (log ω). Normally, such curve, on which the impedance is plotted on the coordinate axis, and log ω—on the abscissa axis, formally, completely corresponds to the entropic nomogram (Fig. 6.3).
2. The fluctuations of biomembrane conductivity (conditioned by random processes) "have the form of Lorentz curve" (p. 99).[18] In this graph, the fluctuation spectral density (ρ) is plotted on the coordinate axis, and the frequency logarithm function (log ω)—on the abscissa axis.

The type of such curve also corresponds to the entropic nomogram in Figure 6.3.

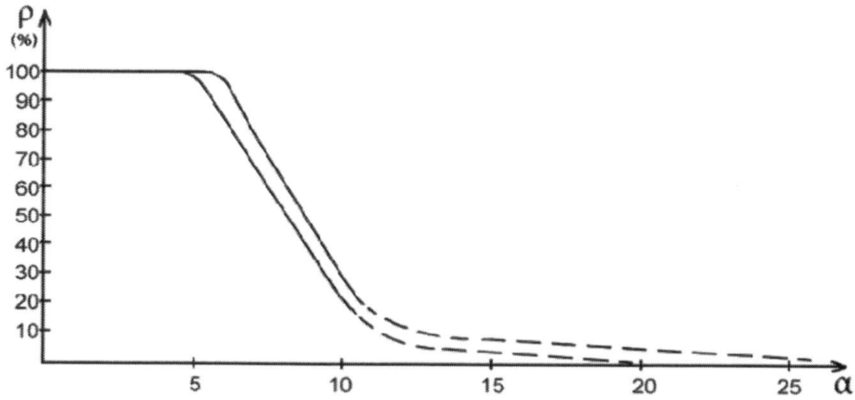

FIGURE 6.3 Nomogram of structural interaction degree dependence (ρ) on coefficient α.

6.8 LORENTZ CURVE OF SPATIAL-TIME DEPENDENCE

In Lorentz curve[19], the space-time graphic dependence (Fig. 6.4) of the velocity parameter (θ) on the velocity itself (β) is given, which completely corresponds to the entropic nomogram in Figure 6.1.

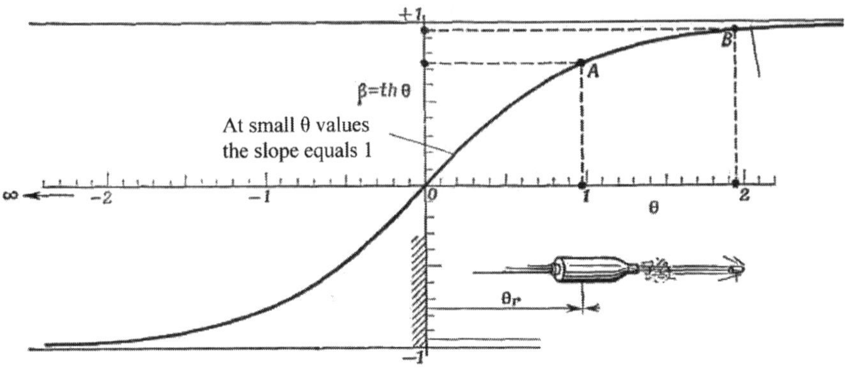

FIGURE 6.4 Connection between the velocity parameter θ and velocity itself $\beta = \text{th}\Theta$.

6.9 ENTROPIC CRITERIA IN BUSINESS AND NATURE

The main properties of free market providing its economic advantages are (1) effective competition and (2) maximal personal interest of each worker.

But on different economy concentration levels these ab initio features function and demonstrate themselves differently. Their greatest efficiency corresponds to small business—when the number of company staff is minimal, the personal interest is stronger and competitive struggle for survival is more active. With companies and productions increase the number of staff goes up, the role of each person gradually decreases, the competition slackens as new opportunities for coordinated actions of various business structures appear. The quality of economic relations in business goes down, that is, the entropy increases. Such process is mostly vivid in monostructures at the largest enterprises of large business (syndicates and cartels).

The concept of thermodynamic probability as a number of microstates corresponding to the given macrostate can be modified as applicable to the processes of economic interactions that directly depend on the parameters of business structures.

A separate business structure can be taken as the system macrostate, and as the number of microstates—number of its workers (N) which is the number of the available most probable states of the given business structure. Thus, it is supposed that such number of workers of the business structure is the analog of thermodynamic probability as applicable to the processes of economic interactions in business.

Therefore, it can be accepted that the total entropy of business quality consists of two entropies characterizing: (1) decrease in the competition efficiency (S_1) and (2) decrease in the personal interest of each worker (S_2), that is, $S = S_1 + S_2$. S_1 is proportional to the number of workers in the company: $S \sim N$, and S_2 has a complex dependence not only on the number of workers in the company but also on the efficiency of its management. It is inversely proportional to the personal interest of each worker. Therefore, it can be accepted that $S_2 = 1/\gamma$, where γ is the coefficient of personal interest of each worker.

By analogy with Boltzmann's equation (6.1), we have:

$$S = (S_1 + S_2) \sim \left[\ln N + \ln\left(\frac{1}{\gamma}\right)\right] \sim \ln\left(\frac{N}{\gamma}\right)$$

or $S = k \ln\left(\frac{N}{\gamma}\right)$

where k is proportionality coefficient.

Here N shows how many times the given business structure is larger than the reference small business structure, at which $N = 1$, that is, this value does not have the name.

For nonthermodynamic systems when we consider not absolute but relative values, we take $k = 1$. Therefore:

$$S = \ln\left(\frac{N}{\gamma}\right) \tag{6.22}$$

In Table 6.2, you can see the approximate calculations of business entropy by Equation (6.22) for three main levels of business: small, medium, and large. At the same time, it is supposed that number N corresponds to some average value from the most probable values.

When calculating the coefficient of personal interest, it is considered that it can change from 1 (one self-employed worker) to 0, if such worker is a deprived slave, and for larger companies, it is accepted as $\gamma = 0.1$–0.01.

Despite the rather approximate accuracy of such averaged calculations, we can make quite a reliable conclusion on the fact that business entropy, with the aggregation of its structures, sharply increases during the transition from the medium to large business as the quality of business processes decreases.

TABLE 6.2 Entropy Growth with the Business Increase.

Structure parameters	Business		
	Small	Average	Large
N_1–N_2	10–50	100–1000	10,000–100,000
γ	0.9–0.8	0.6–0.4	0.1–0.01
S	2.408–4.135	5.116–7.824	11.513–16.118
$\langle S \rangle$	3.271	6.470	13.816

In thermodynamics, it is considered that the uncontrollable entropy growth results in the stop of any macrochanges in the systems, that is, to their death. Therefore, the search of methods of increasing the uncontrollable growth of the entropy in large business is topical. At the same time, the entropy critical figures mainly refer to large business. A simple cut-down of the number of its employees cannot give an actual result of entropy decrease. Thus, the decrease in the number of workers by 10% results in diminishing their entropy only by 0.6%, and this is inevitably followed by the common negative unemployment phenomena.

Therefore, for such supermonostructures controlled neither by the state nor by the society, the demonopolization without optimization (i.e., without decreasing the total number of employees) is more actual to diminish the business entropy.

Comparing the nomogram (Fig. 6.3) with the data from Table 6.2, we can see the additivity of business entropy values (S) with the values of the coefficient of spatial-energy interactions (α), that is, $S = \alpha$.

It is known that the number of atoms in polymeric chain maximally acceptable for a stable system is about 100 units, which is 10^6 in the cubic volume. Then we again have $\lg 10^6 = 6$.

6.10 *S*-CURVES ("LIFE LINES")

Already in the last century some general regularities in the development of some biological systems depending on time (growth in the number of bacteria colonies, population of insects, weight of the developing fetus, etc.) were found.[20] The curves reflecting this growth were similar, first of all, by the fact that three successive stages could be rather vividly emphasized on each of them: slow increase, fast burst-type growth and stabilization (sometimes decrease) of number (or another characteristic). Later it was demonstrated that engineering systems go through similar stages during their development. The curves drawn up in coordinate system where the numerical values of one of the most important operational characteristics (e.g., aircraft speed, electric generator power, etc.) were indicated along the vertical and the "age" of the engineering system or costs of its development along the horizontal, were called *S* curves (by the curve appearance), and they are sometimes also called "life lines."

As an example, the graph of the changes in steel specific strength with time (by years) is demonstrated (Fig. 6.5).[20]

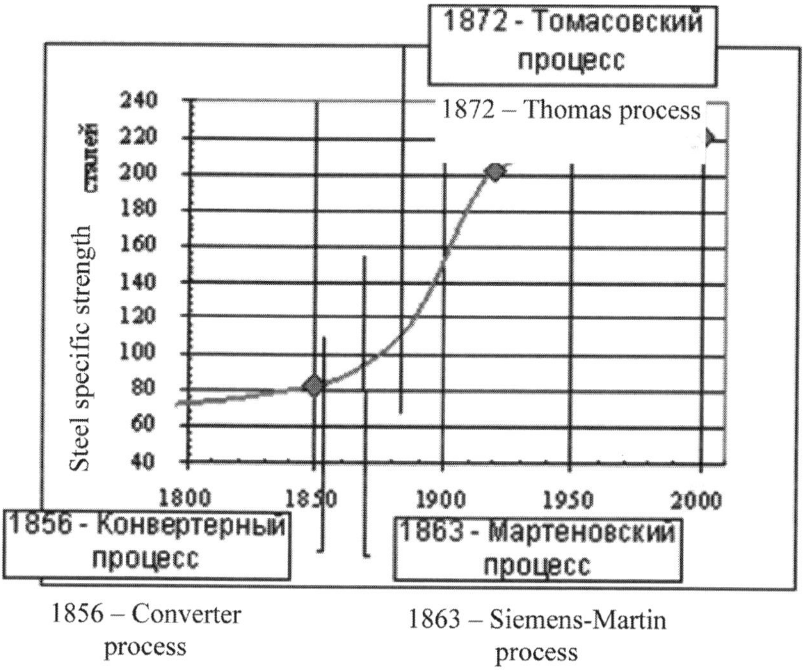

FIGURE 6.5 Dependence of steel specific strength on time.

Thus, the similarity between *S*-curves and entropic nomogram in Figure 6.1 is observed.

And in this case, the same as before, the time dependence (t) is proportional to the entropy reverse value ($1/\alpha$). As applicable to business, such curves characterize the process intensity, for example, sale of the given products.

At the same time, entropic nomograms in accordance with Figure 6.3 assess the business quality (ordinate in such graphs).

It is known that the entropy of isolated systems decreases. The entropy growth in open systems is compensated by the negative entropy due to the interaction with the environment.

All the above systems can be considered as open ones. This also refers to spatial-energy processes, when any changes in quantitative energy characteristics are conditioned by the interaction with external systems.

It is obviously observed in engineering and technological systems, the development of which is followed by additional innovations, modifications, and financial investments.

The entropy in thermodynamics is considered as the measure of nonreversible energy dissipation. From the point of technological and economic principles, the entropy is mainly the measure of irrational energy resource utilization. With the time dependence increase, such processes stabilize in accordance with the nomogram to more optimal values—together with the growth of anti-entropy, that is, the value $1/\alpha = 1/\rho$.

The similar growth with time of rationality of technological, economic, and physical and chemical parameters proves that such nomograms are universal for the majority of main processes in nature, technology, and economy.

6.11 GENERAL CONCLUSION

The idea of entropy is diversified in physical and chemical, economic, engineering, and other natural processes that are confirmed by their nomograms.

KEYWORDS

- **entropy**
- **nomogram**
- **spatial-energy parameter**
- **biophysical processes**
- **business**
- **engineering systems**

REFERENCES

1. Reif, F. *Statistic Physics*; M.: Nauka, 1972; 352 p.
2. Gribov, L. A.; Prokofyeva, N. I. *Basics of Physics*; M.: Vysshaya shkola, 1992; 430 p.
3. Rubin, A. B. *Biophysics. Book 1. Theoretical Biophysics*; M.: Vysshaya shkola, 1987; 319 p.
4. Blokhintsev, D. I. *Basics of Quantum Mechanics*; M.: Vysshaya shkola, 1961; 512 p.
5. Yavorsky, B. M.; Detlaf, A. A. *Reference-book in Physics*; M.: Nauka, 1968; 939 p.
6. Christy, R. W.; Pytte, A. *The Structure of Matter: An Introduction to Modern Physics*. Translated from English; M.: Nauka, 1969; 596 p.
7. Korablev, G. A. *Spatial-Energy Principles of Complex Structures Formation*; Brill Academic Publishers and VSP: Netherlands, 2005; 426 p. (Monograph).
8. Eyring, G.; Walter, J.; Kimball, G. *Quantum Chemistry*; M.: F. L., 1948; 528 p.
9. Fischer, C. F. *Atomic Data* **1972,** *4*, 301–399.
10. Waber, J. T.; Cromer, D. T. *J. Chem. Phys.* **1965,** *42*(12), 4116–4123.
11. Clementi, E.; Raimondi, D. L. Atomic Screening constants from S.C.F. Functions, 1. *J. Chem. Phys.* **1963,** *38*(11), 2686–2689.
12. Clementi, E.; Raimondi, D. L. *J. Chem. Phys.* **1967,** *47*(4), 1300–1307.
13. Gombash, P. *Statistic Theory of an Atom and its Applications*; M.: I.L., 1951; 398 p.
14. Clementi, E. J. B. M. S. *Res. Develop. Suppl.* **1965,** *9*(2), 76.
15. Korablev, G. A.; Zaikov, G. E. *J. Appl. Polym. Sci. U.S.A.* **2006,** *101*(3), 2101–2107.
16. Korablev, G. A.; Zaikov, G. E. *Progress on Chemistry and Biochemistry*; Nova Science Publishers, Inc.: New York, 2009; pp 355–376.
17. Kodolov, V. I.; Khokhriakov, N. V.; Trineeva, V. V.; Blagodatskikh, I. I. Activity of Nanostructures and its Manifestation in Nanoreactors of Polymeric Matrixes and Active Media. *Chem. Phys. Mesoscopy* **2008,** *10*(4), 448–460.
18. Rubin, A. B. *Biophysics. Book 2. Biophysics of Cell Processes*; M.: Vysshaya shkola, 1987; 303 p.
19. Taylor, E.; Wheeler, J. *Spacetime Physics*. Mir Publishers. M., 1987, 320 p.
20. Kynin, A. T.; Lenyashin, V. A. Assessment of the Parameters of Engineering Systems Using the Growth Curves. http://www.metodolog.ru/01428/01428.html.

CHAPTER 7

THE ELASTIC MODULUS CALCULATION OF THE MICRO- AND NANOPARTICLES MATERIAL OF ARBITRARY SHAPE

A. V. VAKHRUSHEV[1], A. A. SHUSHKOV[2], S. N. SYKOV[1], L. L. VAKHRUSHEVA[2*], and V. S. KLEKOVKIN[2]

[1]*Institute of Mechanics, Ural Division, Russian Academy of Sciences, Kalashnikov Izhevsk State Technical University, Izhevsk, Russia*

[2]*Kalashnikov Izhevsk State Technical University, Izhevsk, Russia*

[*]*Corresponding author. E-mail: vakhrushev-a@yandex.ru*

CONTENTS

ABSTRACT

Elastic modulus determination method of the material for the micro and nanometer geometry objects based on interconnected analysis and correlation of experimental indentation results and numerical computer simulation by finite element method is proposed. Results matching is carried out by coincidence of indenter penetration depth into a particle under given constant load in both cases, by varying Young's modulus in the method of finite elements. The method is differed from the previously known that is applicable to micro- and nanoobjects for any arbitrary geometrical shape.

7.1 INTRODUCTION

In recent years, investigations in determining area of the structural, physical and mechanical properties of micro- and nanostructures (nanoparticles, nanowires, nanotubes, etc.) are put in the forefront.[1–4] The study of micro- and nanoobjects properties allows to determine micro- and nanocomposite materials properties that have new improved characteristics.[5–9] These investigations have great potential and quickly find applications in strategic areas such as space and military technics.

Thereupon that nanostructures size is small (less than the wavelength of visible light); it is technically difficult to measure their properties. One main defect of the most widespread method of experimental measurement of mechanical properties of micro- and nanostructures (indentation method) is that the indenter should be pressed only in the plane.[10–13] Under existing conditions, the combined using of experimental indentation method and numerical computer simulation is the most optimal and acceptable method for determining of the nanostructures mechanical properties, it is the purpose of this work.

At present, there are no basic theoretical foundations of determining of the mechanical properties of individual micro- and in particular nanoparticles. Most currently existing methods for determining the elastic modulus and Poisson's ratio of the particles are not direct. The mechanical characteristics of the particles are determined on basis of composite material deforming and subsequent indirect calculation of the particles elastic modulus included in its composition. In particular, at article [14] using Monte Carlo method computer simulation, samples composed of titanium nanoparticles aggregates are investigated. "Equivalent" model of composite material is deformed along one axis. Elastic modulus is calculated from the energy of simulated sample deformation by the finite element method.

Therefore, investigations into the creation of direct calculation methods in combination with the combined using of experimental and numerical methods of the elastic characteristics determination of micro- and nanoparticles are actual. Investigations are continuation of the authors work.[15]

7.2 METHOD

Solution of determining task of the particles elastic modulus of arbitrary shape is divided into three stages: preparatory stage for determining of the numerical experiment inaccuracy, bench testing stage, and the stage of numerical investigations and analysis of their results.

For accuracy evaluation of results obtained by the further numerical experiment of elastic modulus determining, the test problem of coincidence analysis determining of indenter penetration depths h from the load F applied to the plane by the finite element method with the analytical solution of the hard ball indentation into the plane was carried out. The indentation task of absolutely hard ball in the plane has analytic solution.[15]

The values of Poisson's ratio $\mu = 0.33$ and elastic modulus $E = 7.3\times10^{10}$ Pa of the plane were set arbitrarily. The elastic modulus and Poisson's ratio of absolutely hard ball and Berkovich indenter tip are equal $E = 1.141\times10^{12}$ Pa, $\mu = 0.33$. The approximate radius of Berkovich indenter tip curving is $r =$ 200 nm. The values of the plane elastic modulus, Poisson's ratio, the radius of the ball (indenter), loads defined in the analytical solution and the finite element method are equal.

The values of the coordinated relative scaled weighting coefficients can be set[15]: $k_r = 10^{-6}$, $k_F = 10^{-12}$ (in the numerator F^2 and in the denominator r of analytic solution). Thus, the task is solved at the macro level by the finite element method to translate the results of the solution in the nanometer range values of the radius and load must be multiplied on $k_{r,\ P}$, respectively. For example, the real radius of the ball is $r = 0.2\ m$, multiplying it on $k_r = 10^{-6}$ obtain $r = 200$ nm.

By results of finite element methods modeling of the indentation task of the indenter spherical tip into the plane were obtained following results, with transferring in nanoscale (Table 7.1).

TABLE 7.1 Dependences of the Indentation Depths in the Plane from the Applied Load, Obtained by Finite Element Method and Analytical Solution.

N	*F, N*	h_{anal} (m)	h_{FEM} (m)
1	0	0	0
2	1×10^{-7}	0.161×10^{-9}	0.1117×10^{-9}
3	3×10^{-7}	0.335×10^{-9}	0.304×10^{-9}
4	5×10^{-7}	0.4714×10^{-9}	0.439×10^{-9}
5	7.5×10^{-7}	0.618×10^{-9}	0.578×10^{-9}
6	1×10^{-6}	0.751×10^{-9}	0.701×10^{-9}
7	2×10^{-6}	1.188×10^{-9}	1.124×10^{-9}
8	3×10^{-6}	1.553×10^{-9}	1.1478×10^{-9}
9	5×10^{-6}	2.188×10^{-9}	2.074×10^{-9}

The mean-square error is calculated by following formula:

$$\sigma = \sqrt{\frac{1}{n}\sum_{i=1}^{n}\left(h_{\text{anali}} - h_{\text{FEMi}}\right)^2} = 1.54 \times 10^{-10}\ \text{m} \tag{7.1}$$

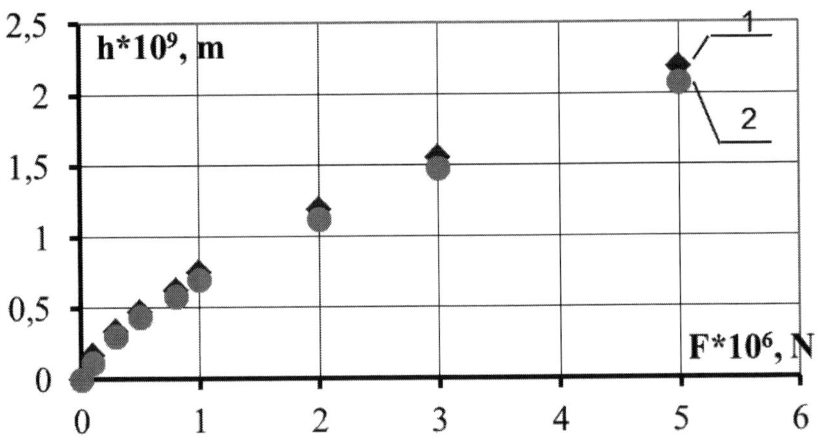

FIGURE 7.1 The dependence of the ball penetration depth into the plane *h*, *m* from the load *F*, *N*: 1—analytical solution and 2—the finite element method.

In Fig. 7.1, the indenter penetration depth dependence (spherical tip) from the applied load is shown. Acceptable consistency of results with the analytical dependence of the indentation of absolutely hard ball into the plane is shown.

7.3 INVESTIGATION RESULTS

During the implementation stage of bench tests high-precision scanning of the glass surface (substrate) was carried out with deposited micro- and nanoparticles at minimum load of scanning—0.01 mN and a established scanning step—0.05 mm, section 20 on 20 μm by the piezoprofilometer of system NanoTest 600. Investigated material of micro- and nanoparticles was aerosol. Investigated particle on considered section was chosen.

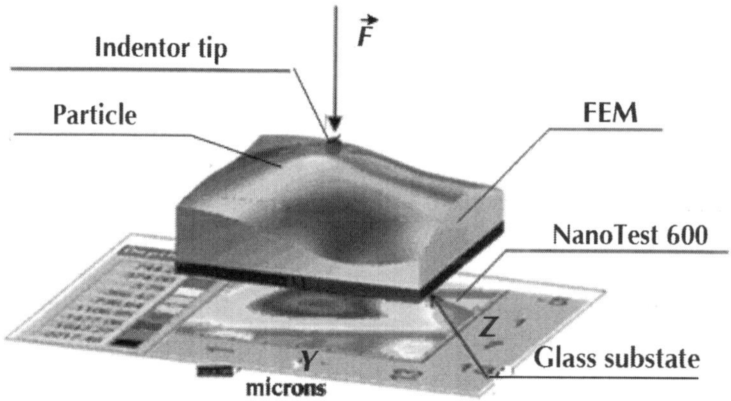

FIGURE 7.2 Finite element method (FEM) modeling of the aerosol particles surface created on basis of the results of scanning procedure by the system NanoTest 600 and indentation procedure with force *F*.

Experimental indentation in the investigated particle (Fig. 7.2) with force $F = 1$ mN is carried out. Indentation in the particle is carried out with the force that is smaller change limit of the elastic deformation into the plastic deformation; thus, all deformation of material is reversible. The depth of the indentation $h_{exp} = 85.421$ nm is defined.

A series of numerical solutions by finite element method of indentation contact task of the indenter tip into the top of the investigated particle (elastic modulus is varied, Poisson's ratio is set $\mu = 0.33$) were carried out.

Dependence of the indenter penetration depth into the particle from the values of elastic modulus is plotted for load $F = 1$ mN (Fig. 7.3). Dependence is plotted $E(\Delta h)$, where $\Delta h = h_{exp} - h_{FEM}$. For investigated particle on reaching $\Delta h = 0$, elastic modulus is determined $\mathring{A}_1 = 5.1 \times 10^{10}$ Pa (Fig. 7.4).

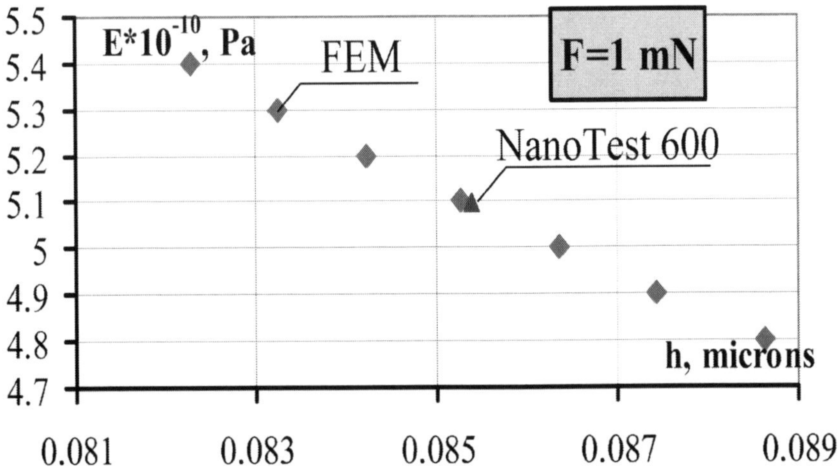

FIGURE 7.3 The dependence of elastic modulus E (Pa) from indenter penetration depth h (μm) into investigated particle with the load $F = 1$ mN. Finite element method modeling, experimental indentation by system NanoTest 600.

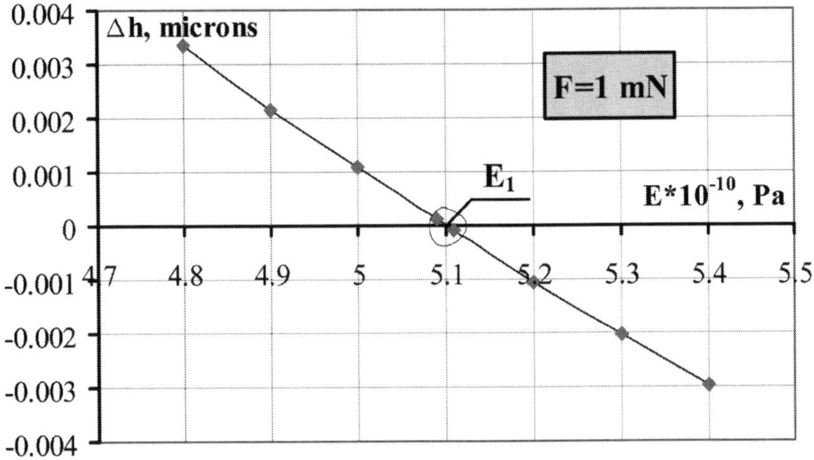

FIGURE 7.4 The dependence $E(\Delta h)$ for the indentation load $F = 1$ mN. At $\Delta h = 0$, the value of elastic modulus E_1 corresponds to the modulus of investigated particle.

7.4 CONCLUSION

A determining method of elastic modulus of micro- and nanoparticles of arbitrary geometric shape is developed based on the matching of the task solving of experimental indentation depth comparison with the depth obtained by the simulation of indentation process into the investigated particle by finite element method, varying the elastic modulus.

The proposed method of elastic modulus calculating of micro- and nanoparticles is applicable only for reversible indentation deformation of diamond tip and is based on an interconnected analysis and correlation of the results of full-scale deformation investigation, performed on an integrated system of determining the physical and mechanical characteristics NanoTest 600 and numerical computer simulation by finite element method.

ACKNOWLEDGMENT

This work is performed as part of the state assignment of Kalashnikov IzhSTU No. 201445-1239.

KEYWORDS

- **elastic modulus**
- **micro- and nanoparticles**
- **finite element method**

REFERENCES

1. Vakhrushev, A. V.; Shushkov, A. A.; Sykov, S. N. Patent of RA on an invention no 2494038, MKP B82Y 35/00, G01N 3/40. Sposob opredeleniya modulya uprugosti Yunga materiala mikro- i nanochastits [A Determination Method of Elastic Young's Modulus of Micro- and Nanoparticles Material], Russia. Zayavl. 20.03.2012, opubl. 27.09.13, Byul. no 27.
2. Vakhrushev, A. V.; Shushkov, A. A. Patent of RA on an Invention No 2297617, MKP G01N 3/08. Sposob opredeleniya modulya uprugosti Yunga i koeffitsienta Puassona materiala mikro i nanochastits [A Determination Method of Elastic Young's Modulus and Poisson's Ratio of Micro and Nanoparticles Material] Russia. Zayavl. 24.10.2005, opubl. 20.04.2007, Byul. no 11.

3. Vakhrushev, A. V.; Shushkov, A. A. Patent of RA on an Invention No. 2292029, MKP G01N 3/08. Sposob opredeleniya modulya uprugosti Yunga materialov [A Determination Method of Elastic Young's Modulus of Materials]. Russia. Zayavl. 06.05.2005, opubl. 20.01.2007, Byul. no 2.
4. Vakhrushev, A. V.; Lipanov, A. M.; Shushkov, S. N. Patent of RA on an Invention No. 2296972, MKP G01N 3/08. Sposob opredeleniya modulya uprugosti Yunga materialov [A Determination Method of Elastic Young's Modulus of Materials]. Russia. Zayavl. 29.07.2005, opubl. 10.04.2007, Byul. no 10.
5. Andrievskiy, R. A.; Kalinnikov, G. V.; et al. Nanoindentirovanie i deformatsionnyie harakteristiki nanostrukturnyih boridonitridnyih plenok [Nanoindentation and Deformation Characteristics of Boride Nitride Films]. *Fizika tverdogo tela* [*Solid State Phys.*] **2000,** *42*(9), 1624–1627.
6. Jung, Y.-G.; Lawn, B. R. "Huang et al. Evaluation of Elastic Modulus and Hardness of thin Films by Nanoindentation", *J. Mater. Res.* **2004,** *19*(10), 3076–3080.
7. Shojaei, O. R.; Karimi, A. Comparison of Mechanical Properties of TiN Thin Films Using Nanoindentation and Bulge Test. *Thin Solid Films* **1998,** *332*(1–2), 202–208.
8. Sklenika, V.; Kucharova, K.; et. al. Mechanical and Creep Properties of Electrodeposited Nickel and its Particle-reinforced Nanocomposite. *Rev. Adv. Mater. Sci.* **2005,** 10, 171–175.
9. Gong, J.; Miao, H.; Peng, Z. A New Function for the Description of the Nanoindentation unloading Data. *Scr. Mater.* **2003,** *49*(1), 93–97.
10. Vakhrushev, A. V.; Shushkov, A. A.; Shushkov, A. V. Eksperimentalnoe issledovanie modulya uprugosti Yunga i tverdosti mikrochastits zheleza metodom indentirovaniya [Experimental Investigation of Elastic Young's Modulus and Hardness of Steel Micro Particles by Indentation]. *Himicheskaya fizika i mezoskopiya* [*Chem. Phys. Mezoskop.*] **2009,** *11*(2), 258–262.
11. Lyahovich, A. M.; Shushkov, A. A.; Lyalina, N. V.; et al. Prochnostnyie svoystva nanorazmernyih polimernyih plenok, poluchennyih v nizkotemperaturnoy plazme benzola [Strength Properties of Nanoscale Polymer Films Prepared in a Low Temperature Benzol Plasma]. *Himicheskaya fizika i mezoskopiya* [*Chem. Phys. Mezoskop.*] **2010,** *12*(2), 243–247.
12. Golovin, Yu. I. Nanoindentirovanie i mehanicheskie svoystva tverdyih tel v submikroob'emah, tonkih pripoverhnostnyih sloyah i plenkah [Nanoindentation and Mechanical Properties of Solid in Submicrovolume, Thin Surface Layers and Films]. *Fizika tverdogo tela* [*Solid State Phys.*] **2008,** *50*(12), 2113–2142.
13. Odegard, G. M.; Clancy, T. C.; Gates, T. S. Modeling of the Mechanical Properties of Nanoparticle/Polymer Composites. *Polymer* **2005,** *46*(2), 553–562.
14. Ogunsola Oluwatosin, A. Synthesis of Porous Films from Nanoparticle Aggregates and Study of their Processing–Structure–Property Relationships. Doctor of Philosophy Dissertation, 2005, 142 p.
15. Vakhrushev, A. V.; Shushkov, A. A.; Sykov, S. N.; Klekovkin, V. S. Opredelenie modulya uprugosti Yunga nanochastits na osnove chislennogo modelirovaniya i eksperimentalnyih issledovaniy. Chast 1. Metodologicheskie osnovyi chislennogo modelirovaniya [Determination of Nanoparticles Young's Modulus on the Basis of Numerical Modeling and Experimental Studies. Part 1: Methodological Bases of Numerical Simulation]. *Himicheskaya fizika i mezoskopiya* [*Chem. Phys. Mezoskop.*] **2014,** *16*(3), 243–247.

CHAPTER 8

SHORT COMMUNICATIONS: TRENDS ON TECHNOLOGICAL CONCEPTS

A. V. VAKHRUSHEV[1,2], A. V. SEVERYUKHIN[1], A. Y. FEDOTOV[1,2], M. A. KOREPANOV[1], O. YU. SEVERYUKHINA[1] and S. A. GRUZD[2]

[1]*Institute of Mechanics, Ural Division, Russian Academy of Sciences, Izhevsk, Russia*

[2]*Kalashnikov Izhevsk State Technical University, Izhevsk, Russia*

CONTENTS

8.1 MODELING PROCESSES OF SPECIAL NANOSTRUCTURED LAYERS IN EPITAXIAL STRUCTURES

The problem of modeling processes of special nanostructured layers for solar cells in the structures solved by molecular dynamics (MD). Depending on the type of building and external forces in the system, the problem will have different accuracies and various thermodynamic parameters. Previous studies proposed semi-empirical approach that combines the benefits of great potentials and embedded atom method (MEAM). In numerical calculations the modern MEAM potentials are used.

An algorithm for the simulation of this particular task is presented below. The silicon substrate is heated to a fixed temperature T_0. This process is described by the equation of motion in the form of Newton, with the initial conditions, where the rate is set according to the Maxwell distribution. More detail the mathematical model described.[1,2] The equations for the resulting system of deposited atoms will be similar to the description of the first stage, but other than that given by the velocity vector for each of the deposited atoms and their direction is opposite to the direction of the axis z, that is, velocities are directed to the substrate. In the next step of modeling on the resulting system epitaxially deposited atoms of the second of this type.

The work was performed as part of the state task IzhSTU them. MT Kalashnikov number 201445-1239 and with the financial support of the Russian Foundation for Basic Research (grant no. 13-08-01072).

8.2 MODELING PROCESSES OF SPECIAL NANOSTRUCTURES FORMATION ON SOLID SURFACE

To develop methods of theoretical modeling of the formation of semiconductor nano hetero structures, such as quantum dots, nano whiskers, etc., and the modeling of their physical properties in order to control and predict their performance is very important.[1] A detailed study of thermo-physical properties of semiconductor nanoheterostructures and the processes taking place in their composition opens up new horizons for their use.

The kinetics of formation of multilayer nanoheterostructures can be described by molecular dynamic model of the process.

The studies have developed methods of mathematical modeling of processes of formation of quantum dots in the surface layers of the silicon substrate with different spatial orientation. The simulation was performed using the apparatus of molecular dynamics using parallelization. As the interaction, potentials used modern many-particle potentials (Modified Embedded-Atom Method [MEAM]).[2]

The work was performed as part of the state task IzhSTU them. MT Kalashnikov number 201445-1239 and with the financial support of the Russian Foundation for Basic Research (grant no. 13-08-01072).

8.3 MATHEMATICAL MODELING OF SUPERSONIC FLOW WITH HOMOGENEOUS CONDENSATION

The classical theory of homogeneous nucleation in accordance with the ideas of Mayer and Frenkel[1] considered that in real gas, along with monoparticles present agglomerates, which form by supercooling supercritical clusters, giving rise to the formation of a new phase.

Monoparticles can form a double particles via reaction I + I = II, triple particles (trimer)—III = I + II, etc. It is assumed that the agglomerates behave as gas particles resulting mixture of monoparticles and agglomerates satisfies the equation of state of an ideal gas.

Thus, the basic question remains determining the concentration or partial pressure of agglomerates. Assuming local chemical equilibrium partial pressures of the agglomerates can be found from the equation:

$$K(g,T) = \frac{p_1^g}{p_g}. \tag{1}$$

The works of Zhukhovitskii[2] have shown that for small clusters determine the equilibrium constant can be used Einstein crystal model. At the same time, for large clusters, the equilibrium constant can be determined from the condition of equality of chemical potentials of the cluster and monogas particles.

The smallest value of the equilibrium constant is used to determine the partial pressure of the agglomerates (Figs. 8.1 and 8.2).

A mathematical model that considers homogeneous condensation in the framework of the Smoluchowski coagulation theory[3] is proposed. Coagulation rate, suggesting that the formation of near-critical clusters is due to the collision of two sufficiently large agglomerates can be written as

$$I_g = kn_{g-1}n_{g-2}, \tag{2}$$

where n_g is the particle concentration and $k = 8\,\pi\,RD$ is the coagulation rate constant, and D is the diffusion coefficient.

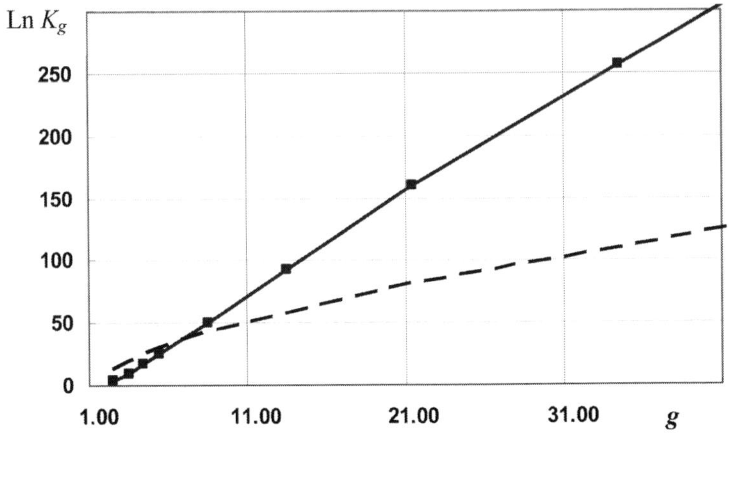

FIGURE 8.2 The partial pressure of agglomerates in the water vapor during supercooling.

The flows of water vapor in the conical nozzle are studied.

Temperature distribution along the length of the supersonic nozzle according to the taken a sufficient amount of supercritical clusters $N = 10^{10}$–10^{14} m^{-3}.

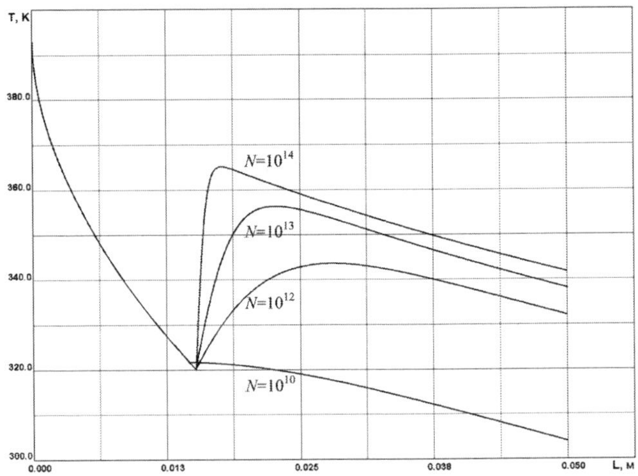

FIGURE 8.3 Temperature distribution along the length of the supersonic nozzle in dependence on supercritical clusters quantities.

By comparison with experimental data the critical value of supercritical clusters is defined $N = 10^{13}$–10^{14} m^{-3}. By calculating of the condensation in the nozzle with different critical concentration supercritical clusters following values of condensation drops were obtained: $r_{cl} = 3.16$ μm by $N = 10^{12}$ m^{-3}, $r_{cl} = 1.55$ μm by $N = 10^{13}$ m^{-3} and $r_{cl} = 0.72$ μm by $N = 10^{14}$ m^{-3}.

Condensing droplet size can be controlled by changing the cooling rate. The results can be used to produce submicron and nanoparticles of desired size.

KEYWORDS

- **homogeneous condensation**
- **supersonic flow**
- **nanoparticles**

REFERENCES

1. Severyukhina, O. Yu.; Vakhrouchev, A. V.; Galkin, N. G.; Severyukhin, A. V. Modeling of Formation of Multilayer Nanoheterostructures with Variable Chemical Bonds. *Chem. Phys. Mezoskop.* **2011,** *12*(1), 53–58.
2. Vakhrouchev, A. V.; Fedotov, A. Y.; Severyukhin, A. V.; Suvorov, S. V. Modeling Processes of Special Nanostructured Layers in Epitaxial Structures for Sophisticated Photovoltaic Cells. *Chem. Phys. Mezoskop.* **1998,** *16*(3), 364–381.
3. Severyukhina, O. Yu.; Vakhrouchev, A. V.; Galkin, N. G.; Severyukhin, A. V. Modeling of Formation of Multilayer Nanoheterostructures with Variable Chemical Bonds. *Chem. Phys. Mezoskop.* **1998,** *12*(1), 53–58.
4. Vukalovich, M. P.; Novikov, I. I. *Uravnenie sostoyania realnykh gasov*; Moscow, 1948; p 340 (in Russian).
5. Zhukhovitskii, D. I. Hot Clusters in Supersaturated Vapor. In *Progress in Physics of Clusters*; Chuev, G. N., Lakhno, V. D., Nefedov, A. P., Eds.; World Scientific Publ.: Singapore, 1998; pp 71–101.
6. Smoluchowski coagulation equation. https://en.wikipedia.org/wiki/Smoluchowski_coagulation_equation.

PART III

Investigations of Structure and Properties of Nanostructures, Nanosystems, and Nanostructured Materials

CHAPTER 9

REVIEW: APPLICATION OF UNIQUE X-RAY ELECTRON MAGNETIC SPECTROMETERS FOR THE INVESTIGATION AND CHEMICAL ANALYSIS OF THE ELECTRONIC STRUCTURE OF THE ULTRATHIN SURFACE LAYERS OF SYSTEMS IN LIQUID AND SOLID STATE

V. A. TRAPEZNIKOV[1], I. N. SHABANOVA[1], V. I. KODOLOV[2*], N. S. TEREBOVA[1], A. V. KHOLZAKOV[3], I. V. MENSHIKOV[3], F. F. CHAUSOV[3], and E. A. NAIMUSHINA[3]

[1]*Physical Technical Institute, Ural Division, Russian Academy of Sciences, Izhevsk, Russia*

[2]*Kalashnikov Izhevsk State Technical University, Basic Research—High Educational Center of Chemical Physics and Mesoscopy, Udmurt Scientific Center, Ural Division, RAS, Izhevsk, Russia*

[3]*Physico-Technical Institute, Ural Division, Russian Academy of Sciences, Izhevsk, Russia*

[*]*Corresponding author. E-mail: vkodol.av@mail.ru*

CONTENTS

ABSTRACT

The review dedicates to the development of unique X-ray photoelectron magnetic spectrometers and their using for the electron structures investigations during the substances transformations with nanostructures participation. The X-ray photoelectron spectroscopy investigations results obtained disclose new possibilities for progress in sciences about materials as well as for growth of electron transfer processes understanding.

9.1 INTRODUCTION

In this review, the results of X-ray photoelectron spectroscopy investigations of different processes as well as the development of unique X-ray photoelectron magnetic spectrometers are discussed. The investigation results of electron structure changes for different substances, including metallic and polymeric materials, in redox processes are considered. The mechanism of minute quantities action on polymeric compositions explained the electron structure changes with properties improvement is proposed.

9.2 UNIQUE X-RAY ELECTRON MAGNETIC SPECTROMETERS FOR THE INVESTIGATION AND CHEMICAL ANALYSIS OF THE ELECTRONIC STRUCTURE OF THE ULTRATHIN SURFACE LAYERS OF SYSTEMS IN LIQUID AND SOLID STATE

The first Russian electron magnetic spectrometers with an automated control system have been developed and built in the Physicotechnical Institute of the Ural Branch of the Russian Academy of Sciences, which are not inferior to the best foreign spectrometers with monochromators (the device resolution is 0.1 eV, luminosity—0.1%, vacuum 10^{-9} Torr) and are much better equipped with technological adapters: a chamber for deposition with resistive evaporators; an impact machine for fracture and impact tests of standard samples in vacuum; a chemical chamber for studying the processes of oxidation, reduction, adsorption, etc. The devices allow acting upon a sample by ions, laser radiation, and electrons. The devices are equipped with adapters for thermal actions in the range of 80–2000 K, sputtering, spalling, cutting, scraping, fracturing, layer-by-layer mechanical stripping (with a specified step) of the surface of samples in vacuum; they are also equipped with manipulators and mass spectrometers (Figs. 9.1–9.4).

FIGURE 9.1 An X-ray electron magnetic spectrometer for low-temperature investigations of electronic structure (device resolution—0.1 eV, luminosity—0.1%, vacuum—10^{-9} Torr).

FIGURE 9.2 An X-ray electron magnetic spectrometer with technological adapters: a machine for impact testing in vacuum, a sputtering adapter and manipulator for transferring samples into the investigation chamber of the spectrometer without deterioration of vacuum (device resolution—0.1 eV, luminosity—0.1%, vacuum—10^{-9} Torr).

FIGURE 9.3 An X-ray electron magnetic spectrometer for the investigation of the electronic structure of metal melts and the comparative studies of amorphous, crystalline, and liquid states at temperatures from room temperature up to 2000 K (device resolution—0.1 eV, luminosity—0.1%, vacuum—10^{-5}—10^{-9} Torr).

FIGURE 9.4 A unique portable small-size magnetic spectrometer with the radius of 10 cm of the electron orbit. The spectrometer is designed for working out technologies for making microelectronics elements by building the spectrometer into a technological line. Specification: the spectrometer resolution—0.1 eV; luminosity—0.1%; sensitivity is to a fraction of a monoatomic layer. The time for determining the concentration of an element is from 10 s to fractions of a second. Sizes are 500 × 500 mm^3; weight—30 kg (device resolution—0.1 eV, luminosity—0.1%, vacuum—10^{-9} Torr).

A unique automated electron magnetic spectrometer for investigating melts up to 1800 K (resolution at start—1.0 eV, luminosity—0.1%) was created. The creation of the device has significantly increased the field of the application of the method of X-ray photoelectron spectroscopy; it allows investigating the electronic structure of amorphous metal melts in solid and liquid state.

An energy analyzer of the magnetic type has an advantage over an electrostatic energy analyzer since it is constructively separated from the spectrometer vacuum chamber.

The created unique 100-cm electron magnetic spectrometer with a large radius of the cyclotron orbit has a very high sensitivity (by two orders of magnitude higher than that for 30-cm spectrometers), which allows investigating small doses of radiation at fast processes.

FIGURE 9.5 A unique supersensitive electron magnetic spectrometer with the radius of 100 cm of the electron orbit for studying fast processes and small doses of radiation from the surface layers of solid and liquid systems, and gaseous state.

9.3 THE DEVELOPMENT OF THE METHOD OF X-RAY ELECTRON SPECTROSCOPY FOR THE INVESTIGATION OF THE NEUTRINO FLUX DENSITY WITH THE USE OF DIFFRACTION AND THE INTERACTION OF DENSE NEUTRINO FLUXES WITH SUBSTANCE PLACED AT AN INDEFINITELY LARGE DISTANCE FROM SOURCES

ABSTRACT

In the present paper, a new method is substantiated for studying the inverse β-decay with the use of Auger- and photoelectron spectra for the registration of electrons, which appear at the interaction of low-energy (~1 MeV and smaller) electron neutrinos (v_e) with a substance. The new method is an alternative to the existing method based on the use of high-energy v_e with the energy of ~10 MeV, which form electrons at the interaction with a substance, the energy of which is sufficient for Cherenkov effect to appear. The offered method was realized using a high-sensitive electron magnetic spectrometer with double focusing by a nonuniform magnetic field $\pi\sqrt{2}$ with the cyclotron orbit radius of 100 cm, which was built at our institute; the first X-ray electron spectra were obtained on it and the constant of the device was determined.

9.3.1 INTRODUCTION

The studies of the inverse β-decay are characterized by a very small cross-section of the interaction of neutrinos with a substance, ~10^{-43}. Naturally, it imposes stricter requirements to the sensitivity of the instruments used for the registration of the interactions, which results in huge sizes of detectors. Thus, in the case of studying the inverse β-decay on the device Superkamiokande (Japan),[1] the neutrino detector is a reservoir containing 50,000 t of fluorescent liquid and thousands of Cerenkov's counters immersed in the reservoir, which record the tracks of electrons formed as the result of the interaction of the high-energy neutrino flux with the liquid. Another device for recording neutrino flux, which is based on the submarine neutrino detector and uses sea-water, requires a significant amount of sea-water that is estimated at 1 km^3.[2]

9.3.2 THEORY (METHODS AND TECHNIQUE)

For the inverse β-decay $\nu_e + d \rightarrow p + p + e^-$ by the charge current (CC), the use of an electron spectroscopy method for recording electron fragments resulting from the interaction of low-energy $\nu_{e,l}$ with a substance is more efficient in comparison with the registration of Cerenkov's electrons using high-energy $\nu_{e,h}$, the amount of which is four orders of magnitude smaller than that of low-energy $\nu_{e,l}$ in the range from 1 to 10 MeV (Fig. 9.1).[3]

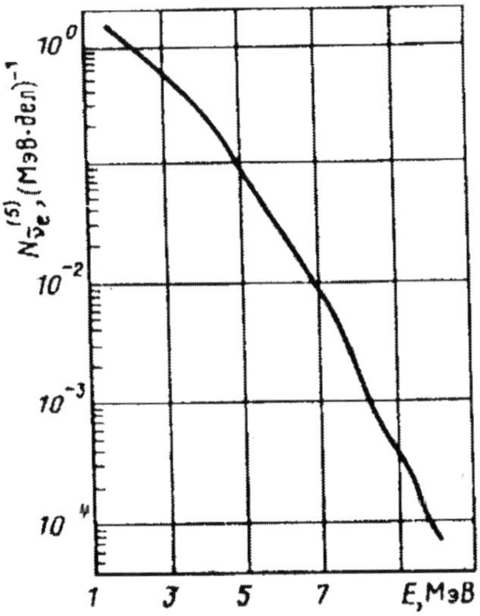

FIGURE 9.1 An anti-neutrino spectrum from fission fragments of ^{235}U.[3]

In addition, the excitation of the electron system by low-energy $\nu_{e,l}$ is by several orders of magnitude more efficient than the excitation by high-energy $\nu_{e,h}$ since high-energy particles with ~MeV energies exceed by several orders of magnitude the potential of ~keV of the electron system excitation.

When the electron system is excited by low-energy $\nu_{e,l}$, in the case of the $\nu_{e,l}$ interaction with the detector substance, consisting of the atoms with a large atomic number Z, the total pulse resulting from the ensemble of Auger-electrons of the detector system will significantly (per number of Auger-transitions in the excited system) exceed the number of electrons recorded by the system for high-energy $\nu_{e,h}$, where only one Cerenkov electron is recorded per one interaction. The entire ensemble of Auger-transitions will take place in the case of high-energy $\nu_{e,h}$ as well, however, Auger-electrons have much smaller energy than that which is necessary for producing Cerenkov effect, consequently they are not registered by Cherenkov's counters.

The inverse β-decay allows to solve actual problems in the sphere of practical use of neutrinos: evaluation of the power of a reactor, the variation of the composition of the fissile region, etc.[4] Here, the information carrier is the density of states of neutrinos from the energy $N\overline{\nu}_e(E)$; the hypothesis on the difference in the ν_e spectra obtained from the fission products of ^{235}U and ^{239}Pu was first suggested at the Institute of Atomic Energy (Russia)[5], and it was experimentally confirmed.[6] The method can be used for the evaluation the operation regime of a reactor (the evaluation of the reactor power, the amount of the plutonium used, etc.).

In the present work, the use of the electron spectroscopy method for the solution of the problems similar to those of nuclear physics is considered because the method is more available and more accurate since the density of the electron states $N_e(E)$ is dependent of the electrons in the range of energies much smaller (by orders of magnitude) than the energies in the range corresponding to the density of states of electron neutrinos $N\nu_e(E)$, which results in better resolution, and finally in significantly smaller cost. Large resolution of the method of electron spectroscopy of $N_e(E)$ in contrast to the neutrino method for the evaluation of the density of states of $N\nu_e(E)$ allows to increase the effectiveness of the signal detection from the reactor when the distance is increased.

It is of interest to evaluate the possibilities of the use of linearly polarized low-energy electron neutrinos for the identification of neutrino signals coming from different sources. By analogy with the X-ray linear polarization, in which the diffraction is obtained with the use of crystals at Bragg–Wulf angle of 45° depending on the interplanar spacing, the neutrino energy is determined unambiguously according to Bragg–Wulf equation:

$$n\lambda = 2d\sin\theta,$$

where n is the order of reflection, λ is the length of neutrino wave, d is the interplanar spacing in a crystal, and θ is the reflection angle.

For example, when an aluminum crystal is used, which has the maximal intensity of reflection of 140 from the octahedron plane (III), at $d = 2.33$ and the angle sin 45° the neutrino energy is 3.7 keV. When tritium is selected as the neutrino source, the electrons formed at the β-decay will have the energy of 14.9 eV, since the maximal energy of neutrinos is 18.6 keV.

9.3.3 RESULTS AND DISCUSSION

For recording neutrino flux from the Sun, nuclear reactors, and other sources,[7] it is offered to use a supersensitive electron magnetic spectrometer with double focusing by nonuniform magnetic field $\pi\sqrt{2}$.[4] The spectrometer was first used by K. Seigbahn and K. Edvardson at the University in Uppsala; it has the largest possible radius of the cyclotron orbit as the sensitivity of such instruments varies proportionally with the square of the radius. The selection of focusing in our device, which can simultaneously register electrons of different energies due to the presence of the focal plane, makes it possible to use the set of microchannel plates without limiting the number of channels, which is important for increasing the sensitivity of the device.

One of the largest in the world,[8] electron iron-free magnetic spectrometer with double focusing by nonuniform magnetic field $\pi\sqrt{2}$ with the radius of the cyclotron orbit of 100 cm (see Fig. 9.2) was developed and built by mutual efforts of the Udmurt State University and Physicotechnical Institute of the Ural Branch, RAS. The interval of the spectra studied is from 100 eV to 4 MeV.[9]

FIGURE 9.2 The 100-cm electron spectrometer (front view).

The direct detector of neutrinos is a chamber containing lead oxide in the superconducting state, which is necessary for increasing the length of the free path of the excited electrons and due to this their number registered by the spectrometer also increases, since the electrons appearing as the result of the ν_e interaction with lead are directed by the quadrupole lens to the slit of the electron spectrometer. The quadrupole lens was designed at the Institute of Analytical Instrument-making of RAS.[10] The electron detector is a micro-channel plate produced by the company "Baspic," Vladikavkaz, Russia. When polarized neutrinos are used, a crystal holder with the crystal bending according to Johann[11] is placed in front of the chamber (Fig. 9.3a) for the case when the radiation is coming from the outside of the focal circle[12] (Fig. 9.3b) and for the case with neutrinos as in work (Fig. 9.4).[13]

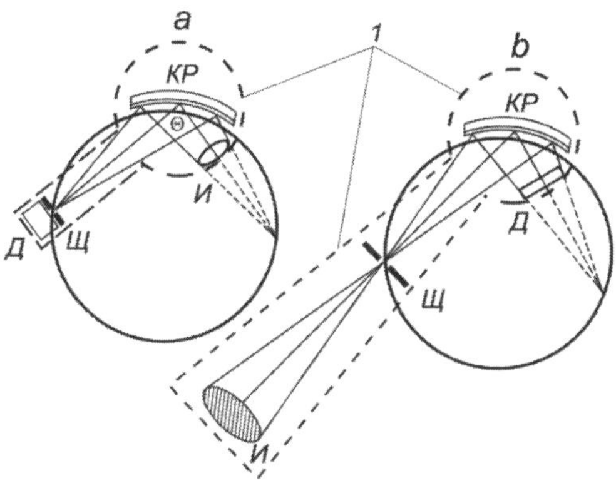

FIGURE 9.3 The scheme of the X-ray path according to Johann method: (a) normal variant: D is a detector, S is a slit, CR is a crystal, and S is a radiation source; (b) the variant with an X-ray generator placed outside the focal circle of the path of rays according to Johann method.

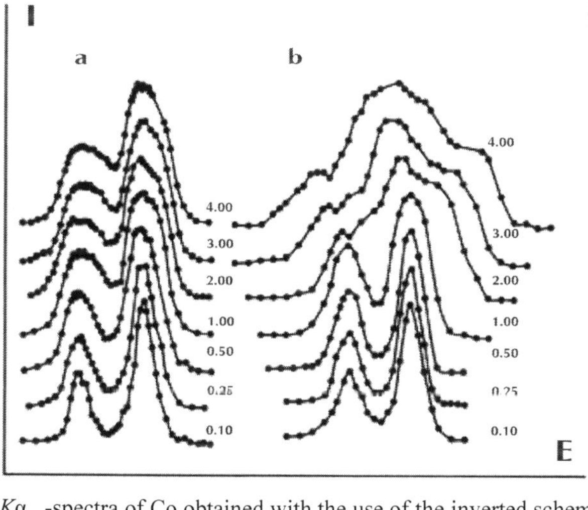

FIGURE 9.4 $K\alpha_{1,2}$-spectra of Co obtained with the use of the inverted scheme: (a) primary spectra; (b) secondary spectra.

The spectrometer can rotate over the axis perpendicular to the plane of the magnetic meridian, which leads to the width dependence and allows to take into account the angle of ecliptic at investigating solar antineutrino flux in the daytime and at night. In addition to the neutrino processes, it is possible to investigate fast processes using the device, for example, to investigate surface layers of solids for one impulse of impact load lasting for 10^{-5} s. Small time of the registration of photoelectrons allows to investigate the electronic structure of the surface of components of machines and mechanisms operating in cyclic mode (compressor and turbine blades, air-engines, gascompressor units, valves, rings, and cylinders of internal-combustion engines, springs, barrels of automatic guns, teeth of cutters and gear cutters, etc.) during one cycle. The device provides the possibility to investigate short-lived free radicals, which is important for understanding biological processes at the molecular and atomic levels. Some problems were refined in connection with new data received on the investigation of electron spectra of small doses of radiation.[14] We believe that the task discussed in the present paper concerning the application of the electron spectroscopy for studying the inverse β-decay is rather new, since neither in the keynote article of D. A. Shirley, B. L. Peterson, and M. P. Howels "Electron spectroscopy in the 21st century" nor in the review[16] nothing is mentioned of such statement of the task[12].

The experiment on obtaining electron spectra, which reflect the density of states of v_e, consists of three successive steps:

First, the whole spectrum, which fits into the aperture of the microchannel plates occupying the entire plane of the spectrometer, and the channels of which are connected to one anode, which will show that there is the interaction of neutrinos with the detector substance.

Second, the set of channels of the microchannel plates is divided into equal amounts, which are intended for equal energy parts of the focal plane, and each group of channels is connected to a separate anode; this will show a rough spectrum of the distribution of the neutrino density of states by energy.

Finally, the partition of each large region of channels into narrow spectral intervals with a separate anode for each of them in order to reveal the fine structure of the neutrino spectrum. It is necessary to note that in all the cases, the whole spectrum will be obtained simultaneously. On the spectra under study, the variation of the neutrino density of states will be reflected in the process of the variation of the neutrino radiation mode in time or during the variation of either the distance or the angle dependence between the neutrino source and the electron spectrometer.

9.3.4 CONCLUSION

In the present paper, a scheme of a device is offered allowing to increase the density of neutrino flux, wherein diffraction on a bent crystal is used, which allows to increase the density of the total neutrino flux by an order of magnitude.

A brief description of the main components of the 100-cm electron magnetic spectrometer is given, which has a focal plane, the rotation axis of an energy analyzer that is perpendicular to the magnetic meridian plane. The first Auger- and photoelectron spectra are obtained. The constant of the instrument is determined equal to c = 0.4 Ga cm/mA. Initial projects have been programed.

A scheme of a device is offered allowing to increase the neutrino flux density due to the diffraction on bent crystals, the number of which can be considerable, and each crystal can be of the size as large as possible, which allows to increase the density of the total neutrino flux by 9–10 orders of magnitude. The device can be used as a neutrino telescope. The general task of the present investigation on the increasing of the neutrino flux density is obtaining of a new source of restorable energy, that is, neutrino energy.

Work is supported by a basic unit of the state task of the Ministry of Education and Science.

ACKNOWLEDGMENTS

The authors thank their colleagues A. Ye. Kazantsev, A. V. Kholzakov, N. S Terebova, B. A. Budenkov, A. A. Voronchikhin, V. V. Pelenev, A. D. Frolov, and L. V. Rude for their assistance in obtaining the curve of the dependence of the focusing field on the radius of the cyclotron orbit of the energy analyzer and the first spectra.

KEYWORDS

- **inverse β-decay**
- **Cherenkov effect**
- **neutrino detector**
- **neutrino polarization**

- **cross-section of neutrino and substance interaction**
- **100-cm electron magnetic spectrometer**
- **focal plane**
- **quadrupole lens, electron spectra**
- **device constant**

REFERENCES

1. Totsuka, Y. University of Tokio Report NJCCR—Report 227-90-20, 1987.
2. Presti, D. L. Low Power Electronics for a Submarine Neutrinos Detector. *Nucl. Phys. B (Proc. Suppl.)* **2000,** *87*, 523–524.
3. Borovoy, A. A.; Khakimov, S. K. *Neutrino Experiments on Nuclear Reactors*; M.: Energoatomizdat, 1990, 149 p.
4. Kozlov, Yu. V.; Martemyanov, V. P.; Mukhin, K. N. The Problems of Neutrino Mass in Modern Neutrino Physics. *UFN* **1997,** *167*(8), 849–885.
5. Borovoy, A. A.; Micaelyan, L. A. Preprint IAE-2763, Moscow, 1976.
6. Borovoy, A. A.; et al. Measurement of Electron Spectra from Fission Products ^{235}U and ^{239}Pu by Heat Neutrons. *YaF* **1983,** *37*(6), 1345–1350.
7. Siegbahn, K.; Edvardson, K. Beta-ray spectroscopy in the precision range of $1{:}10^5$. *Nucl. Phys.* **1956,** *1*, 137–149.
8. Graham, R. L.; Evan, J. T.; Geiger, T. S. A One Meter Radius Iron-free Double Focusing $\pi/\sqrt{2}$ Spectrometer for β-ray Spectroscopy with a precision of $1{:}10^5$. *Nucl. Instr. Meth.* **1960,** *9*(3), 245–286.
9. Trapeznikov, V. A.; Shabanova, I. N.; Zhuravlev, V. A. The Development of the 100-cm Electron Magnetic Spectrometer with Double Focusing. *Vest. Udm. Univer., Izhevsk* **1993,** *5*(1), 111–121.
10. Gall, L. N.; Gall, R. N.; Stepanova, M. S. Ion-optic System with Rigid Focusing for Mass-Spectrometer. Part 1. The Formation of Quasi-parallel Beams. *ZhTF* **1968,** *38*, 1043–1047; Part 2. Gall, L. N.; Gall, R. N.; Demidova, V. A.; Stepanova, M. S. The Formation of Ribbon Beams. *ZhTF* **1968,** *38*, 1048–1051.
11. Johann, H. H. Die Erzeugung lichtstarker Röntgenspektren mit Hilfe von Konkavkristallen. *Z. Phys. A*, **1931,** *69*(3–4), 185–206.
12. Trapeznikov, V. A.; Sapozhnikov, V. P. The Use of the X-ray Spectrometer with a Bent Crystal for Analyzing the Radiation Outside the Focal Circle. *PTE* **1970,** *3*, 227–232.
13. Trapeznikov, V. A. *The Use of the Diffraction Phenomenon for Increasing the Density of the Neutrino Flux*; Nauka i tehnologii, RAS: Moskva, 2007; pp 375–382.
14. Trapeznikov, V. A. Electron Spectroscopy of Small Doses of Radiation. *UFN* **1998,** *7*, 793–799.
15. Shirley, D. A.; Petersen, B. L.; Howells, M. R. Electron Spectroscopy into the Twenty-first Century. *J. Electron Spectrosc. Relat. Phenom.* **1995,** *76*, 1–7.
16. Gershtein, S. S.; Kuznetsov, E. P.; Ryabov, V. A. The Nature of Neutrino Mass and Neutrino Oscillations. *UFN* **1997,** *167*(8), 811–848 (in Russian).

9.4 XPS INVESTIGATION OF THE COMPOSITION OF THE SURFACE LAYERS OF NI-BASED BINARY MELTS

ABSTRACT

The XPS method has been used for investigating the composition of the surface layers of the $Ni_{84}P_{16}$, $Ni_{81}P_{19}$, $Ni_{78}P_{22}$, and $Ni_{82}B_{18}$ alloys at temperature variation and isothermal holding in the liquid state. When the temperature is changed, jump-like changes in the composition of the melt surface layers are observed, which are treated as structural transformations in the liquid state. A nonmonotonic behavior of the composition of the melts surface layers is found under isothermal holdings. The maximal melt instability is observed in the structural transformation region. The transition to the equilibrium state is observed only for the $Ni_{81}P_{19}$ melt surface layers, which can be due to the appearance of strong bonds between the alloy elements at the formation of the melt cluster structure in contrast to the $Ni_{84}P_{16}$, $Ni_{78}P_{22}$, and $Ni_{82}B_{18}$ alloys.

9.4.1 INTRODUCTION

The appearance of new technological processes of the alloy production requires a special investigation of the processes taking place in metal melts, because in most cases, liquid phase is a phase preceding solid state. Nowadays, the production of special alloys with the use of vacuum metallurgy is especially important.

In the investigation of metal melts, three stages can be singled out.[1] At the *first* stage, it was established that multicomponent metal melt was a much more complex object than ideal solution, and its design–theoretical schemes are more complicated as well. At the *second* stage of the investigations, it became obvious that polytherms of properties of some pure liquid metals, and, a fortiori, alloys, had bends, jumps, and other anomalies. The latter indicate that there are jump-like structural changes alongside with smooth ones, which are somewhat similar to phase transformations. The investigation of metal melts behavior and their properties during long isothermal holdings should be referred to the *third* series of the experimental works on metal melts. The obtained **time** dependences showed many different specific features; unfortunately, they still cannot be systematized and comprehensively and explicitly explained.

Binary alloys are the simplest alloys; thus, they can be considered as model alloys. The investigation of the model alloys can reveal general regularities of the formation of the structure of complex (multicomponent) metal melts.

As a rule, the structure of metal melts is considered from the viewpoint of the generation of different formations in melts, that is, microgroups, microinhomogeneities, microheterogeneous regions, clusters, etc.[2–5] According to some evaluations, the above microgroups can have sizes from 10 to 100 nm and larger than that; however, there are no reliable data on the structure and sizes of the clusters. The difficulty, in the first place, is connected with the structure of metal melts themselves, and with the limited number of investigation methods suitable for investigating metal melts.

Methods characterizing bulk properties are mainly used. Such methods are viscosity, diffractometry, various kinds of thermometry. Investigations of the surface of metal melts are quite few due to the limited number of experimental methods. These methods mainly deal with measuring the surface tension of metal melts. In the present work, the investigations have been conducted with the use of the method of X-ray photoelectron spectroscopy (XPS). The XPS method is a method for studying surface. The method is nondestructive and allows investigating the chemical structure of the surface layers of melts. The chemical structure of surface layers is understood as the composition of a studied layer (3–5 nm), the chemical bond of the alloy elements, and electronic structure as well (the distribution of the valence electrons of alloy melt by energy is not discussed in the present work, since it is a separate object for investigation).

9.4.2 *EXPERIMENTAL*

The chemical structure and nearest surroundings of atoms in the liquid state of binary alloys $Ni_{84}P_{16}$, $Ni_{81}P_{19}$, $Ni_{78}P_{22}$, and $Ni_{82}B_{18}$ were investigated "in situ" during one experiment by the method of XPS.[6]

The experiment was conducted on a unique X-ray electron magnetic spectrometer for investigating both solid samples and their melts during long time at different temperature of melt superheating (up to 1500°C).[7] The specific feature of the above spectrometer is that the investigation chamber and the energy analyzer are spatially separated. Such devise construction allows acting upon a sample, heating a sample in particular, without disturbing the energy-analyzer focusing properties. Figure 9.1 shows the schematic view of the X-ray electron magnetic spectrometer for studying metal melts.

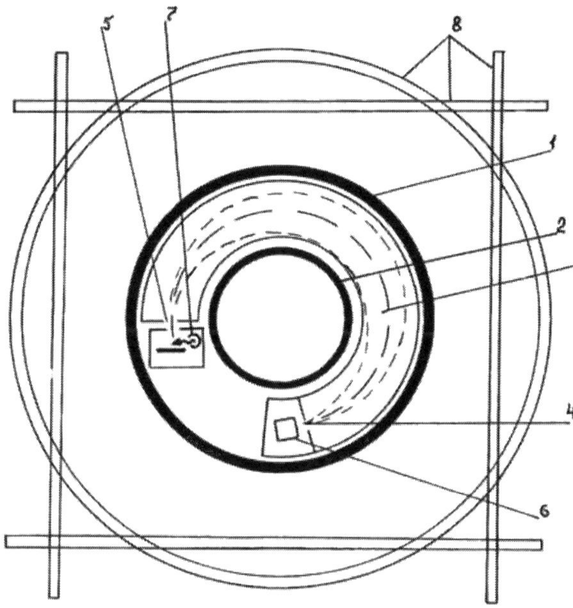

FIGURE 9.1 The schematic view of the X-ray electron magnetic spectrometer for studying metal melts.

The spectrometer consists of the following: 1 and 2—magnetic energy-analyzer; 3—chamber of electron transit (vacuum of 10^{-3} Pa); 4 and 5—entrance and exit slits; 6—electron detector; 7—X-ray tube (exciting radiation AlK$\alpha_{1,2}$ (hn = 1486.6 eV)), and 8—the system for compensating external magnetic fields. The spectrometer operates as follows. Under the X-ray radiation, photoelectrons are knocked out from a sample; their kinetic energy and energy of binding are described with the photoeffect expression $h\nu = E_{\text{kinetic}} + E_{\text{binding}} + \varphi$. Electrons get into the spectrometer chamber through the entrance slit. After that, under the action of the magnetic field of a certain configuration, electrons with the energy of interest pass along the cyclotron orbit and get to the recording path. By changing the current strength in the coils of the energy-analyzer, it is possible to obtain an entire spectrum of the obtained photoelectrons (or scan separate regions of the spectrum). Figure 9.2 shows a Mo survey spectrum as an example, which was obtained during the spectrometer testing. In the present work, the measurements were conducted as follows. At the first step, the required heating of a sample was set. After that, the measurements at isothermal holding were conducted, and the spectra $P2p_{3/2}$, $Ni2p_{3/2}$ and B1s, $Ni2p_{3/2}$ for the corresponding alloys were scanned by turns (the time of signal accumulation at the point is 10 s). Then, the above spectra were scanned during longer time (30–60 s at the point) for constructing the temperature dependence of the concentration, *C*(*T*). During the experiment, the core level spectra for all the elements present in the alloys were analyzed. At isothermal holdings, the ratio of the parameters (areas, relative intensity) of the most intensive spectra was analyzed.

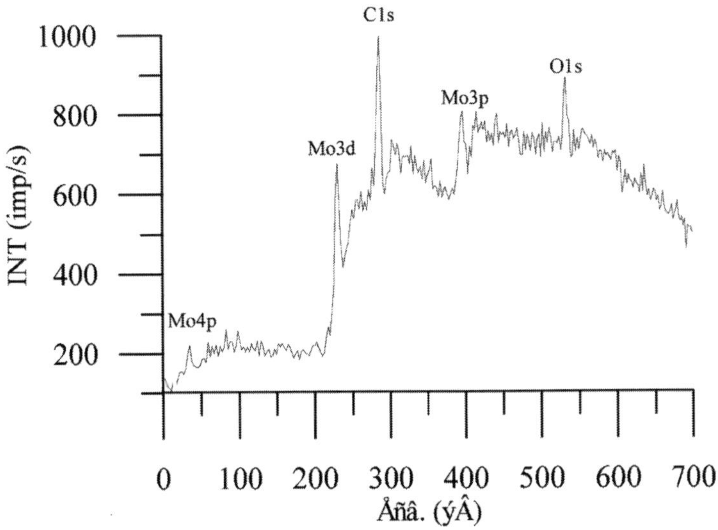

FIGURE 9.2 The Mo X-ray photoelectron spectrum.

The XPS method allows to control the surface state directly during the experiment. The presence of impurities and oxidation of the surface layers were controlled by the C1s and O1s spectra directly during the experiment.

The transition from the crystalline state to the liquid state was controlled by the variation of the oxygen content in the alloy surface layers. The oxygen content dropped almost to zero[8] as the result of melting.

During the entire experiment, the vacuum in the spectrometer chamber was 10^{-3} Pa. All the measurements were conducted without disturbing the vacuum. Such vacuum is not ultrahigh vacuum; consequently, in such residual medium there are elements, though in a significantly smaller amount, which are present in the atmosphere.

Since in the crystalline state the metal type of bond prevails, it is natural that on the surface, the processes of the oxidation of the elements are taking place. Due to melting, the situation on the surface changes significantly. We believe that the formation of a cluster structure takes place, and the presence of oxygen in the surface layers indicates the change of the type of bond. In the crystalline state the metal type of bond prevails, whereas in the melt, atomic groups (clusters) predominantly with the covalent type of bond prevail on the surface.

Such type of bond between the cluster elements is the required condition for the formation of clusters, because in the opposite case, any "open" bonds will be used by oxygen of the residual gaseous medium for the formation of oxides in the surface layers of melts. In the alloys under study, such situation has not been observed.

9.4.3 RESULTS AND DISCUSSION

First of all, let us consider the change of the composition of the surface layers of the binary eutectic melts $Ni_{81}P_{19}$ and $Ni_{82}B_{18}$ and non-eutectic alloys $Ni_{84}P_{16}$ and $Ni_{78}P_{22}$ as the temperature increases.

In the $Ni_{81}P_{19}$, in the crystalline state the ratio of the concentration $C(P)/C(Ni)$ on the surface makes up 0.2. Nickel is in the oxidized state. The process of the alloy melting is controlled by the oxygen content in the surface layers. Sharp decrease of the oxygen content in the alloy surface layers indicates the melting of the sample.[8] Melting leads to the formation of the cluster structure in the melt surface layers; at the same time, the type of the chemical bond changes. In the crystalline state the metal type of bond prevails, whereas in the melt there is the covalent bond between the elements in the clusters.

Figure 9.3 presents the change of the composition of the $Ni_{81}P_{19}$ alloy surface layers as the temperature increases from T_{melt} to 1100°C. The chosen criterion is the concentration ratio $C(P)/C(Ni)$. In the melt, two characteristic temperatures 920 and 1000°C can be singled out, at which a jump-like change in the composition of the surface layers takes place. Based on the knowledge about the cluster structure of metal melts, it can be stated that at the above-mentioned temperatures, the change of the surface layers composition is due to the change of the type of the clusters. In other words, in the clusters, the nearest environment of the atoms is changed. At each of the above-mentioned changes, the number of Ni atoms increases in the environment of phosphorus atoms. The XPS data do not allow to determine what specific type of clusters prevails on the melt surface at a particular temperature; however, the fact that the type of clusters changes is undoubtedly important for understanding the processes of the formation of the structure of metal liquids surface. Thus, the jump-like changes of the composition of the $Ni_{81}P_{19}$ melt surface are connected with the change of the cluster composition, which can be interpreted as structural transformations in the liquid state. In the $Ni_{82}B_{18}$ alloy, in the crystalline state, the concentration ratio C(B)/C(Ni) makes up 0.4. Nickel is in the oxidized state. As with the $Ni_{81}P_{19}$ alloy, the process of the alloy melting is controlled by the content of oxygen in the surface layers. The decrease of the oxygen content almost to zero in the alloy surface layers indicates the melting of the sample.

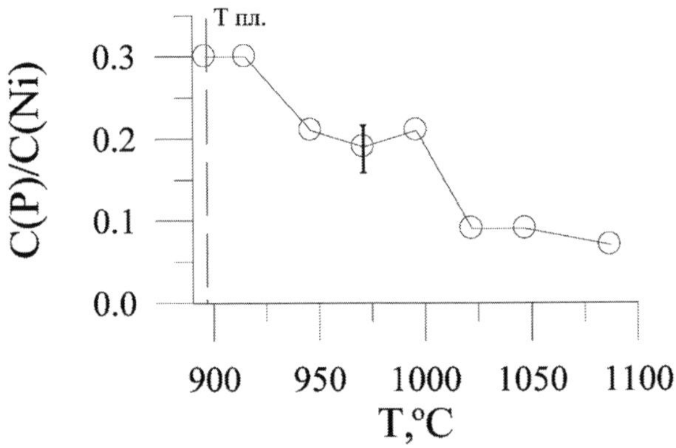

FIGURE 9.3 The change of the concentration ratio C(P)/C(Ni) in the surface layers of the $Ni_{81}P_{19}$ melt.

Figure 9.4 shows the change of the composition of the surface layers of the $Ni_{82}B_{18}$ alloy as the temperature increases from 1050 to 1400°C. The chosen criterion is the concentration ratio C(B)/C(Ni). The behavior of the change of the concentration ratio C(B)/C(Ni) differs from that of the change of the concentration ratio C(P)/C(Ni) for the $Ni_{81}P_{19}$ alloy. When the temperature increases, the growth of the boron concentration is observed in the surface layers. At the temperature 1270°C, the jump-like variation of the composition of the melt surface layers is observed. We believe that such change is connected with a change in the nearest surrounding of atoms in the clusters present on the melt surface. Two temperature regions, that is, from T_{melt} to 1270°C and from 1270 to 1400°C, can be singled out. In the first region, the Ni–B clusters with a relatively small content of boron are stable, and in the second region, the Ni–B clusters with a higher content of boron are stable. The change of the cluster composition of the surface layers of the $Ni_{82}B_{18}$ melt at the temperature 1270°C is considered as a structural transformation in the liquid state.

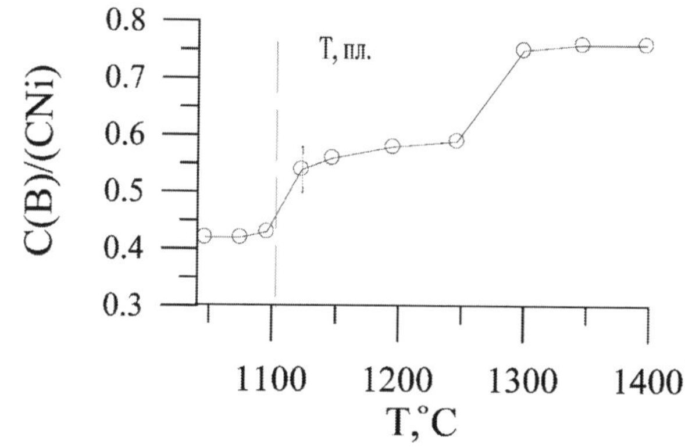

FIGURE 9.4 The change of the concentration ratio *C*(B)/C(Ni) in the surface layers of the $Ni_{82}B_{18}$ melt.

Similar investigations have been conducted for the alloys $Ni_{84}P_{16}$ and $Ni_{78}P_{22}$. Figures 9.5 and 9.6 present the changes of the composition of the surface layers of the $Ni_{84}P_{16}$ hypoeutectic alloy as the temperature increases from T_{melt} to 1400°C and the $Ni_{78}P_{22}$ hypereutectic alloy at the temperature increase from T_{melt} to 1200°C.

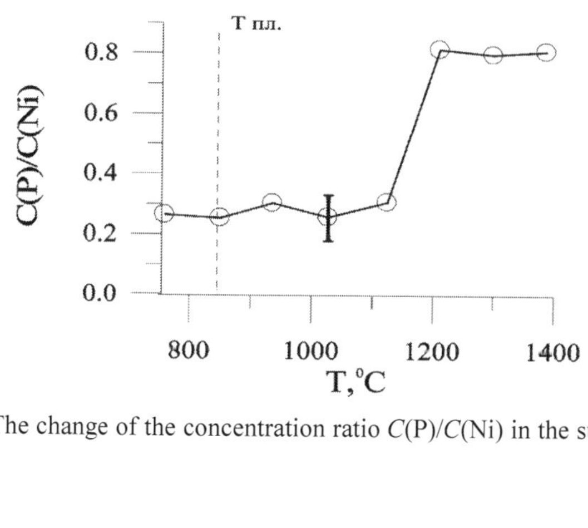

FIGURE 9.5 The change of the concentration ratio *C*(P)/*C*(Ni) in the surface layers of the $Ni_{84}P_{16}$ melt.

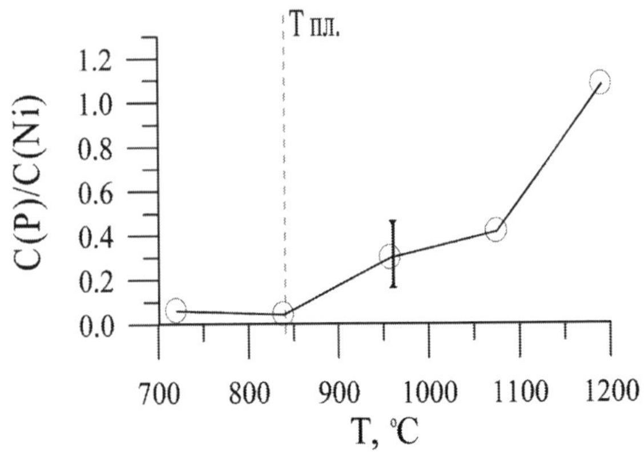

FIGURE 9.6 The change of the concentration ratio *C*(P)/*C*(Ni) in the surface layers of the $Ni_{78}P_{22}$ melt.

At heating from 850 to 1200°C (the $Ni_{84}P_{16}$ melt), the relative composition of the surface layers changes insignificantly, within the experimental error. The change of the melt surface composition occurs in a jump-like manner at the temperature of 1200°C; however, in contrast to the $Ni_{89}P_{19}$ system, the concentration of metalloid on the surface increases, the concentration ratio of phosphorus and nickel grows from 0.3 to 0.8. In the temperature region from 1200°C and above, the Ni–P and P–P clusters come out upon the surface, which is indicated by the beginning of oxidation and the appearance of additional maxima at the distance of 3 and 5 eV from the main peak in the Ni3p and P2p spectra. When the temperature increases, phosphorus starts evaporating.

The temperature investigation of the $Ni_{78}P_{22}$ melt shows the similar dependence of the variation of the relative chemical composition of the surface layers at heating (Fig. 9.6). The increase of the number of clusters with the increase in the phosphorus content on the melt surface is observed at the temperature of 1180°C. At further heating, the melt surface becomes unstable and the phosphorus selective evaporation takes place.

We think that for the formation of strong covalent bonds in the nearest surrounding of nickel atoms, a certain number of phosphorus atoms are required. This number is determined by the spatial and energy overlapping of the wave functions of the valence electrons of both the alloy components. In the Ni–P alloy, the required number of P atoms in the nearest surrounding of Ni atoms should be more than 2. Such composition on the liquid state surface is only characteristic of the eutectic alloy $Ni_{81}P_{19}$. At the deviation of the composition from the eutectic one, less strong clusters are formed on the melt surface, and at heating they are destructed.

The results of the investigation of the composition of the $Ni_{84}P_{16}$, $Ni_{81}P_{19}$, $Ni_{78}P_{22}$, and $Ni_{82}B_{18}$ melt surface layers at isothermal holdings are of special interest. The concentration ratios *C*(P)/*C*(Ni) and *C*(B)/*C*(Ni) for corresponding alloys are chosen as a criterion.

Figure 9.7 shows the change of the concentration ratio *C*(P)/*C*(Ni) with time. It can be seen, that the value of *C*(P)/*C*(Ni) decreases with time from 5.5 to 3.5; the scatter of points relative to the approximation line decreases and is within the experimental error. Such behavior of the surface layers composition corresponds to the transition to the equilibrium state. In the given case, the equilibrium state is understood as the absence of changes of the composition of the melt surface layers with time. We think that the process of the redistribution of the clusters on the surface and the clusters in the near-surface region of the melt takes place due to the presence of the temperature gradient in the melt and the necessity to minimize the melt surface energy. As the result of the action of the above factors, a nonmonotonic variation of the composition of the surface layers is observed.

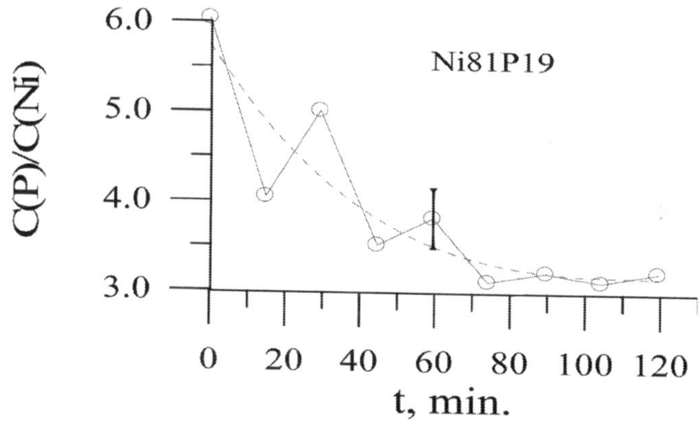

FIGURE 9.7 The change of the concentration ratio *C*(P)/*C*(Ni) in the surface layers of the $Ni_{81}P_{19}$ melt at the isothermal holding at 980°C.

For the $Ni_{82}B_{18}$ melt (Fig. 9.8), a different pattern of the change of the surface layers composition is observed at the isothermal holding. The changes of the surface composition are nonmonotonic (the criterion is *C*(B)/*C*(Ni). In contrast to the case with $Ni_{81}P_{19}$, during 200 min, the transition of the melt to the equilibrium state is not observed. Such difference in the behavior of the surface layers of $Ni_{82}B_{18}$ and $Ni_{81}P_{19}$ can be due to that the clusters Ni–B have less strong bonds inside a cluster in contrast to the Ni–P clusters.

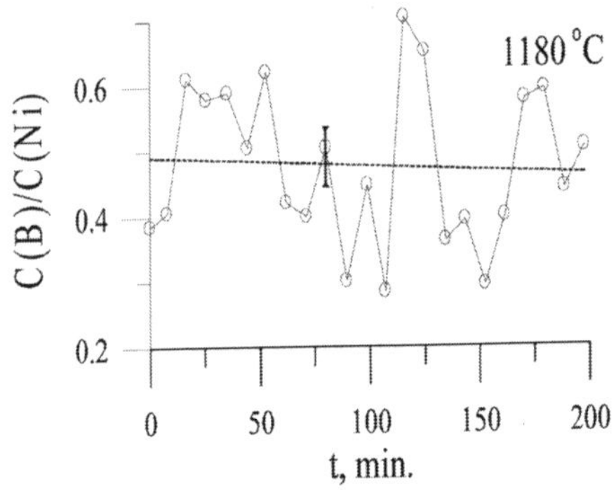

FIGURE 9.8 The change of the concentration ratio $C(B)/C(Ni)$ in the surface layers of the $Ni_{82}B_{18}$ melt at the isothermal holding at 1180°C.

Similar measurements have been conducted for the melts $Ni_{84}P_{16}$ and $Ni_{78}P_{22}$. As an example, in Figures 9.9 and 9.10, the changes of the concentration ratio $C(P)/C(Ni)$ in the $Ni_{84}P_{16}$ and $Ni_{78}P_{22}$ melt surface layers are shown at the isothermal holdings. The equilibrium state has been reached at none of the temperatures.

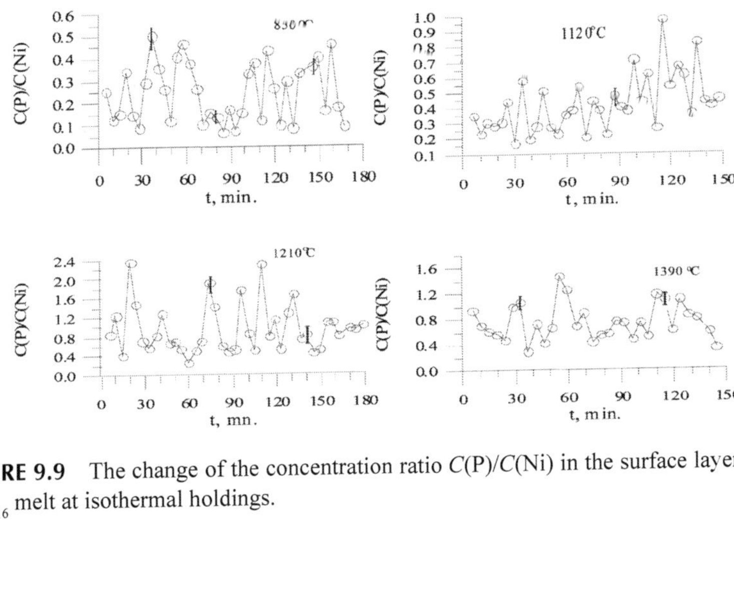

FIGURE 9.9 The change of the concentration ratio $C(P)/C(Ni)$ in the surface layers of the $Ni^{84}P_{16}$ melt at isothermal holdings.

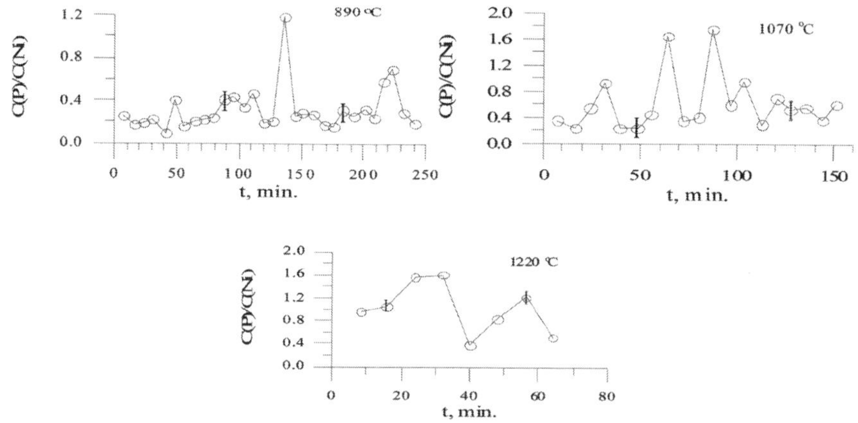

FIGURE 9.10 The change of the concentration ratio $C(P)/C(Ni)$ in the surface layers of the $Ni_{78}P_{22}$ melt at isothermal holdings.

The studied changes of the composition of the surface layers indicate complex character of the surface change of the studied melts, which allows to state that the surface of the melts cannot be ascribed to equilibrium systems.

For the quantitative evaluation of the instability of the surface layers of the studied melts, the quantity $\xi = 1/\bar{a}\sqrt{\sum(\bar{a}-a)^2/N(N-1)}$ is offered, where a – $C(P)/C(Ni)$ for the melts $Ni_{84}P_{16}$, $Ni_{81}P_{19}$, $Ni_{78}P_{22}$; a – $C(B)/C(Ni)$ for $Ni_{82}B_{18}$; N is the number of measurements at each of the temperatures. Figure 9.11 shows the instability x for the surface layers of the melts $Ni_{84}P_{16}$, $Ni_{81}P_{19}$, $Ni_{78}P_{22}$, and $Ni_{82}B_{18}$. In the figure, it can be seen that the least instability is observed for the melts with the eutectic composition. For the alloys with the noneutectic composition, the increase in the instability of the composition of the surface layers of the melts $Ni_{84}P_{16}$ and $Ni_{78}P_{22}$ is observed. Such behavior of the surface layers of the melts can be due to the processes of mixing the surface clusters with the near-surface region clusters and due to the formation of stronger bonds in the clusters of the eutectic composition elements of $Ni_{81}P_{19}$ and $Ni_{82}B_{18}$ than those in the clusters of the noneutectic composition of the $Ni_{84}P_{16}$ and $Ni_{78}P_{22}$ alloys.

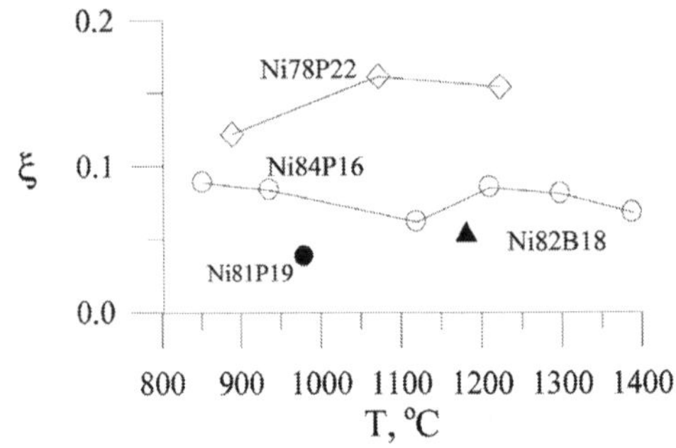

FIGURE 9.11 Instability ξ for the surface layers of the melts $Ni_{84}P_{16}$, $Ni_{81}P_{19}$, $Ni_{78}P_{22}$, and $Ni_{82}B_{18}$.

9.4.4 CONCLUSION

In the liquid state of the alloys, $Ni_{81}P_{19}$, $Ni_{82}B_{18}$, $Ni_{84}P_{16}$, and $Ni_{78}P_{22}$, jump-like variations of the composition of the surface layers have been found which are due to the change in the nearest surrounding of the atoms in the clusters. Such variations can be interpreted as structural transformations in the liquid state.

At isothermal holdings, the change of the composition of the surface layers is of nonmonotonic character, which can be due to the redistribution of the cluster composition of the surface and near-surface layers of the melt and which is determined by the strength of the chemical bond in the clusters forming the melt surface. The transition to the equilibrium state is only observed for the $Ni_{81}P_{19}$ eutectic composition.

The evaluation of the instability of the surface layers of the melts has been conducted. It is shown that the least instability is observed for the eutectic compositions, which allows to state that the surface of the melts cannot be referred to equilibrium systems.

KEYWORDS

- **metal melts**
- **Clusters**
- **chemical bond**
- **X-ray photoelectron spectroscopy (XPS).**

REFERENCES

1. Zamyatin, V. M.; Baum, B. A.; Mezenin, A. A.; Shmakova, K. Yu. Time Dependences of the Melt Properties: Their Meaning, Peculiarities, and Explanations. *Rasplavy* **2010,** *5*, 19–31.
2. Yelansky, G. N.; Yelansky, D. G. Structure and Properties of Metal Melts. M.: MGVMI, **2006,** 228 p.
3. Vatolin, N. A.; Pastukhov, E. A.; Kern, E. M. The Influence of Temperature on the Structure of Melted Iron, Nickel, Palladium and Silicon. *Doklady Akademii Nauk* **1974,** *217*(1), 127–130.
4. Popel, P. S. Metastable Microheterogeneity of the Melts in the Systems with Eutectic and Monotectic and its Influence on the Alloy Structure after Solidification. *Rasplavy* **2005,** *1*, 22–48.
5. Regel, A. R.; Glazov, V. M. Physical Properties of Electron Melts. M.: Nauka, 1980, p 296.
6. Siegbahn, K.; Nordling, K.; Falman, A.; Nordberg, R.; Hamrin, K.; Hedman, J. Electron Spectroscopy. M.: Mir, 1971, p 493.
7. Trapeznikov, V. A.; Shabanova, I. N.; Varganov, D. V.; et al. The Creation of an Automated Electron Magnetic Spectrometer for Investigating Melts: Report. VNTI Centre, No. 12880067297, 1989.
8. Kholzakov, A. V.; Isupov, N. Yu.; Ponomarev, A. G. Structural Transformations in Amorphous and Liquid States of Fe- and Co-based Systems. *Khim. Fiz. Mezosk.* **2011,** *13*(4), 516–519.

9.5 THE INFLUENCE OF THE CARBON METAL-CONTAINING NANOSTRUCTURES QUANTITIES ON THE DEGREE OF THE POLYMER MODIFICATION

In the present paper, the influence of minute additions of carbon metal-containing nanoforms is studied on the polymer structure for polycarbonate, polymethylmethacrylate and polyvinyl alcohol which have different content of oxygen. In this case the structure of the nanomodified polymers changes and becomes similar to the structure of the nanoforms. The degree of the modification depends on the amount of oxygen atoms in the polymer structure. The larger is the number of oxygen atoms, the smaller is the content of the nanoforms necessary for the polymer modification; in polycarbonate, the modification starts when the content of the nanostructures is 10^{-5}%, in polymethylmethacrylate, the modification is observed at the nanostructure content of 10^{-4}%, and for the polyvinyl alcohol modification the minimal content of 10^{-3}% is necessary.

9.5.1 *INTRODUCTION*

High structure-forming activity of nanostructures allows their use for modification of materials. It is known that the modification of different materials with minute amounts of nanostructures improves material performance characteristics. At the same time, the mechanism of such influence of nanoforms on the structure and properties of materials has not been fully clarified yet. In the present paper, the dependence of the nanomodification of polymer systems on the content of oxygen in their structure is studied, and the explanations of the process are presented based on the XPS results using the modification of polymer systems as an example.

9.5.2 *EXPERIMENT*

Carbon metal-containing nanostructures are multilayered nanotubes formed in a nanoreactor of a polymer matrix in the presence of the 3D-metal systems. Nanocarbon structures are prepared by an original procedure in the conditions of low temperatures (not higher than 400°C) which is pioneer investigations.[1]

The method for the modification of polymer with carbon metal-containing nanostructures includes the preparation of fine-dispersed suspension (FDS) based on the solution of polymer in methylene chloride. Then, carbon metal-containing nanostructures are added into the prepared FDS. For the refinement and uniform distribution of carbon metal-containing nanostructures, the prepared mixture is treated with ultrasound. To prepare a film, the solvent is evaporated from the mixture at heating to 100°C.

The XPS investigations were conducted on an X-ray electron magnetic spectrometer with the resolution 10^{-4}, luminosity 0.085% at the excitation by the Al$K\alpha$ line 1486.5 eV in vacuum 10^{-8}–10^{-10} Pa. In comparison with an electrostatic spectrometer, a magnetic spectrometer has a number of a advantages connected with the construction capabilities of X-ray electron magnetic spectrometers which are the constancy of luminosity and resolution independent of the energy of electrons, high contrast of spectra and the possibility of external actions on a sample during measurements.[2]

The study of the variations of the C1s-spectrum shape lies in the basis of the investigations of the change of the polymer structure during nanomodification.

The identification of the C1s-spectra with the use of their satellite structure has been developed allowing the determination of the chemical bond of elements, the nearest surrounding of atoms and the type of sp-hybridization of the valence electrons of carbon in nanoclusters and materials modified with them.[3–5]

Since the surface of nanostructures has low reactivity, for increasing the nanostructure surface activity functionalization is used, that is the attachment of certain atoms of sp- or d-elements to the atoms on the nanostructure surface and the formation of a covalent bond between them; the functionalization results in the formation of an interlink between the atoms of the nanostructure surface and the atoms of material.

The XPS study has shown[6] that the formation of the covalent bond between the atoms of the functional sp-groups and the atoms on the nanostructure surface is influenced by the electronegativity of the atoms of the components and the closeness of their covalent radii. Thus, it is most probable that the functionalization of nanostructures leads to the formation of the bond between phosphorus atoms and nanostructure d-metal atoms (Fe, Ni, Cu) and between nitrogen or fluorine atoms and carbon atoms of the nanostructures.

9.5.3 RESULTS AND DISCUSSION

The changes of the structure of the organic glass, polycarbonate, and polyvinyl alcohol modified with different amounts (10^{-5}%, 10^{-4}%, 10^{-3}%, 10^{-2}%, 10^{-1}%) of carbon copper-containing nanostructures have been studied.

Figure 9.1 presents the C1s-spectrum of the carbon copper-containing nanostructure, which consists of three components C–C (sp^2)—284 eV, C–H—285 eV, and C–C (sp^3)—286.2 eV.

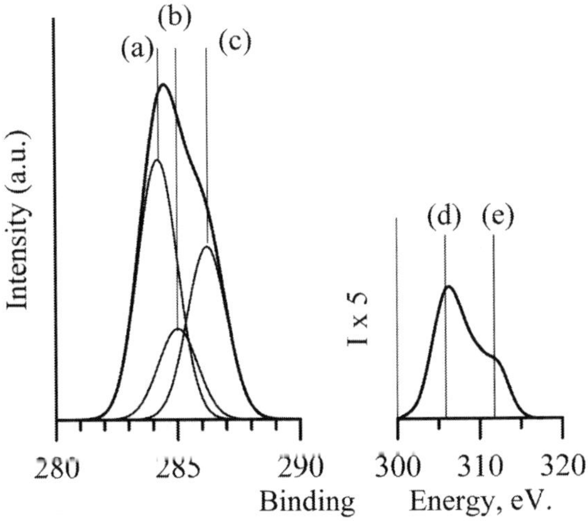

FIGURE 9.1 The XPS C1s-spectrum of carbon copper-containing nanostructures consisting of three components: (a) C–C (sp^2)—284 eV; (b) C–H—285 eV; (c) C–C (sp^3)—286.2 eV, and the satellite structure (d) satellite (sp^2); and (e) satellite (sp^3).

The presence of a small amount of the C–H component indicates an incomplete synthesis of nanostructures from the polymer matrix. The ratio of the maxima intensities of C–C (sp^2) and C–C (sp^3) depends on the dimension of the nanostructure: the larger is the surface area compared to the volume, the higher is the C1s-spectrum component with the sp^3-hybridization of valence electrons.[3]

Figure 9.2 shows the C1s-spectra of polymethylmethacrylate (chemical formula $[CH_2C(CH_3)(CO_2CH_3)]_n$).

In the reference sample, which is the film of organic glass, the bonds C–H and C–O (287 eV) prevail characterizing the organic glass structure. With increasing content of carbon copper-containing nanostructures in organic glass, starting from 10^{-4}%, the C1s-spectrum structure changes and the structure characteristic of carbon copper-containing nanostructure appears, namely, C–C (sp^2) and C–C (sp^3). The C–H component characterizes the remnants of the organic glass structure. Note that the C–O component is absent in the C1s-spectrum. During the organic glass modification, with increasing concentration of the nanostructure up to 10^{-3}%, in the C1s-spectrum the C–H component is decreasing and the C–C (sp^2) and C–C (sp^3) components characteristic of nanostructures are growing, that is, the degree of the polymer modification is growing.

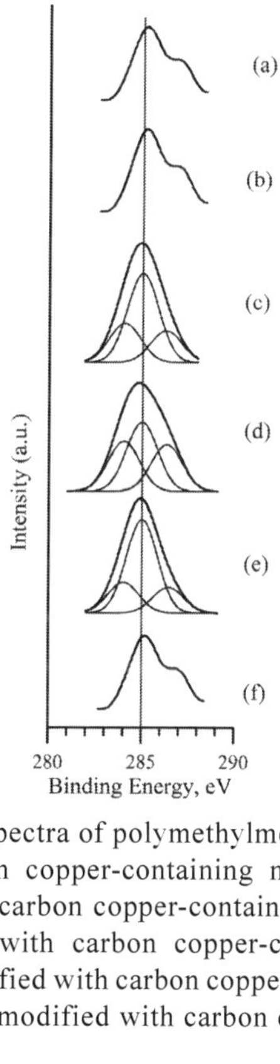

FIGURE 9.2 The XPS C1s-spectra of polymethylmethacrylate: (a) reference sample; (b) nanomodified with carbon copper-containing nanostructures in the amount of 10^{-5}%; (c) nanomodified with carbon copper-containing nanostructures in the amount of 10^{-4}%; (d) nanomodified with carbon copper-containing nanostructures in the amount of 10^{-3}%; (e) nanomodified with carbon copper-containing nanostructures in the amount of 10^{-2}%; and (f) nanomodified with carbon copper-containing nanostructures in the amount of 10^{-1}%.

With a further increase in the nanostructure concentration (10^{-2}%), the degree of the polymer modification decreases and at the nanostructure content of 10^{-1}% the modification is absent. The structure of the C1s-spectrum becomes similar to that of the C1s-spectrum of the unmodified organic glass. This can be explained by the processes of the nanostructure coagulation taking place when the nanostructure content in the polymer is high. Consequently, it is possible to judge about the degree of the polymer nanomodification by the ratio of C–C bonds to C–H bond in the C1s-spectrum. Maximal modification takes place at the nanostructure content of 10^{-3}%.

Similar results have been obtained for the nanomodification of polycarbonate (chemical formula –[–OArOC(O)OR–}$_n$–, Ar–C_6H_4–; R—(CH_2)). The C1s-spectrum of the reference sample of polycarbonate containing a large amount of oxygen also consists of two components C–H (285 eV) and C–O (287 eV); however, the relative intensity of the C–O component is significantly larger than that observed for organic glass. When the content of nanostructures is in the range of 10^{-5}–10^{-2}%, in the C1s-spectrum the structure characteristic of carbon copper-containing nanoform appears. In contrast to organic glass, in polycarbonate the change of the structure starts at the nanoform content of 10^{-5}%. In polycarbonate, at the nanostructure content of 10^{-1}%, no changes are observed in the polymer structure.

Figure 9.3 shows the C1s-spectra of polyvinyl alcohol (chemical formula [$CH_2CH(OH)]_n$) having in its structure the least content of oxygen in comparison with the other polymers under study.

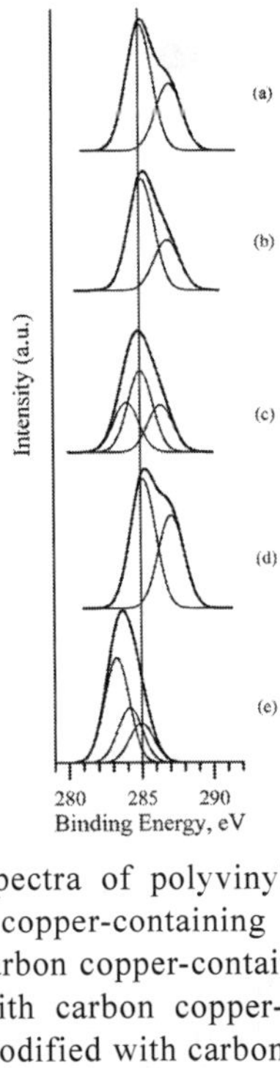

FIGURE 9.3 The XPS C1s-spectra of polyvinyl alcohol. (a) Reference sample; (b) nanomodified with carbon copper-containing nanostructures in the amount of 10^{-4}%; (c) nanomodified with carbon copper-containing nanostructures in the amount of 10^{-3}%; (d) nanomodified with carbon copper-containing nanostructures in the amount of 10^{-1}%; and (e) nanomodified with carbon copper-containing nanostructures in the amount of 1%.

Similar to the reference samples of polycarbonate and organic glass, the C1s-spectrum of the reference sample of PVA (a PVA film) contains two components C–H and C–O; however, the C–O component is less intensive compared to C–H. The change of the PVA structure is observed only when it is modified with carbon copper-containing nanostructures, the content of which is in the range of 10^{-3}–10^{-2}%. When nanoforms are added into the PVA solution, in the C1s-spectrum structure the components appear which are characteristic of the C1s-spectrum of carbon copper-containing nanostructure (Fig. 9.1). When in PVA, the nanostructure content is 10^{-1}%, no changes are observed in the polymer structure similar to the above-mentioned polymers.

At the higher content of the nanostructures (1%) in the PVA polymer nanomodification is also absent; however, in contrast to the samples with the content of nanostructures of 10^{-1}%, the absence of oxygen atoms (there is no component C–O in the C1s-spectrum) and the decrease of the component characterizing the C–H bond are observed. Consequently, the PVA carbonization takes place in the presence of the nanostructures as catalysts of the reaction in which the bonds between carbon and oxygen and hydrogen are broken and deoxygenation and dehydrogenization and the formation of C–C bonds occur. The C1s-spectrum consists of three components C–C (283.2 eV), C–C with sp^2-hybridization (284.2 eV), and C–H (285 eV). The presence of a small amount of the C–H component in addition to the C–C component indicates incomplete carbonization. Since the binding energy of the C–C component (283.2 eV) is smaller than that of the C–C component with the sp^2-hybridization of the valence electrons (284.2 eV), the formation of the C–C component with sp-hybridization valence electrons of the atoms carbon can be suggested.

A quite definite amount of the nanostructures is required for the modification of each of the studied polymers, that is, for changing their structure. The XPS studies show that the smaller is the number of oxygen atoms bound to carbon in the polymer, the larger is the amount of nanoparticles necessary for changing the polymer structure. For polycarbonate, it is 10^{-5}%, for organic glass—10^{-4}%, and for polyvinyl alcohol—10^{-3}%. It is associated with different reactivity of the C–O bond in polymers. When in the polymer the amount of nanoparticles is in the range from minimal up to 10^{-1}%, the C1s-spectrum changes; the component C–O (277.0 eV) disappears and the components C–C (sp^2) and C–C (sp^3) characteristic of a carbon metal-containing nanoform appear. The comparison of the structure of the studied polymers show high reactivity of the C–O bond in the CO_3 group in polycarbonate; the C–O bond reactivity decreases in the CO_2 group in organic glass, and its further decrease is observed in the CO group in polyvinyl alcohol. The disappearance of the component characterizing the C–O bond in the C1s-spectrum indicates the possibility of the replacement of this group of atoms by nanoparticles, that is, the formation of strong bonds between the polymer atoms and the atoms of the nanostructure surface. In this case, the polymer gains the structure-forming activity of nanostructures which are the centers of a new appearing structure. The breakage of the chemical bond of the C–O group with the nearest environment of the polymer atoms due to a different reactivity of the C–O bond takes place at different minimal contents of nanoparticles in the polymers.

The investigation of vaseline oil (chemical formula $[CH_2]_m$) that does not contain oxygen atoms in its structure shows the absence of the vaseline oil modification at any content of nanoforms. The C1s spectrum has the only C–H component similar to the reference sample.

Based on the results obtained it can be suggested that the larger is the content of oxygen atoms in the bond with carbon in the studied polymers, the larger is a change in the polymer structure and the larger is the formation of the regions in the polymer structure, which have a similar structure to that of carbon metal-containing nanoforms. The smaller is the number of oxygen atoms in the starting polymer, the larger is the amount of nanoforms necessary for the polymer structurization and the transformation of the polymer structure into the one similar to the nanoform structure during the modification.

The experimental data show that the modification of the polymers occurs in the range from room temperature up to 100°C; at higher temperatures the modification is absent because the polymer decomposition starts.

9.5.4 CONCLUSION

In the present paper, it is shown that the degree of the nanostructure influence on the interaction with a polymer is determined by the content of nanostructures and their activity in this medium. The temperature growth blocks the development of self-organization in the medium. Thus, for describing the medium structurization process under the nanostructure influence it is necessary to enter some critical parameters, namely, the content and activity of nanostructures and critical temperature.

The change of the polymer structure is accompanied by the change of their technological properties: the tensile strength of the studied films improves by 13%, the surface electrical resistance decreases by a factor of

3.3, and the transmission density of the films increases in the region close to an infrared one, which leads to an increase in the polymer heat capacity.

Thus, the X-ray studies show that

1. For obtaining the maximal degree of the modification of the studied polymers, it is necessary that the content of carbon copper-containing nanoforms in them would be ~10^{-3}%. In this case, the structure of the nanomodified polymers changes and becomes similar to the structure of the nanoform.
2. The larger is the content of oxygen atoms bound to carbon atoms in the polymers, the larger is the degree of the polymer modification and the smaller is the percentage of nanoforms required for the beginning of modification.
3. High percentage of nanostructures can lead to the nanostructure coagulation and the absence of the polymer modification. In the studied polymers, it is observed at the nanostructure content of 10^{-1}%.
4. When oxygen atoms bound to carbon atoms are absent in the polymer structure, modification is not observed at any content of nanostructures.
5. At the higher content of nanostructures (1%), in the studied polymers the process of partial carbonization is observed, that is, the removal of O and H atoms and the formation of the C–C bonds with sp-hybridization of the valence electrons.

KEYWORDS

- **polymethylmethacrylate (organic glass, PMMA)**
- **polycarbonate**
- **polyvinyl alcohol (PVA)**
- **carbon copper-containing nanostructures**
- **X-ray photoelectron spectroscopy (XPS)**
- **satellite structure of C1s-spectra**
- **functionalization**
- **modification**

REFERENCES

1. Kodolov, V. I.; Khokhryakov, N. V. *Chemical Physics of the Processes of the Formation and Transformations of Nanostructures and Nanosystems*. Iz-vo IzhGSHA, Izhevsk, 2009. In two volumes. Vol 1, 360 p. Vol 2, 415 p.
2. Shabanova, I. N.; Dobysheva, L. V.; Varganov, D. V.; Karpov, V. G.; Kovner, L. G.; Klushnikov, O. I.; Manakov, Yu. G.; Makhonin, E. A.; Khaidarov, A. V.; Trapeznikov, V. A. New Automated X-ray Electron Magnetic Spectrometer: Spectrometer with Technological Manipulators and Spectrometer for Investigation of Melts. *Izv. AN SSSR. Ser. fiz.* **1986,** *50*(9), 1677–1682.
3. Makarova, L. G.; Shabanova, I. N.; Terebova, N. S. Application of X-ray Photoelectron Spectroscopy to Study the Chemical Structure of Carbon Nanostructures. Zavodskaya laboratoria. *Diagn. Mater.* **2005,** *71*(5), 26–28.
4. Makarova, L. G.; Shabanova, I. N.; Kodolov, V. I.; Kuznetsov, A. P. X-ray Photoelectron Spectroscopy as a Method to Control the Received Metal–Carbon Nanostructures. *J. Electr. Spectr. Rel. Phen.* **2004,** *137–140*, 239–242.
5. Shabanova, I. N.; Kodolov, V. I.; Terebova, N. S.; Trineeva, V. V. X-ray Photoelectron Spectroscopy in the Investigation of Carbon Metal-containing Nanosystems and Nanostructured Materials. Iz-vo "Udmurtski Universitet": Izhevsk, 2012, 250 p.
6. Shabanova, I. N.; Terebova, N. S. Dependence of the Value of the Atomic Magnetic Moment of d-Metals on the Chemical Structure of Nanoforms. *J. Nanosci. Nanotechnol.* **2012,** *12*(11), 8841–8844.

9.6 INVESTIGATION OF THE INTERATOMIC INTERACTION IN THE IMMUNOGLOBULIN G FRAGMENTS BY THE METHOD OF X-RAY PHOTOELECTRON SPECTROSCOPY

ABSTRACT

A vaccine based on the Fc fragment of immunoglobulin G has been developed for treating a human for rheumatoid arthritis; its biological activity depends on the purity of the separation of the fragments Fc and Fab. In the present work, the chemical structure of the immunoglobulin fragments Fc and Fab has been studied by the method of X-ray photoelectron spectroscopy, which has allowed to determine the purity of the separation of the fragments.

9.6.1 INTRODUCTION

Earlier, we have shown that Fc fragments of rat's immunoglobulin G can contain antigenic determinants for rheumatoid factor inhibiting autoimmune reactions.[1] A vaccine has been developed for treating a human for rheumatoid arthritis, which contains Fc fragments of homologous immunoglobulin as an active principle.[2] However, in the process of separation and purification of the immunoglobulin fragments Fc and Fab, an uncontrolled change in the composition of the fragments can take place, which, in its turn, leads to the loss of the vaccine biological activity. In this connection, it becomes necessary to control the purity of the separation of the Fc and Fab fragments in the process of their obtaining and purifying.

Chemical and biological methods are not capable of determining the purity of the immunoglobulin fragments' separation with a specified accuracy. Therefore, it was necessary to develop other effective methods to control the purity of the separation of the immunoglobulin fragments. First of all, the method of X-ray photoelectron spectroscopy (XPS) has drawn interest as it is nondestructive and the most informative as far as the investigation of interatomic interaction is concerned. The possibility of using the XPS method has been evaluated for determining the chemical state of the functional groups in the Fc- and Fab-fragments of immunoglobulin of a human and a rat.

9.6.2 EXPERIMENTAL

Immunoglobulin IgG was separated from serum of rats by sedimentation with ammonium sulfate; after that, it was purified from impurities by the method of anion-exchange chromatography on a DEAE-sefarose. The purity was analyzed in polyacrylamide gel in the presence of SDS by electrophoresis. The Fc- and Fab-fragments of immunoglobulin IgG were obtained by papain hydrolysis in the presence of a reducing agent.[3,4] The separation of the Fc- and Fab-fragments was conducted on a protein G-sefarose in the mode of sorption of the Fab-fragments. The obtained Fc-fragments were analyzed by electrophoresis.[3]

The XPS investigations were conducted on an X-ray electron magnetic spectrometer with the resolution 10^{-4}; the device luminosity was 0.085% at the excitation by Al*K*α-line 1486.5 eV, in vacuum 10^{-8}–10^{-10} Pa. In comparison with an electrostatic spectrometer, the magnetic spectrometer has a number of advantages due to constructive possibilities of X-ray electron magnetic spectrometers, which are the constancy of luminosity and resolution independent of the electron energy, high contrast of spectra, and the possibility of external acting upon a sample during measurements.[5] The investigations of changes in the immunoglobulin structure are based on studying changes in the shape of C1s, N1s, and S2p electron spectra.

9.6.3 RESULTS AND DISCUSSION

The C1s, O1s, N1s, and S2p spectra of the core levels of samples of the native and split into fragments immunoglobulin were studied at the temperatures from room temperature to 473 K. To study the state of atoms of carbon, oxygen, sulfur, and nitrogen, the investigation of the reference samples of amino acids (glycine, histidine, and albumin) was conducted.[6] Based on the results of the XPS investigations, the spectra parameters were established, which allowed to determine the quality of the immunoglobulin fragmentation.

Figures 9.1–9.3 show the X-ray photoelectron 1s-spectra of carbon and nitrogen and 2p-spectra of sulfur at room temperature and at heating.

At room temperature, the immunoglobulin C1s-spectrum (Fig. 9.1a) consists of four components due to different surrounding of carbon atoms: C–C with the sp^2-hybridization of the valence electrons (283.7 eV), C–H (285 eV), C–N(H) (286.5 eV), and COOH (289.1 eV). A satellite at 306 eV[7] also indicates the presence of the C–C bond. The immunoglobulin

N1s-spectrum (Fig. 9.1a) consists of the component indicating the bonds of nitrogen with hydrogen (N–H). The S2p-spectrum consists of one component which is close to NH_2–S–S–NH_2 by the energy of binding.

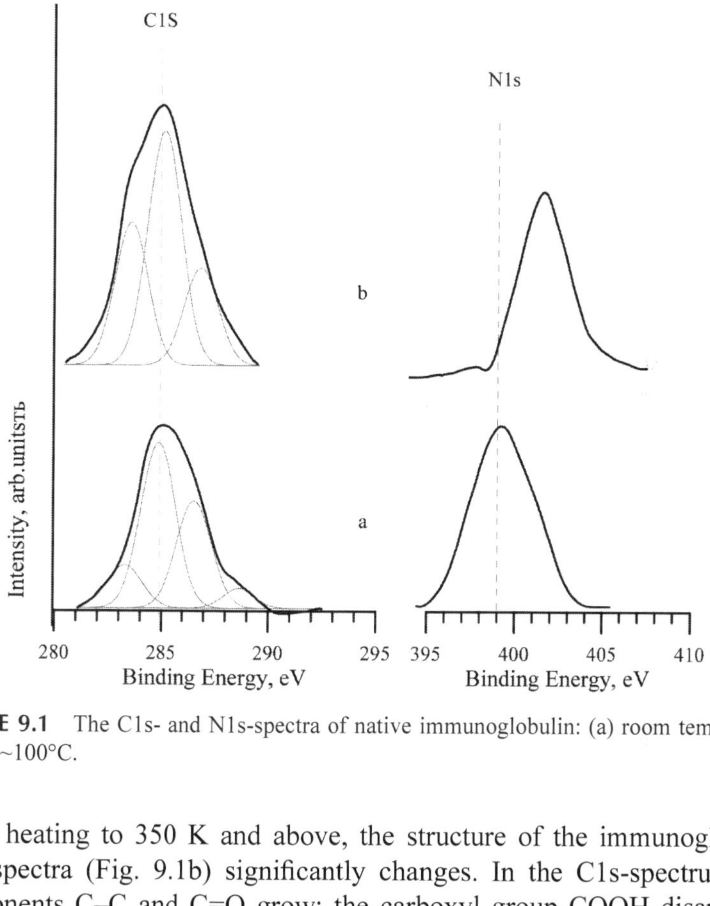

FIGURE 9.1 The C1s- and N1s-spectra of native immunoglobulin: (a) room temperature and (b) ~100°C.

At heating to 350 K and above, the structure of the immunoglobulin XPS spectra (Fig. 9.1b) significantly changes. In the C1s-spectrum, the components C–C and C=O grow; the carboxyl group COOH disappears; the components C–C, CH, and carbonyl group C=O (287 eV) remain. In the N1s-spectrum, the component NH disappears, and oxidized nitrogen N–O (401 eV) appears. In the S2p-spectrum, the contribution of the NH_2–S–S–NH_2 component decreases, and the component S–O (168 eV) appears. Consequently, the oxidation of the immunoglobulin leads to its decomposition at the increase of temperature. The concentration of the carbonyl groups relative to carboxyl ones indicates the degree of the immunoglobulin decomposition.[6] It should be noted that at heating, the C–C bonds grow, which indicates the partial break of C–H bonds. Thus, we have determined the parameters of the XPS spectra characterizing the immunoglobulin structure. The components N–H in the N1s-spectrum, C–N(H) in the C1s-spectrum, and NH_2–S–S–NH_2 in the S2p-spectrum are responsible for the native form of the immunoglobulin. The absence of the above components in the C1s-, N1s-, and S2p-spectra and the growth of the N–O, S–O, and C=O bonds indicate the oxidative immunoglobulin decomposition. Thus, the method for determining the spectra parameters characterizing the immunoglobulin structure has been worked out, and the criteria of the immunoglobulin decomposition have been established.

The structure of the XPS C1s-, N1s-, S2p-, and O1s-spectra of the Fab and Fc fragments of the rat and human immunoglobulin has been investigated.

Figure 9.2 shows the C1s-, N1s-, and S2p-spectra of the Fab fragment of the human immunoglobulin G.

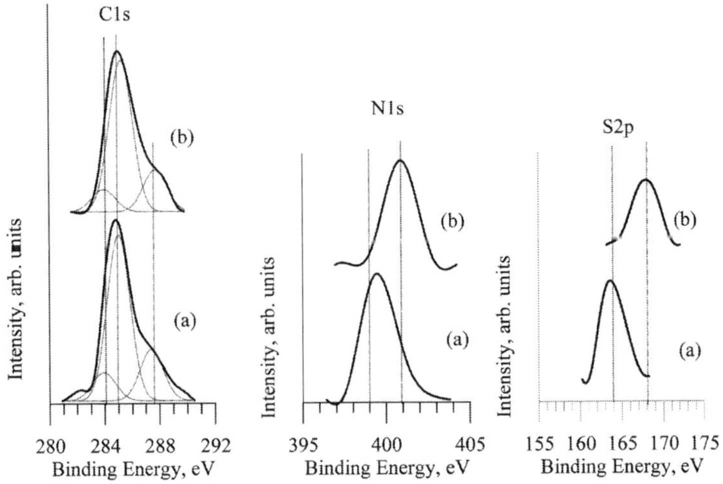

FIGURE 9.2 The C1s, N1s, S2p-spectra of the Fab-fragment: (a) room temperature and (b) ~100°C.

The C1s-spectrum of the immunoglobulin Fab fragment (Fig. 9.2a) consists of three components C–C (284 eV), C–H (285 eV), and C–O (288 eV).

At room temperature, in the N1s-spectrum there are bonds N–H and N–O. The chemical bond N–H is weak. At heating above 30–40°C it breaks, and the N–O bond only remains.

In the S2p-spectrum, at room temperature, the S–S component is observed for sulfur atoms. At small heating to 40°C, in the spectrum, the

S–O component appears. Thus, in the Fab fragment, atoms N and S dot not form strong bonds with the surrounding atoms, and the Fab fragment easily decomposes.

The investigation of the Fc fragment of the human immunoglobulin shows that in the C1s-spectrum of the Fc fragment (Fig. 9.3), in addition to the C–H and C–O bond components, there is C–N(H) bond component. In the Fc fragment, the bond C–N(H) is stronger than N–H in the Fab fragment, which is determined by the hybridization of the valence p electrons of nitrogen and carbon atoms. At heating to 100°C, the chemical bond between the atoms of nitrogen and carbon remains; at further heating, the C–N(H) starts breaking, and the oxidation of nitrogen is observed. In the sulfur S2p-spectrum, the S–S bond remains up to 100°C; at heating above 100°C, it is not observed. Such difference in the strength of the chemical bond of the atoms of nitrogen and sulfur in the Fa and Fc fragments apparently is due to the difference in their atomic structures, most probable due to the difference in the distance between the atoms of N, S, and C and their surroundings, and, consequently, due to the overlapping of the wave functions of their valence electrons.

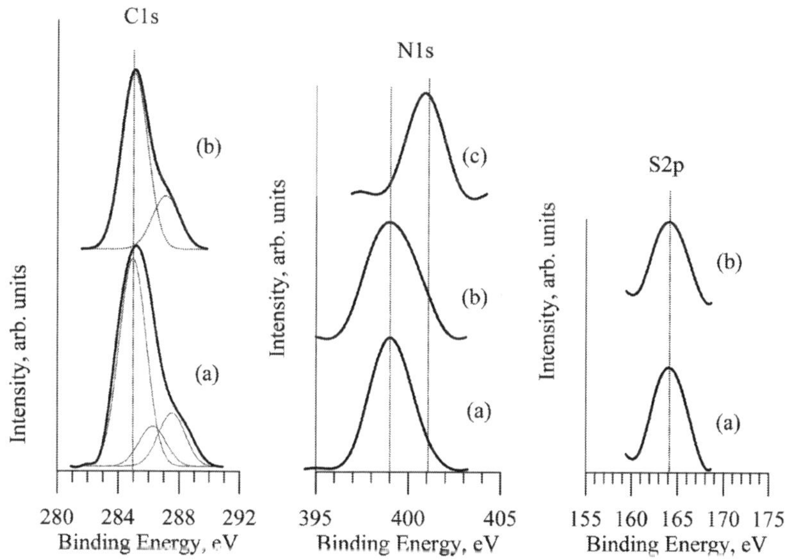

FIGURE 9.3 The C1s N1s, S2p-spectra of the Fc-fragment: (a) room temperature and (b) ~100°C.

For the rat immunoglobulin, there are difficulties in the separation of the Fc and Fab fragments. The comparative analysis of the XPS spectra of three samples of Fc-fragments, which were obtained by the same method for the Fc and Fab separation (hydrolysis), has been conducted. The XPS spectra C1s and N1s of all the three samples of Fc fragments differ from one another by the energy of binding and components, which indicates different nearest surrounding of the component atoms. Only for one sample of the Fc-fragment of the rat immunoglobulin, one can observe the results similar to those for the Fc-fragment of the human immunoglobulin. For the other two samples of the Fc fragment of the rat immunoglobulin, oxidative decomposition and the absence of the sulfur S2p-spectrum are observed.

9.6.4 CONCLUSION

The results obtained allow to draw the following conclusions:

In the immunoglobulin Fc fragment, a strong chemical bond of atoms of nitrogen and sulfur with carbon atoms is formed due to the hybridization of their valence electrons with p electrons of carbon atoms.

In the Fab fragment, atoms of nitrogen and sulfur form weak bonds with carbon atoms, which are easily broken at external actions.

The separation of the rat immunoglobulin G into the Fc and Fab fragments by papain hydrolysis does not guarantee 100%-purity of the fragments.

Thus, the results of the present investigation show that the method of X-ray photoelectron spectroscopy allows controlling the purity of the separation of the immunoglobulin into fragments, which is required for the creation of a vaccine; the method can be used for the control over the purity of the obtained Fc fragments.

KEYWORDS

- **X-ray photoelectron spectroscopy (XPS)**
- **interatomic interaction**
- **immunoglobulin (IgG)**
- **immunoglobulin fragments Fc and Fab**
- **method of papain hydrolysis**

REFERENCES

1. Beduleva, L.; Menshikov, I.; Stolyarova, E.; Fomina, K.; Lobanova, O.; Ivanov, P.; Terentiev, A. Rheumatoid Factor in Idiotypic Regulation of Autoimmunity. *Int. J. Rheum. Dis.* **2014.** DOI:10.1111/1756-185X.12335. [Epub ahead of print]
2. Menshikov, I. V.; Beduleva, L. V. The Use of Fc Fragments of Immunoglobulin G as an Antigen for Treating for Rheumatoid Arthritis: Means and Way of Treatment. Patent No. 2385164. 2010.
3. Terentyev, A. S.; Sidorov, A. Yu.; Beduleva, L. V.; Stolyarova, I. V.; Menshikov, I. V. Vestnik UdGU. *Biologia. Nauka o zemle* **2012,** *4*, 91.
4. Utsumi, S. Stepwise. *Biochem. J.* **1969,** *112*, 343.
5. Trapeznikov, V. A.; Shabanova, I. N.; Varganov, D. V.; Karpov, V. G.; Kovner, L. G.; Klushnikov, O. I.; Manakov, Yu. G.; Makhonin, E. A.; et al. *Izv. AN SSSR. Ser. fiz.* **1986,** *50*(9), 1677.
6. Shabanova, I. N.; Terebova, N. S.; Naimushina, E. A. *J. Electron Spectrosc. Related Phenom.* **2012,** *185*, 609.
7. Makarova, L. G.;. Shabanova, I. N.; Terebova, N. S. Zavodskaya laboratoria. *Diagnos. Mater.* **2005,** *71*(5), 26.

9.7 THE COMPARATIVE ELECTROCHEMICAL AND XPS INVESTIGATION ON PROTECTIVE LAYERS OF CORROSION INHIBITORS—THE ZINC COORDINATION COMPLEXES WITH NITRILO TRIS–(METHYLENEPHOSPHONIC) ACID

ABSTRACT

The protective properties and electronic structure of the surface layers of two corrosion inhibitors—the zinc coordination complexes with nitrilo tris–(methylenephosphonic)—having different atomic-molecular structure on carbon steel are investigated. It is shown that they possess different protective properties. The bridging complex $[Zn(H_2O)_3\mu\text{-}N(CH_2PO_3)_3H_4]$, where a Zn atom is octahedrally coordinated by six oxygen atoms, forms low-spin complexes with iron, the structure of which is almost similar to that of the complexes formed by the free $N(CH_2PO_3)_3H_6$ ligand. The chelate complex $[ZnN(CH_2PO_3)_3]Na_4\cdot 13H_2O$, where a Zn atom is coordinated in the trigonal pyramid configuration containing a nitrogen atom and four oxygen atoms, forms a layer of surface complexes containing Fe atoms in a high-spin state, which determines better protective properties of $Na_4[ZnN(CH_2PO_3)_3]\cdot 13H_2O$ as a corrosion inhibitor. The electronic spectra parameters correlating with the inhibitors protective properties are determined.

9.7.1 INTRODUCTION

It is known that when organic-phosphonic acids interact with metals, mainly with *d*-elements, products are formed capable of inhibiting corrosion of metals, particularly, of steel.[1] The anticorrosion effectiveness of the products of the interaction between different organic-phosphonic acids and different metals was thoroughly studied by Kuznetsov.[2,3] The main mechanisms of the influence of zinc-phosphonic adducts on the kinetics of steel electrochemical corrosion were revealed: (1) precipitation of zinc hydroxocomplexes shielding the surface of metals; (2) oxygen-free passivation of the steel surface with the layer of the surface iron–zinc-phosphonate complexes; (3) modification of the crystal lattice of a magnetite protective film; (4) alloying of magnetite protective film by the complex-forming metal ions. Kuznetsov et al. studied the influence of the nature of a metal[4,5] and organic-phosphonic ligand[6] on the anticorrosion effectiveness of the obtained adducts. Later, particularly in Refs. [7,9], a number of investigations were conducted on the protective action of zinc-phosphonic adducts confirming Kuznetsov's conclusions in general. Some of zinc–phosphonic adducts are produced as corrosion inhibitors in industry.

It is very strange, that the products of the interaction of zinc and organic-phosphonic acids have not been isolated as separate substances and studied for a long time as the conventional approach in chemistry requires.

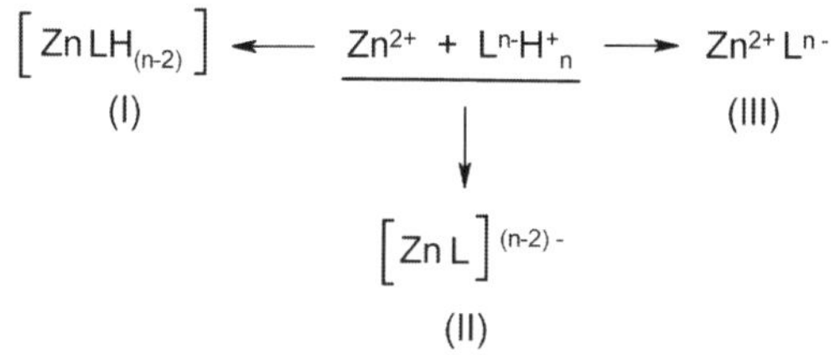

FIGURE 9.1 The scheme of the interactions of Zn^{2+} ions with organic-phosphonic acid ($L^{n-}H^{+}_{n}$).

Interactions of metal ions with organic-phosphonic acids (Fig. 9.1) lead to the formation of protonated coordination complexes (I) and deprotonated coordination complexes (II) with a different degree of the inner-sphere hydration and the formation of outer-sphere of ionic compounds (III)[10,11] as well. The quantitative ratio of the reaction products strongly varies at the change of pH and concentration of reagents; however, in practice, the mixture of different products is always formed. The results of the investigations on the anticorrosion influence of an obtained mixture do not allow making any conclusions concerning inhibiting properties of particular products of the reaction and this diminishes the scientific value of the respective studies.

The interaction of the Zn^{2+} ions with nitrilo tris–(methylenephosphonic) acid $N(CH_2PO_3)_3H_6$ (NTP) can lead to the formation of a nonelectrolytic protonated bridging three-aqua zinc nitrilo tris–(methylenephosphonate) compound $[Zn(H_2O)_3\mu\text{-}N(CH_2PO_3)_3H_4]$ (Zn_ONTP). The compound was first obtained and studied by Demadis et al.[12] The structure of the compound (CCDC 257224) is displayed in Fig. 9.2(a). The gravimetric corrosion experiment showed the inhibition coefficient of oxygen-type corrosion of carbon steel in neutral water environment equal to 270%, or $\gamma = V/V(i) = 2.7$, where V is the corrosion rate in the absence of the inhibitor (in the blank experiment) and $V(i)$ was the corrosion rate in the presence of the inhibitor. The value $\gamma = V/V(i) = 2.7$ corresponded to the degree of protection $Z = [V - V(i)]/V = 1 - \gamma^{-1} = 0.63$.

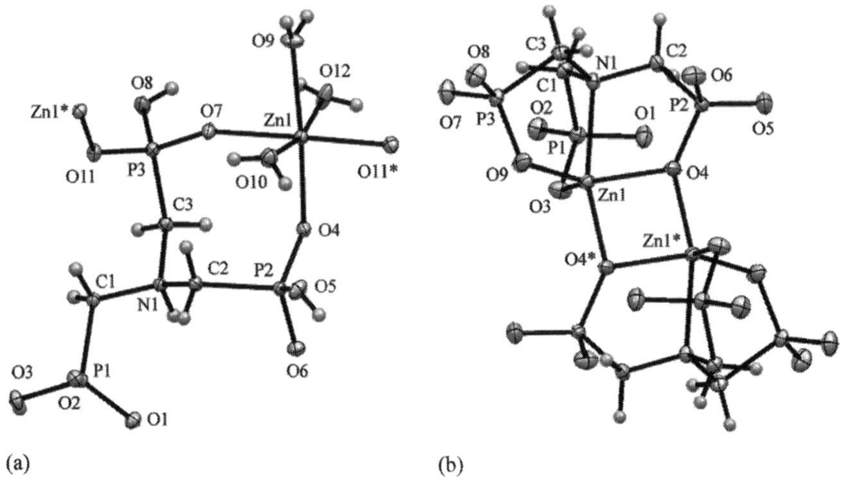

FIGURE 9.2 The structure of the products of the interaction of Zn^{2+} ions with nitrilo tris–(methylenephosphonic) acid: (a) Zn_ONTP; (b) $Zn_{N,O}$NTP (inner coordination sphere).

The interaction of Zn^{2+} with NTP in the alkaline environment (in the presence of NaOH) leads to the formation of the sodium salt of the full deprotonated chelate nitrilo tris–(methylenephosphonato) zincate $[ZnN(CH_2PO_3)_3]Na_4 \cdot 3H_2O$ ($Zn_{N,O}$NTP) first obtained and studied by Somov and Chausov.[13] $Zn_{N,O}$NTP is an electrolyte. The structure of the inner coordination sphere of $Zn_{N,O}$NTP (CCDC 919565) is shown in Fig. 9.2(b); the zinc atom is coordinated in the configuration of the trigonal pyramid, at one top of which there is a nitrogen atom and at the other top and in the base plane there are oxygen atoms. The inner coordination sphere is stable due to the closing of three 5-atom cycles with the shared Zn–N bond. The outer coordination sphere is represented by sodium ions in the octahedral and trigonal- bipyramidal surrounding of crystallization water molecules.[14] The gravimetric experiment showed $\gamma = 14.3$, or $Z = 0.93$.

A great difference is observed in the anticorrosion efficiency of the coordination complexes with a similar or almost similar composition but a different structure. This difference together with a strong change in the quantitative ratio of the obtained products due to a change in the conditions of the interaction of the reagents, leads to a significant variation of the properties of zinc-phosphonate corrosion inhibitors produced by different manufacturers and in different batches of inhibitors produced by one manufacturer.[15] It is one of the reasons which cause disagreements in the evaluation of the efficiency of such inhibitors.[16]

The products of the interaction of Zn^{2+} with nitrilo tris–(methylenephosphonic) acid (Zn_0NTP and $Zn_{N,O}NTP$) are separated as individual substances, which gives the possibility of investigating an inhibiting effect and mechanism of the effect of each of the substances. In the present paper, the results of the comparative study on the anticorrosion efficiency and electronic structure of the Zn_ONTP and $Zn_{N,O}NTP$ protective layers on the carbon steel surface are given.

9.7.2 INVESTIGATION TECHNIQUE

The synthesis of Zn_0NTP was conducted according to the method from Ref. [12]. At room temperature, 0.02 mol NTP aqueous solution was added drop by drop into 0.02 mol $ZnSO_4 \cdot 7H_2O$ aqueous solution at constant stirring. During 24 h, Zn_0NTP crystals were precipitating, which were then washed, separated by filtering and dried.

The synthesis of $Zn_{N,O}NTP$ was conducted in accordance with the method from Ref. [13]. At 70—80°C, 0.45-mol NaOH aqueous solution was added into 0.1-mol ZnO aqueous suspension; after that 0.1-mol NTP aqueous solution was added. The obtained clear solution was evaporated and held for 10 days at room temperature for crystallization. Precipitated $Zn_{N,O}NTP$ crystals were washed, separated by filtering and dried.

Synthesized Zn_ONTP and $Zn_{N,O}NTP$ were identified by the XRD method. Crystalline products were milled to particles of size 3–5 μm. In Fig. 9.3, the experimental X-ray patterns and the X-ray patterns calculated with the use of the CCDC 257224 and CCDC 919565 data are presented for comparison, and they show good agreement. For further electrochemical and XPS experiments, all the solutions were prepared from separated pure crystalline Zn_ONTP and $Zn_{N,O}NTP$, the crystal structure of which had been identified.

The samples for the corrosion experiments were prepared from hot-rolled carbon steel 20 (0.24% C, 0.48% Mn, 0.21% Si, <0.02% P, <0.02% S). The samples of 10 × 10 × 4 and 15 × 18 × 0.5 mm in size were milled and polished to the surface average roughness $R_a = 1–2$ μm.

The electrochemical investigations were conducted in the borate buffer solution[17] having a pH = 7.4 in the conditions of natural aeration of the environment. The measurements were conducted in the background solution (the check experiment) in the presence of 0.5 g/l NTP, Zn_ONTP and $Zn_{N,O}NTP$ and in the same environments with the addition of 0.01 mol/l NaCl. Corrosion potentials were measured relative to the saturated Ag|AgCl electrode by the thin-layer method[18] using a millivoltmeter DT890B. The polarization curves were taken in a standard measuring cell relative to the saturated Ag|AgCl electrode by the potentiodynamic method using a potentiostat IPC-ProL. Before registering the anodic branch of the polarization curve, the sample had been held at the cathode potential for destructing oxides layers and forming the metal native surface. It allowed observing the process of the formation of a protective layer on the metal surface at the anode potentials. The results of the electrochemical experiments were converted into the potential values relative to the standard hydrogen electrode.

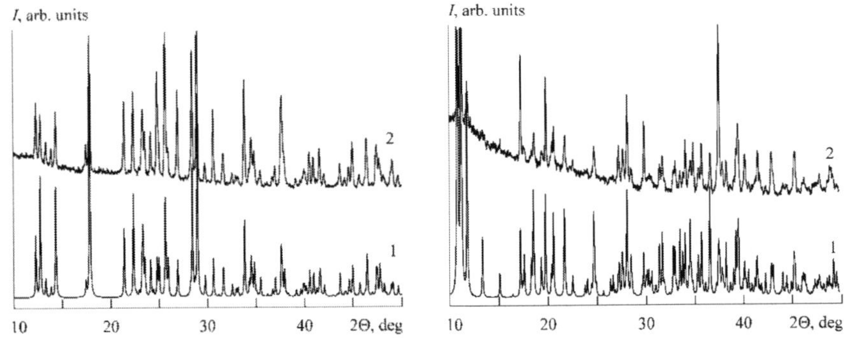

(a) (b)

FIGURE 9.3 The experimental (2) Zn_ONTP (a) and $Zn_{N,O}NTP$ (b) X-ray patterns (a diffractometer DRON-6, powder specimen, shooting with rotation, Co*K*α radiation) in comparison with the calculated (1) X-ray patterns based on the data for the structures CCDC 257224 (a) and CCDC 919565 (b) (Mercury 2.4, Build RC5).

The protective layers on the steel sample surface obtained in the above corrosion experiments were studied by XPS. An X-ray electron magnetic spectrometer with double focusing by the magnetic field $H \sim r^{-1/2}$ was used.[19] The device provided the constant luminosity of 0.15% and resolution 10^{-4} in the electron kinetic energy range of 0–1000 eV. The excitation by Al*K*α-radiation ($h\nu$ = 1486.6 eV) was used. The design of the sample holder and the energy analyzer placed outside of the spectrometer working vacuum chamber allowed obtaining spectra at the temperature action upon a sample in the range of 50–400°C in situ.

9.7.3 RESULTS AND DISCUSSION

The corrosion potentials of carbon steel in the studied environments are presented in Table 9.1. It can be seen that the addition of NTP into the corrosion environment leads to the decrease of the corrosion potential of steel, which, probably, indicates the prevalence of the cathodic reaction slowdown.

TABLE 9.1 The Stationary Electrode Potentials of Carbon Steel in the Borate and Borate-chloride Solutions in the Presence of 0.5 g/l Zn_ONTP and $Zn_{N,O}$NTP Inhibitors.

Equipment*	Inhibitor	Corrosion potential (V)
BBS7.4	None**	–0.48 ± 0.01
BBS7.4	NTP	–0.76 ± 0.01
BBS7.4	Zn_ONTP	–0.46 ± 0.01
BBS7.4	$Zn_{N,O}$NTP	–0.41 ± 0.01
BBS7.4 + 0.01 mol/l NaCl	None**	–0.51 ± 0.01
BBS7.4 + 0.01 mol/l NaCl	NTP	–0.77 ± 0.01
BBS7.4 + 0.01 mol/l NaCl	Zn_ONTP	–0.53 ± 0.01
BBS7.4 + 0.01 mol/l NaCl	$Zn_{N,O}$NTP	–0.49 ± 0.01

*BBS7.4 is borate buffer solution with a pH = 7.4; **check experiment.

The Zn_ONTP addition insignificantly changes the corrosion potential by slowing down almost equally both the anodic and the cathodic reaction. The $Zn_{N,O}$NTP addition into the corrosion environments leads to the increase of the corrosion potential, which indicates the predominant anodic reaction slowing-down. In the presence of chlorides the $Zn_{N,O}$NTP effect is less than in the pure borate buffer solution due to the competition between the chloride and phosphonate surface complexation of iron.

The anodic polarization curves of carbon steel for the studied environment are shown in Figure 9.4. The addition of the three NTP, Zn_ONTP, and $Zn_{N,O}$NTP inhibitors into the corrosion environments decreases the corrosion current both in the area of active dissolution and in the area of passive state. The addition of the above inhibitors increases the transpassivity potential in both the borate and borate-chloride environments. In this case, the protective action of each inhibitor is expressed differently. The less decrease of the corrosion current in both the borate and borate-chloride environments is caused by NTP. The addition of Zn_ONTP in the same amount leads to a large decrease in the corrosion current. The largest decrease of the corrosion current is reached by the addition of $Zn_{N,O}$NTP. In the presence of $Zn_{N,O}$NTP in the borate environment, a change in the polarity of the corrosion current (the cathode passivation peak) is observed, which has earlier been observed at the autopassivation both of chrome-nickel austenitic steels[20] and of carbon steel.[21] The formation of the cathode passivation peak is indicative of the spontaneous formation of a passive layer on the metal surface in the $Zn_{N,O}$NTP presence.

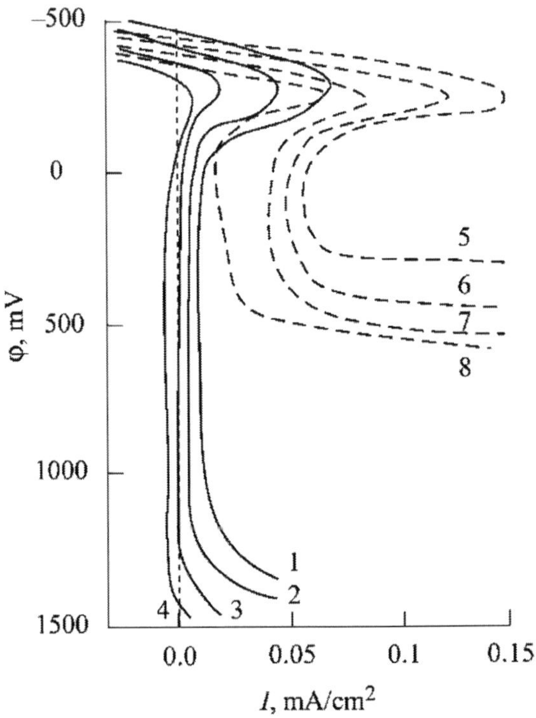

FIGURE 9.4 The anodic polarization curves of carbon steel in the borate buffer solution having a pH = 7.4 (1–4) and said buffer solution with the addition of 0.01 mol/l NaCl (5–8); 1, 5—the check experiment (one inhibitor); 2, 6—with 0.5 g/l NTP; 3, 7—with 0.5 g/l Zn_ONTP; 4, 8 with 0.5 g/l $Zn_{N,O}$NTP.

The $Fe2p_{3/2}$ electron spectra of the surface layers of carbon steel coated with protective layers formed by the NTP,[22] Zn_ONTP and $Zn_{N,O}$NTP inhibitors are displayed in Figure 9.5. Both at 100°C and at 250°C, the $Fe2p_{3/2}$ spectra of steel with the Zn_ONTP protective layer show a narrow maximum near 711 eV with a relatively weakly expressed structure of the spectra of the Fe oxidized states and a characteristic satellite structure. Having regard to the differences in the resolution of the used spectrometers, the spectra, in general, are similar to those of the steel surface with the NTP protective layer.[22]

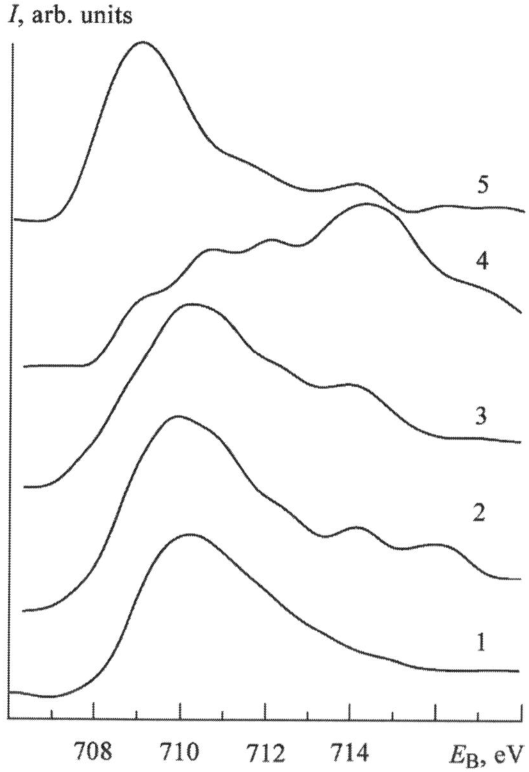

FIGURE 9.5 The Fe2p electron spectra of the carbon steel surface layers coated with the inhibitor layers: 1—NTP (room temperature,[22]); 2—Zn_ONTP (T = 100°C); 3—Zn_ONTP (T = 250°C); 4—$Zn_{N,O}$NTP (T = 100°C); and 5—$Zn_{N,O}$NTP (T = 250°C).

The $Fe2p_{3/2}$ spectrum of steel with the $Zn_{N,O}$NTP protective layer has a different character at 100°C. Its structure is characteristic of atoms of Fe in a high-spin state and has a high intensity of characteristic losses due to an increase in the number of noncompensated Fe3d-electrons, which is spectroscopically revealed as the appearance of an intensive satellite structure in the region of high binding energies.[22–25] The above effect indicates an increase in the number of noncompensated 3d-electrons of Fe atoms of the steel surface layer at their interaction with atoms of the $Zn_{N,O}$NTP inhibitor. At 250°C, the above effect disappears, apparently, due to the destruction of the protective layer, and the spectrum becomes identical to that of pure iron.

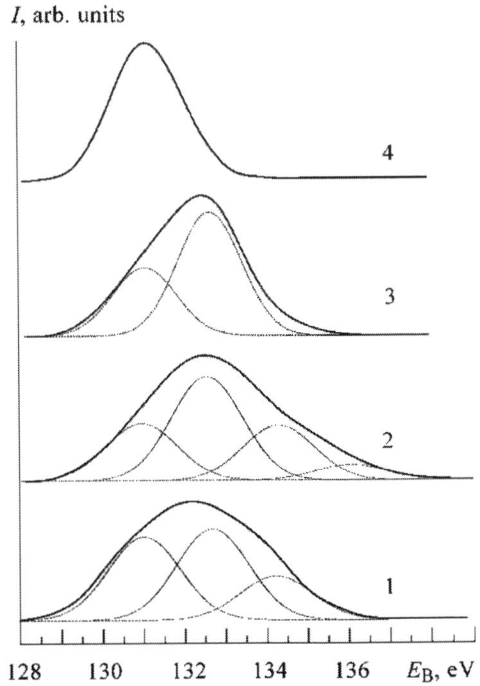

FIGURE 9.6 The P2p electron spectra of free inhibitors and the protective layers formed by them: 1—free Zn_ONTP; 2—the protective layer formed by Zn_ONTP on the carbon steel surface; 3—free $Zn_{N,O}$NTP; 4—the protective layer formed by $Zn_{N,O}$NTP on the carbon steel surface.

The electron P2p spectra of the free Zn_ONTP and $Zn_{N,O}$NTP inhibitors and the protective layers formed by them on the carbon steel surface are shown in Figure 9.6. The free Zn_ONTP inhibitor has a broad P2p spectrum containing three components. The component with a maximum at E_B = 131.2 eV corresponds to the P–O bonds participating in the coordination of Zn atoms. The component with a maximum at E_B = 132.8 eV corresponds to the P–O bonds which do not participate in the Zn atoms coordination. The component with a maximum at E_B = 134.3 eV corresponds to the P–O(H) bonds. The P2p spectrum of the Zn_ONTP adsorption layer contains the three components mentioned above and a component with a maximum at E_B = 136.3 eV corresponding to the oxidation of phosphorus to PO_4 due to the interaction with the surface layers of oxides. It is found that not all the inhibitor PO groups participate in the protective layer–metal bond; the intensity of the respective component (E_B = 131.2 eV) is less than that in the spectrum of the free inhibitor.

The P2p spectrum of the free $Zn_{N,O}$NTP inhibitor contains two components. The component with a maximum at E_B = 131.2 eV corresponds to the participation of the P–O bonds in the Zn-atoms coordination. The component with a maximum at E_B = 132.8 eV corresponds to the P–O bonds which do not participate in the coordination of Zn atoms. In the P2p spectrum of the $Zn_{N,O}$NTP protective layer, there is one component with a maximum at E_B = 131.2 eV, which indicates that all the PO_3 groups of $Zn_{N,O}$NTP participate in the coordination of both Zn atoms and Fe atoms of the steel surface layer, thereby providing more stable bonding of the protective layer with the iron surface layer and increasing the protective effect.

Similar changes can also be observed in the Zn3d spectra of the protective layers formed by the Zn_ONTP and $Zn_{N,O}$NTP inhibitors on the steel surface.

The results obtained indicate that Zn_ONTP and $Zn_{N,O}$NTP essentially differ in their mechanism of the protective layer formation and inhibiting effect.

The mechanism of the Zn_ONTP inhibiting effect is close to that of NTP described in Ref. [22] and it is due to the formation of a weakly bonded layer on the steel surface. The approximately equal slowdown of the anodic and cathodic reaction indicated by a little change in the corrosion potential is apparently due to the formation of zinc hydroxo-complexes because of the interaction of the inhibitor with alkali products of the cathodic reaction.

The mechanism of the $Zn_{N,O}$NTP effect is due to the formation on the steel surface of an adsorption film formed by the $Zn_{N,O}$NTP-complexes, which is strongly bonded with Fe. An increase in the number of noncompensated Fe electrons leads to the formation of the Fe–O–P covalent bonds; all the PO groups, which do not participate in the Zn atoms coordination, are involved into the bond with Fe atoms. The slowdown of the anodic reaction indicated by an increase in the corrosion potential is due to the blocking of the metal surface by the surface complexes and, probably, the inhibition of the Fe^{2+} ions transfer by the electrostatic field of the surface complexes (the anodic ψ_1-effect[26]).

9.7.4 CONCLUSION

A new approach is developed in investigating the mechanism of the inhibiting effect of the phosphonate metal-based complexes, wherein the structure of the surface layers formed by the selectively synthesized and separated coordination complexes of Zn^{2+} with NTP having different structure are investigated by the comparative electrochemical and XPS methods.

It is shown that in similar conditions, the inhibitors Zn_ONTP (non-electrolytic with polymeric bridging structure) and $Zn_{N,O}$NTP (electrolytic with chelate structure) demonstrate different effect of the inhibition of the oxygen-type corrosion of carbon steel in water environments. The Zn_ONTP inhibitor slows down the anodic and cathodic reactions and slightly decreases the corrosion current. The $Zn_{N,O}$NTP inhibitor slows down mainly the anodic reaction, increases the corrosion potential, and provides a strong decrease in the corrosion current and steel surface auto-passivation.

Some differences are found both in the structure of the protective layers formed by the Zn_ONTP and $Zn_{N,O}$NTP inhibitors and in the spin state of iron atoms in the surface layer under the protective layers formed by the above inhibitors. The transition of iron atoms into a high-spin state and the increase of the number of noncompensated 3d-electrons leads to the formation of strong covalent bonds Fe–O–P between the steel surface layer and the $Zn_{N,O}$NTP protective layer.

The structure of the characteristic losses in the Fe2p-spectrum and the contribution of the phosphorus oxidized states in the P2p-spectrum of the protective layer correlate with the kinetic characteristics of corrosion processes and can serve as anticorrosive efficiency indices of the protective layer.

ACKNOWLEDGMENTS

The work is financially supported by the Russian Fund of Fundamental Research and the Government of the Udmurt Republic (project 13-02-96007).

KEYWORDS

- **corrosion inhibitors**
- **structure of coordination compounds**
- **chelate complexes**
- **bridging complexes**
- **nitrilo tris–(methylenephosphonates)**
- **zinc**
- **XPS**

- **the spin state of iron in coordination complexes**
- **electrochemical examination**
- **carbon steel**

REFERENCES

1. Hwa, C. M. US Patent 3431217. **1969**.
2. Kuznetsov, Y. I. *Protect. Met. Phys. Chem. Surf.* **2002,** *38*, 103.
3. Kuznetsov, Y. I. *Russ. Chem. Rev.* **2004,** *73*, 75.
4. Kuznetsov, Y. I.; Raskolnikov, A. F. *Protect. Met.* **1992,** *28*, 249. (in Russian)
5. Kuznetsov, Y. I.; Isaev, V. A.; Trunov, E. A. *Protect. Met.* **1990,** *26*, 798. (in Russian)
6. Kuznetsov, Y. I.; Raskolnikov, A. F. *Protect. Met.* **1992,** *28*, 707. (in Russian)
7. Rajendran, S.; Apparao, B. V.; Palaniswamy, N. *Anti-Corrosion Methods Mater.* **2000,** *47*(2), 83.
8. Rajendran, S.; Apparao, B. V.; Palaniswamy, N.; Amalraj, A. J.; Sundaravadivelu, M. *Anti-Corros. Methods Mater.* **2002,** *49*(3), 205.
9. Awad, H. S. *Anti-Corrosion Methods Mater.* **2005,** *52*(1), 22.
10. Demadis, K. D.; Mantzaridis, C.; Raptis, R. G.; Mezei, G. *Inorg. Chem.* **2005,** *44*, 4469.
11. Demadis, K. D.; Barouda, E.; Zhao, H.; Raptis, R. G. *Polyhedron* **2009,** *28*, 3361.
12. Demadis, K. D.; Katarachia, S. D.; Koutmos, M. *Inorg. Chem. Commun.* **2005,** *8*, 254.
13. Somov, N. V.; Chausov, F. F. *Crystallogr. Rep.* **2014,** *59, 71.*
14. Shabanova, I. N.; Naimushina, E. A.; Chausov, F. F.; Somov, N. V. *Surface Interface Anal.* **2014,** *12.*
15. Chausov, F. F. *Ecol. Ind. Russ.* **2008,** *9, 28.* (in Russian)
16. Balaban-Irmenin, Y. V. *Ener. Saving Water Treat.* **2011,** *3*, 20. (in Russian)
17. Schwabe, K. *Fortschritte der pH-Meßtechnik.* Verlag Technik: Berlin, 1958.
18. Rosenfeld, I. L.; Pavlitskaya, T. P. In *New Methods of the Physical Chemistry Investigations 2.* Academy of Sciences of USSR: Moscow, **1957**; pp 56–68. (in Russian)
19. Trapeznikov, V. A.; Shabanova, I. N.; Zhuravlev, V. A. *J. Electr. Spectrosc. Relat. Phenom.* **2004,** *137–140*, 731.
20. Tomashov, N. D.; Chernova, G. P.; Markova, O. N. *Russ. J. Appl. Chem.* **1960,** *33,* 1324. (in Russian)
21. Rosenfeld, I. L.; Kuznetsov, Y. I.; Kerbeleva, I. Y.; Persiantseva, V. P. *Protec. Met.* **1975,** *11*, 612. (in Russian)
22. Labjar, N.; El Hajjaji, S.; Lebrini, M.; Serghini Idrissi, M.; Jama, C.; Bentiss, F. *J. Mater. Environ. Sci.* **2011,** *2*(4), 309.
23. Shabanova, I. N.; Trapeznikov, V. A. *JETP Lett.* **1973,** *18*(9), 339.
24. Gupta, R. P.; Sen, S. K. *Phys. Rev. B.* **1974,** *10*, 71.
25. Grosvenor, A. P.; Kobe, B. A.; Biesinger, M. C.; McIntyre, N. S. *Surf. Interface Anal.* **2004,** *36, 1564.*
26. Timashev, S. F. *Russ. J. Electrochem.* **1979,** *15, 333.* (in Russian)

9.8 GENERAL CONCLUSION

Thus, X-ray photoelectron spectroscopy with using unique X-ray photoelectron spectrometers is perspective method for the investigations of nano- and mesoscopic objects and processes with their participation.

CHAPTER 10

NANOSTRUCTURED POWDERS OF COBALT–NICKEL SOLID SOLUTION, THEIR ANALYSIS BY OPTICAL EMISSION SPECTROMETRY WITH INDUCTIVELY COUPLED PLASMA AND USE AS REFERENCE SAMPLES FOR THE CALIBRATION OF SPECTROMETERS WITH SOLID SAMPLING

R. KOLMYKOV

Kemerovo State University, Faculty of Chemistry, Kemerovo, Russia

CONTENTS

ABSTRACT

In this work, nanostructured powders of solid-cobalt–nickel are synthesized. Solid solution provides maximum homogeneity of the elemental composition in the bulk powders. Nickel and cobalt contents in the powders studied by optical emission spectrometry with inductively coupled plasma in the classical liquid performance, as well as the use of laser sampling.

10.1 INTRODUCTION

The methods for the preparation, validation, and application of nanoscale and nanostructured metal powders and their multicomponent systems is interested to develop for modern materials science.[1–9]

It is necessary to investigate these highly complex, both in composition and structure, objects need to use the entire methods of X-ray diffraction, spectroscopic, and other physical methods. A number of papers[1,3–9] are devoted to the synthesis of nanosized particles of nickel and cobalt, and their binary system in redox reactions in aqueous solutions. Along with the issues of synthesis of the acute problem of certification obtained nanomaterials.[2,5,7,9] One of the certification issues of such objects is their chemical composition, the determination of which is devoted to this work.

10.2 THEORY (METHODS AND TECHNIQUE)

The chemical reactor Readley with the thermostat Lauda ECO 4, with the stirring device Heidolf RZR 2102 Control is used for the synthesis of nanostructured metallic powders.[5]

The optical emission spectrometer with inductively coupled plasma iCAP 6500 DUO with laser ablation UP 266 MACRO is used for elemental composition of nanostructured metallic powders.

State standard samples 7265-96 Ni(II) and 7268-96 Co(II) are used to preparation standard 10 and 20 ppm cobalt and nickel solutions. The investigated samples are prepared by dissolution in the chemical pure nitric acid.

The chemical pure nickel (technical condition 3-272-10) and cobalt (technical condition 2-313-11) by Component Reactive Ltd. are used for the calibration of the spectrometer with laser sampling. The investigated samples are prepared by the pressing at a pressure of 400 MPa using an electric hydraulic press PGM-100MG4.[8]

Due to the difference of the samples (pressed powder) and standard samples (bulk metal), it was selected the calibration method related to the basis of concentrations.

In this study nanostructured powders of solid-cobalt–nickel was selected,[5,9] which, because of its solid solution in the nanoscale and nanostructured state extremely homogeneous throughout.

Selection of analytical wavelengths used in the work was carried out in order to avoid spectral overlap nickel and cobalt main components.

Spectrometer parameters are shown in Table 10.1.

TABLE 10.1 Spectrometer Parameters.

Measurement parameters	
Plasma power (W)	1150
Argon nebulizing flow rate (L/min)	0.6
Cooling argon flow rate (L/min)	12
Auxiliary argon flow rate (L/min)	0.5
Plasma observation	Radial
Signal recording (s)	10
Flow rate of solution delivery (mL/min)	2

10.3 RESULTS AND DISCUSSION

Figure 10.1 shows the calibration curves for the analysis of sample solutions. The correlation coefficients tend to 1, which indicates the high quality of the resulting standard solutions.

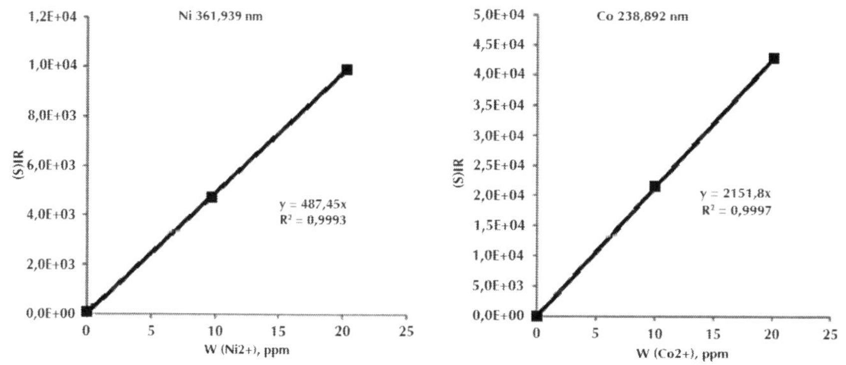

FIGURE 10.1 Calibration curves for the determination of nickel and cobalt in the samples.

The elemental analysis of the test system Co–Ni was carried out using the calibration data. The results of this analysis are shown in Table 10.2.

$Co_{40}Ni_{60}$ was predetermined in the synthesis of ratio of 40%:60%, respectively. Cobalt content of 40.1 wt% and nickel—59.8 wt% were determined by the elemental analysis of the nanostructured powder. For the sample, there is a slight understatement of the nickel content in comparison with the predetermined down.

TABLE 10.2 The Results of Elemental Analysis of the Test Samples Co–Ni from the Solutions.

The name of the sample	Element	Analytical line λ (nm)	W (%)
$Co_{40}Ni_{60}$	Co	238,892	40.1 ± 0.1
	Ni	361,939	59.8 ± 0.1
		In total	99.9 ± 0.2
$Co_{30}Ni_{70}$	Co	238,892	30.2 ± 0.1
	Ni	361,939	69.7 ± 0.2
		In total	99.9 ± 0.3
$Co_{20}Ni_{80}$	Co	238,892	19.8 ± 0.2
	Ni	361,939	80.1 ± 0.2
		In total	99.9 ± 0.4

$Co_{30}Ni_{70}$ was predetermined in the synthesis of ratio of 30%:70%, respectively. Cobalt content of 30.2 wt% and nickel—69.7 wt% were determined by the elemental analysis of the nanostructured powder. For a given sample, as for the previous content occurs too low the nickel component.

$Co_{20}Ni_{80}$ was predetermined with the synthesis ratio of 20%:80%, respectively. Cobalt content of 19.8 wt% and nickel—80.1 wt% were determined by the elemental analysis of the nanostructured powder. The nickel content is a little bit higher than with a predetermined.

Thus, the powders obtained by the elemental composition satisfy the pledged in the synthesis.

The next stage of this work is the study of Co–Ni-solid solution nanostructured powders possibility to use for the calibration of the spectrometer using laser sampling. For this compact, powder materials are manufactured in the form of tablets. Micrograph of the tablet surface is shown in Figure 10.2. This micrograph confirms the stability of the particles of investigated systems to deformation during compaction pressure is sufficiently high.

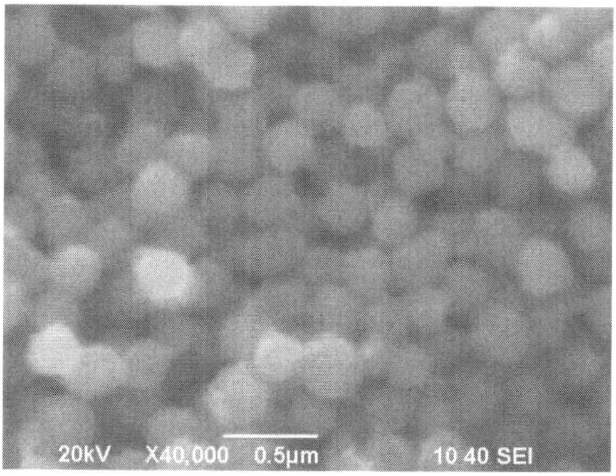

FIGURE 10.2 Electron-microscopic picture of cobalt–nickel compacted pressure of 400 MPa.[8]

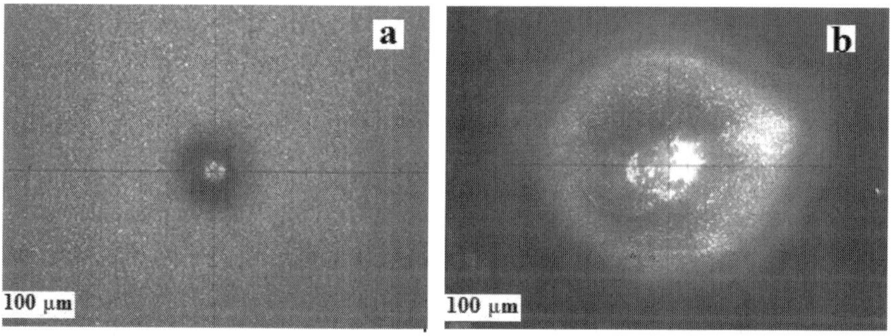

FIGURE 10.3 Microscopic photographs of craters test samples during and after the laser sampling, the diameter of the laser beam 100 μm: (a) $Co_{40}Ni_{60}$ and (b) $Co_{40}Ni_{60}$ during the laser pulses with the expanding cloud of aerosol.

It was selected a method of calibrating a spectrometer based on assigned to concentrations in this paper to neutralize the possible increase of the intensity of the analytical signals from powder materials compared to bulk metal. This method corresponds to the following equation:

$$\frac{C_{Co}}{C_{Ni}} = \frac{100\%}{C_{Ni}} - 1 \tag{1}$$

where C_{Co} and C_{Ni} are the mass fractions of cobalt and nickel (wt%), respectively.

Microscopic photographs of the surface of compact powder metals studied are presented in Figure 10.3. Figure 10.3 shows a photograph of the crater formed on the surface of the pressed tablet from the $Co_{40}Ni_{60}$ sample, and Figure 10.3b fixed the expanding cloud of nanoparticles aerosol of this sample. After, the aerosol carrier gas is transported from the laser set-top box to the torch of a spectrometer where the spectrum is excited. Samples are taken from the sample surface at four points form a square. The results obtained were averaged.

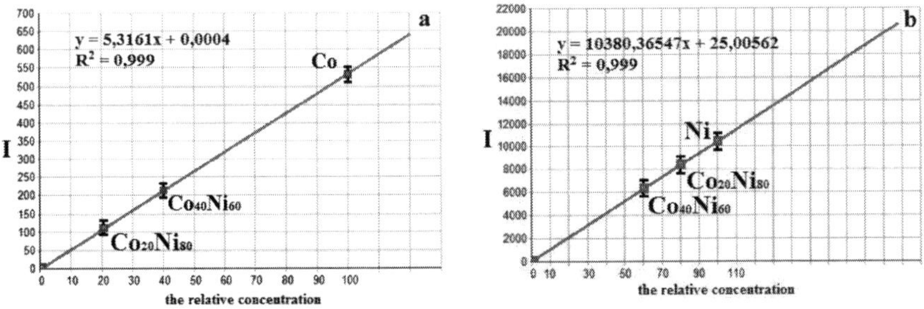

FIGURE 10.4 The calibration curve for the determination of cobalt (238.982 nm) and nickel (361.939 nm) in the samples.

TABLE 10.3 Results of Elemental Analysis of the Samples by Laser Sampling

The name of the sample	Element	Analytical line λ (nm)	*W* (%)
	Co	238,892	39.7 ± 0.6
$Co_{40}Ni_{60}$	Ni	361,939	59.8 ± 0.5
		In total	99.5 ± 1.2
	Co	238,892	30.2 ± 0.2
$Co_{30}Ni_{70}$	Ni	361,939	69.5 ± 0.3
		In total	99.7 ± 0.5
	Co	238,892	19.8 ± 0.3
$Co_{20}Ni_{80}$	Ni	361,939	79.7 ± 0.4
		In total	99.5 ± 0.7

To construct the calibration curves of the analytical wavelengths radiation intensity of the mass concentration of the test components were chosen the same analytical wavelengths as for the liquid-phase analysis: 238.892 nm and 361.939 nm cobalt for nickel. As reference, samples selected were chemically pure metals (nickel and cobalt) and their binary systems, synthesized for the experiment. If the synthesized system is suitable for calibration of the spectrometer, the value of the intensity of their analytical signals will be on the calibration line. If the points on one line will not lay down, it will mean that the test samples are not enough homogeneous under these experimental conditions.

The result of the experiment were obtained from the direct calibration correlation coefficient close to 1 (Fig. 10.4) using Equation (1), indicating that sufficient homogeneity test samples for use in such a sampling. The results of the determination of the basic system components are shown in Table 10.3.

Figure 10.5 is a photomicrograph of a crater on a massive erosion chemically pure nickel (a) and compacted system Co–Ni. The shape and size of the craters are matched, and thus the amount of material evaporated from the surface under the same conditions is approximately the same. Thus, it is another proof of the suitability of a compact nanostructured Co–Ni system suitable for use as a reference sample, along with the massive chemically pure metallic nickel and cobalt.

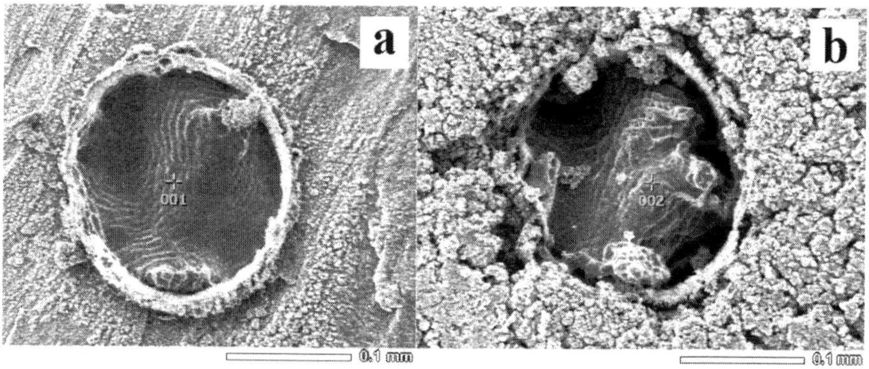

FIGURE 10.5 Electron microscopic photographs of craters on the surface of a bulk nickel (a) and pressed nanostructured system cobalt–nickel.

10.4 CONCLUSION

Used in work the synthesis method allows to obtain a binary composition of sufficient purity and uniform as a result of the experiments.

The results obtained suggest that the sufficient purity binary system used in the method is suitable as a reference material in analytical practice.

This work was supported by the Russian Federation Ministry of Education order No 2014/64.

KEYWORDS

- **cobalt**
- **nickel**
- **nanostructured powders**
- **optical emission**

REFERENCES

1. Dzidziguri, E. L.; Levina, V. V.; et al. Formation of Metallic Nanopowders Upon Chemical Reduction. *Phys. Met. Metallogr.* **2001,** 91(6), pp 583–589.
2. Sidorova, E. N.; Dzidziguri, E. L.; et al. Dispersion Characteristics of a Nickel Nanopowder. *Russ. Metall. (Metally)* **2008,** *2008*(6), 513–517.
3. Zakharov, U. A.; et al. Sintez i svoyistva nanorazmernyh poroshkov metallov gruppyi zheleza i ih vzaimnyh sistem [Synthesis and Properties of Nano-sized Powders of Metals of the Iron Group and their Binary Systems]. *Perspect. Mater.* **2008,** *6*, 249.
4. Zakharov, U. A.; Kolmykov, R. P. Poluchenie nanorazmernyh poroshkov nikelya i kobal'ta dlya sovremennoi promyshlennosti [Preparation of Nanosized Powders of Nickel and Cobalt for Modern Industry]. *Polzunovski vestnik* **2008,** *3*, 137–140.
5. Kolmykov, R. P. *Poluchenie i izuchenie nanoporoshkov nikelya, kobal'ta i ih vzaimnoi sistemyi* [Preparation and study of the Properties of Nanopowders of Nickel, Cobalt, and their Binary System] Thesis for the degree of Candidate of Chemical Sciences, Kemerovo State University. Kemerovo, 2011; 160 p.
6. Pugachev, V. M.; Dodonov, V. G.; et al. Fazovyi sostav I nekotorye svoistva nanorazmernyh poroshkov Ni–Co i Ni–Cu [The Phase Composition and Some Properties of Nanoscale Powders Ni–Co and Ni–Cu] *Perspect. Mater.* **2011,** *11*, 156–163.
7. Dodonov, V. G.; et al. Osobennosti opredelenija razmerov kristallicheskih nanochastic perehodnyh metallov po rentgenograficheskim dannym [Features of Determining the Size of the Crystalline Transition Metal Nanoparticles by X-ray Diffraction Data]. *Polzunovsky Vestnik.* **2008,** *3*, 134–136.
8. Kolmykov, R. P.; Ivanov, A. V. Kompaktirovanie, spekanie i yelektrofizicheskie svoistva nanokristallicheskih nikelja i kobal'ta [Compacting, Sintering and Electrical Properties of Nanocrystalline Nickel and Cobalt]. *Polzunovsky Vestnik* **2009,** *3*, 266–270.
9. Zakharov, U. A.; Kolmykov, R. P.; et al. Fazovyi sostav i morfologija nanoporoshkov sistemy kobal't-nikel' [The Phase Composition and Morphology of Nano Cobalt–Nickel]. *Vestnik Kemerovskogo gosudarstvennogo universiteta* **2014,** *3*(59), 194–200.

CHAPTER 11

EFFECT OF SYNTHESIS CONDITIONS ON LUMINESCENT PROPERTIES OF ALUMINUM OXIDE NANOSTRUCTURED CERAMICS

V. S. KORTOV*, S. V. ZVONAREV, D. V. ANANCHENKO, K. A. PETROVYKH, A. N. KIRYAKOV, and A. M. LUKMANOVA

Ural Federal University, Institute of Physics and Technology, Mira St. 21, Yekaterinburg, Russia

Corresponding author. E-mail: v.s.kortov@urfu.ru

CONTENTS

ABSTRACT

Nanostructured ceramic samples were synthesized from commercial high-purity α-Al_2O_3 nanopowder with 50–70-nm particles. The samples were subjected to static pressure and vacuum annealing in the presence of carbon to create oxygen vacancies forming luminescent centers.

Synthesis temperatures varied from 1500 to 1700°C with annealing time ranging from 30 min to 3 h. To define dependence of TL peak parameters on the absorbed dose, the ceramic samples were preliminarily exposed to the doses of 0.32–500 Gy from the $^{90}Y/^{90}Sr$ β-source with the dose rate of 32 mGy/min.

Scanning electron microscopy results showed that large agglomerates on the surface of samples are subject to destruction with further formation of small size nanoparticles. This phenomenon may be related to oxide stoichiometric composition disorder during high temperature annealing. It was found that high-temperature annealing of Al_2O_3 nanopowder compacts in a vacuum (reducing medium in the presence of carbon) leads to oxygen vacancies generation in the oxide, which is confirmed by pulsed cathodoluminescence (PCL) results for the samples under study.

PCL spectrum of nanostructured ceramics has a band at 420 nm caused by the glow of F-centers. The low-intensity luminescence in the nanostructured samples synthesized at 1500°C (as compared to a single crystal) was caused by a lower concentration of the above-mentioned centers in ceramics. An increase in annealing temperature leads to luminescence centers concentration growth within the band at 400–450 nm. It was observed that TL peak intensity at 410 K of nanostructured aluminum oxide depended on the dose of β-irradiation.

11.1 INTRODUCTION

The potential of nanoscale dielectrics application in micro- and optoelectronics to create phosphors with high light output has been intensively studied in recent years. Physical, chemical, and energy characteristics of nanomaterials have a significant effect on their luminescent properties.

The results of such effects have not been fully investigated yet. Dimensional, electronic, and quantum effects extended surface are the factors that determine unique properties of nanomaterials. Nanostructured ceramics with a complex of high mechanical and luminescent properties is a promising functional material.[1] Thus, the aim of this work is to study the effects of synthesis conditions on the luminescent properties of alumina ceramic.

11.2 EXPERIMENTAL

The aluminum oxide ceramic samples under study were synthesized from commercial α-Al_2O_3 nanopowder. The powder with the particle size of 50–70 nm was obtained using the alcoholate method. The samples were subjected to 8–9 kg/cm^2 static pressure. The following synthesis was carried out in a vacuum electric furnace with temperatures varying from 1500 to 1700°C and the annealing time ranging from 30 min to 3 h. Vacuum is a good reducing medium, which helps to obtain oxygen-deficient samples of alumina ceramics.

The surface structure of the obtained ceramics was studied with a SIGMAVP scanning electron microscope in high vacuum with a 5 kV accelerating voltage. A Clinker C7 analyzer of solid fragment microstructure was used to estimate particle size distributions. Luminescent properties of nanostructured ceramics were compared with the properties of single-crystalline anion-defective Al_2O_3, grown in highly reducing conditions.[2]

The registration of pulsed cathodoluminescence (PCL) spectra was carried out on the spectrometer "KLAVI" in the range of 350–750 nm. PCL in the ceramics was excited with an electron beam with the pulse length of 2 ns, electron energy of 130 ± 10 keV, and the flux density of 60 A/cm^2.

The measurement of the thermoluminescence (TL) was carried out in the temperature range of 50–550°C on the experimental stand with linear heating rate −2°C/s. To define the parameters of the dependence of the TL peak on the absorbed dose, ceramic samples were preliminarily exposed to the doses of 0.32–500 Gy from the $^{90}Y/^{90}Sr$ β-source with the dose rate of 32 mGy/min.

11.3 RESULTS AND DISCUSSION

Scanning electron microscopy (SEM) results showed that ceramics synthesis at 1500°C within 1 h leads to agglomerates with an average size of about 432 nm and high porosity of samples (Fig. 11.1a). Thus, there is fraction of particles with a size less than 100 nm, a portion of which has a size of 10–40 nm, less than the initial nanopowder. This phenomenon may be related to large agglomerates destruction with the formation of small-size nanoparticles.

Figure 11.1b presents quantitative data on the size distributions of the particles obtained in alumina ceramics synthesized at 1500°C within 1 h. The largest share of the particles in the sample (75%) has a small size up to 150 nm, with the amount of particles larger than 150 nm not exceeding 26%.

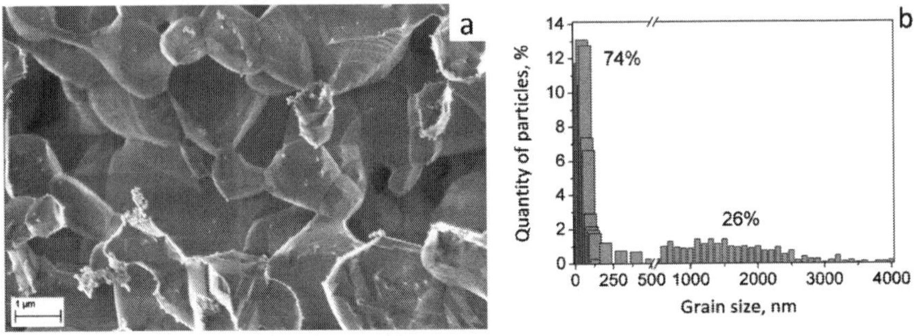

FIGURE 11.1 SEM-images (a) and particle size distribution (b) of nanocrystalline Al_2O_3, ceramics synthesized by annealing at 1500°C within 1 h.

Luminescent properties of the synthesized ceramics were compared with anion-defective single crystal α-Al_2O_3 properties to evaluate the presence of emission centers. In a single crystal alumina oxide, these centers form a complex defect that creates aggregate center, consisting of charged oxygen vacancies and impurity ions.[3] The PCL experimental spectra (Fig. 11.2) show that single-crystal and nanostructured aluminum oxide ceramics synthesized at 1500°C (within 3 h) display luminescence in 400–450-nm band. Luminescence in the band of 695 nm for single-crystal Al_2O_3 was caused by the presence of impurity Cr^{3+} ions and corresponds to the transition $^2E \rightarrow {}^4A_2$[4] Emission in the 400-nm band is due to triplet–singlet transitions $^3P \rightarrow {}^1S_0$ in F-centers (oxygen vacancies with two electrons trapped).

Low-intensity luminescence in nanostructured samples synthesized at 1500°C, as compared to a single crystal, was caused by a lower concentration of these centers in ceramics. The temperature of synthesis reaction was increased to increase the concentration of luminescent centers.

Figure 11.3 shows TL glow curves of the ceramic samples synthesized at different parameters under study exposed to 180 Gy of β-radiation. TL glow curves feature several peaks: namely, a low-temperature peak at 410 K and a high-temperature peak at 600 K. The high-temperature peak is not as intense, but its TL depends on the synthesis mode, similarly to the low-temperature peak. It proved that the samples annealed in a vacuum at high temperatures, display maximum TL intensity, which is caused by an increased concentration of oxygen vacancies forming F luminescence center.

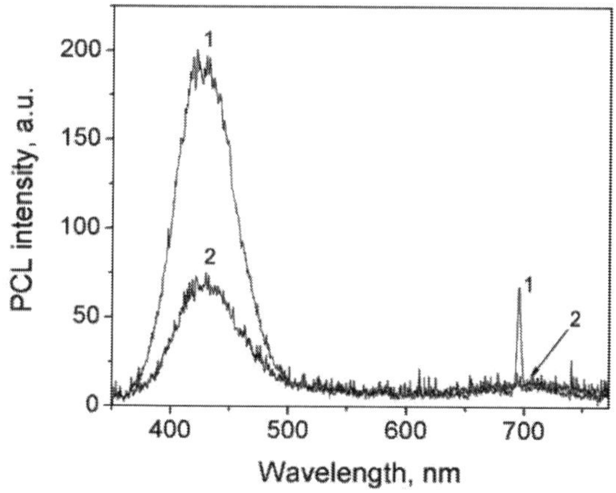

FIGURE 11.2 PCL spectrum of anion-defective single crystal Al_2O_3 (1) and nanocrystalline Al_2O_3 ceramics, synthesized by annealing at 1500°C for 3 h (2).

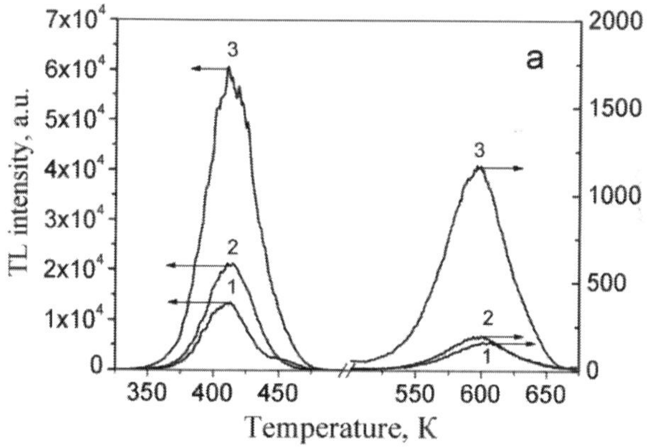

FIGURE 11.3 TL nanocrystalline Al_2O_3 ceramics, synthesized at different temperatures: 1500°C, 3 h (1); 1600°C, 1 h (2); 1700°C, 30 min (3).

Figure 11.4 shows TL glow curves of Al_2O_3 nanostructured ceramics at the peak 410 K for different doses of β-radiation. It is obvious that TL intensity increases with the growth of radiation dose; there is a linear dependence at the range of 10–100 Gy. It is expected that synthesized ceramics would be a promising material for detection of ionizing radiation doses, which exceed the operating dose range of single crystal α-Al_2O_3 detectors be more than one order.[5]

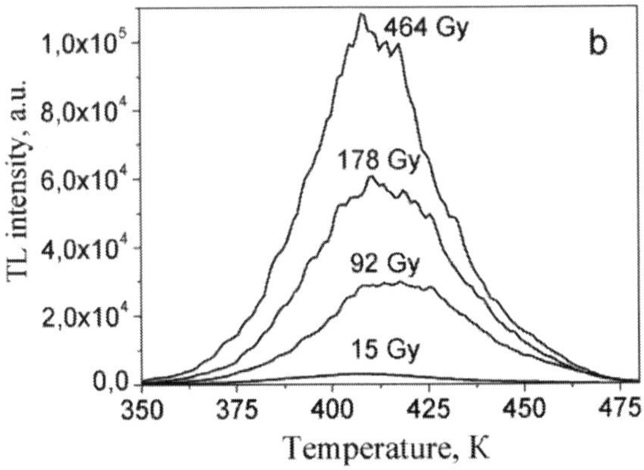

FIGURE 11.4 Dose response of the TL peak at 410 K of the Al_2O_3 ceramics annealed at 1700°C, 30 min.

11.4 CONCLUSION

In this paper, the effect of the synthesis conditions on the nanostructured alumina ceramics luminescence properties was studied. It was found that high-temperature annealing in a reducing atmosphere (vacuum, the presence of carbon) of Al_2O_3 compacts leads to oxygen vacancies formation in the oxide. An increase in annealing temperature causes the growth in the concentration of luminescence centers in the band of 400–450 nm, which are related with oxygen vacancies. It was observed that TL peak intensity at 410 K of nanostructured aluminum oxide depends on the dose of β-irradiation.

ACKNOWLEDGMENTS

This research project was performed as part of state order of The Ministry of Education and Sciences of RF and was supported by Ural Federal University as a part of the Program of Development through the "Young Scientists of UrFU" competition.

KEYWORDS

- **aluminum oxide**
- **thermoluminescence**
- **pulsed cathodoluminescence**
- **nanostructured ceramics**
- **synthesis conditions**
- **oxygen vacancies**

REFERENCES

1. Kortov, V. S.; Nikiforov, S. V.; Moiseykin, E. V.; Vohmincev, A. S.; Simanov, A. S. Luminescent and Dosimetric Properties of Nanostructured Ceramics Based on Aluminum Oxide. *Phys. Solid State* **2013,** *55*(10), 2088–2093.
2. Akselrod, M. S.; Kortov, V. S.; Gorelova, E. A. Preparation and Properties of α-Al_2O_3. *Radiat. Protect. Dosimet.* **1993,** *47*(1/4), 159–164.
3. Kortov, V. S.; Pustovarov, V. A.; Spiridonova, T. V.; Zvonarev, S. V. Photoluminescence of Ultradisperse Alumina Ceramics Under VUV Excitation. *J. Appl. Spectrosc.* **2013,** *80*(6), 835–840.
4. Kulinkin, A. B.; Feofilov, S. P.; Zakharchenya, R. I. Luminescence of Impurity 3d and 4f Metal Ions in Different Crystalline Forms of Al_2O_3. *Phys. Solid State* **2000,** *42*(5), 835.
5. Kortov, V. S. Materials for Thermoluminescent Dosimetry: Current Status and Future Trends. *Radiat. Maesur.* **2007,** *42*(4/5), 576–581.

PART IV

Producing of Nanostructured Materials and Investigations of their Properties

CHAPTER 12

ELECTROSPUN NANOFIBROUS STRUCTURE

G. E. ZAIKOV

Emanuel Institute of Biochemical Physics, Russian Academy of Sciences, Moscow, Russia

CONTENTS

ABSTRACT

Electrospinning is a comparatively simple method of producing nanofibers. Nanofibers produced by this method are widely utilized for varied applications like drug delivery and tissue scaffolding.

Symptoms	Definitions
V_c	critical voltage
H	distance between the capillary exit and the ground
L	length of the capillary
R	radius
γ	surface tension of the liquid/solution
nm	nanometer

12.1 A BRIEF PERSPECTIVE INTO ELECTROSPUN POLYMER NANOFIBERS

The polymer nanofibers have been known with nanoscale diameters because of the unique properties like high surface area and high porosity. They are useful in many important and varied applications such as tissue engineering scaffolds.[1,2]

Researcher and industry can produce them from a variation range of polymers.[3–5] Nowadays, electrospinning of these materials has gained much attention mainly because of cheapest and simplest feature of this method.[6–10]

Additionally, the significantly growing number of publications and patents in this field become important in the recent years.[11–13] The result of it is shown in Figure 12.1.

Electrospinning process requires an interaction between two forces, as shown in Figure 12.2. In this procedure, the polymer solution receives electrical charges from a high voltage supply. These charges are carried by ions through the fluid. If the repulsive force between the charged ions overcomes the fluid's surface tension, an electrified liquid jet could be formed and elongated toward the collector. With evaporation of solvent, nanofibers jet are collected on the surface of screen.[14,15] A schematic of electrospinning are shown in Figure 12.3.

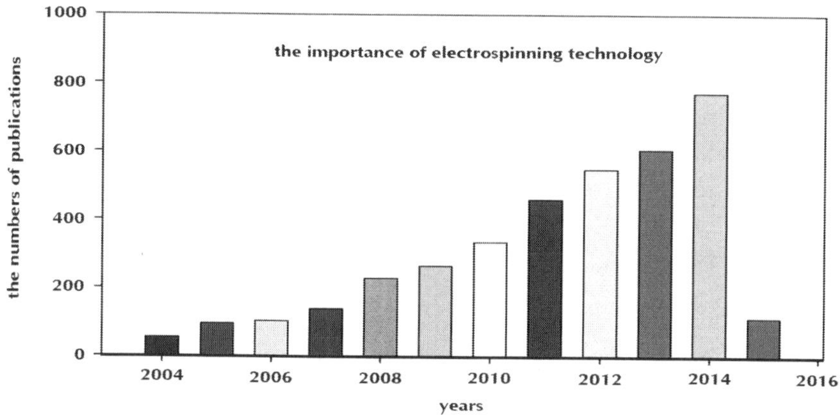

FIGURE 12.1 Numbers of publications about electrospinning.

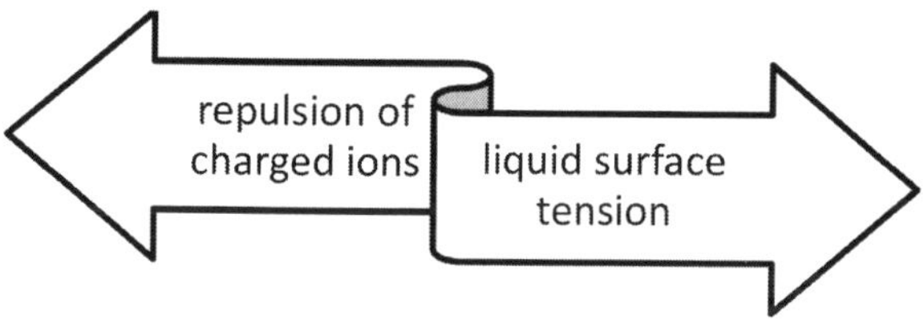

FIGURE 12.2 Forces affected electrospinning process.

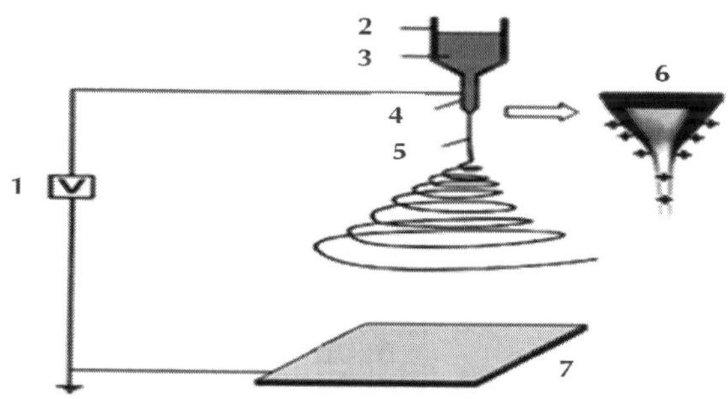

FIGURE 12.3 Sections of electrospinning process: (1) high voltage, (2) polymer solution, (3) syringe, (4) needle, (5) whipping instability, (6) Taylor cone, and (7) collector.

The most significant challenge in this process is to attain uniform nanofibers consistently and reproducibly.[11,16–18] Depending on several solution parameters, different results can be obtained using the same polymer and electrospinning setup.[19]

Factors that are studied to have a primary effect on the formation of uniform fibers are the process parameters, environmental parameters, and solution parameters.[11,20–23] These data are presented in Figure 12.4. In addition, many researcher studies effect of these parameters on final fibers. A summary of most important parameters are presented in Table 12.1.

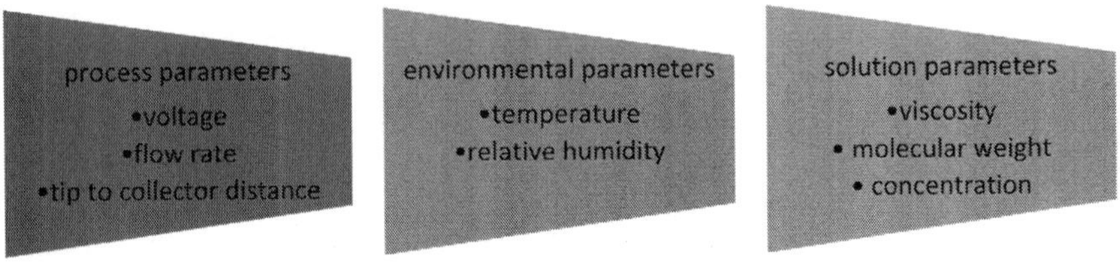

FIGURE 12.4 Parameters affect the morphology and size of electrospun nanofibers.

TABLE 12.1 The Effects of Some Parameters on Electrospinning Nanofibers in Researcher's Studies.

	Parameters	Year	Name of researcher	Effect	Reference
Process parameters	Needle to collector distance	1999	Fong et al.	Inversely proportional to bead formation density	[24]
				Inverse to the electric field strength	
		2003	Gupta & Wilkes	Inversely proportional to bead formation density	[25]
				Inversely proportional to fiber diameter	
		2004	Theron et al.	Exponentially inverse to the volume charge density	[26]
				Inverse to the electric field strength	
	Flow rate	2004	Theron et al.	Directly proportional to the electric current	[26]
				Inversely related to surface charge density	
				Inversely related to volume charge density	
		2005	Sawicka et al.	Directly proportional to the fiber diameter	[27]

TABLE 12.1 *(Continued)*

	Parameters	Year	Name of researcher	Effect	Reference
	Voltage	2001	Deitzel et al.	Direct effect on bead formation	[28]
		2003	Gupta & Wilkes	Inversely related to fiber diameter	[25]
		2004	Theron et al.	Inversely proportional to surface charge density	[26]
		2004	Kessick et al.	AC potential improved fiber uniformity	[29]
Solution parameters	Concen-tration of polymer	2001	Deitzel et al	Power law relation to the fiber diameter	[28]
		2002	Demir et al.	Cube of polymer concentration propor-tional to diameter	[30]
		2003	Gupta & Wilkes	Directly proportional to the fiber diameter	[25]
		2004	Hsu & Shivkumar	Parabolic-upper and lower limit relation to diameter	[31]
	Viscosity	2004	Hsu & Shivkumar	Parabolic relation to diameter and spinning ability	[31]
Environ-mental parameters	Temperature	2002	Demir et al.	Inversely proportional to viscosity	[30]
				Uniform fibers with less beading	

The mechanics of this process deserving a specific attention and necessary to predictive tools or way for better understanding and optimization and controlling process.[32] Also, many researchers have studied the behavior of the jet during formation.[33] For this case, we must to know more about parts of electrospinning process. In next section, we investigate this process with details.

12.2 INVESTIGATED FORMATION PARTS OF ELECTROSPINNING PROCESS

As mentioned above, many parameters affect electrospun. Alike, this process distinguished by four main sections.[34] So, we can separate process into four sections as shown in Figure 12.5. A description of these stages is as follows:

FIGURE 12.5 Different parts of electrospinning process.

12.2.1 *FIRST STEP: FORMATION OF TAYLOR CONE*

An electrospinning solution is usually an ionic solution that contains charged ions. The amounts of positive and negative charged particles are equal; therefore, the solution is electrically neutral.[14]

When an electrical potential difference is given between needle and collector, a hemispherical surface of the polymeric droplet at the orifice of needle is gradually expanded. When potential came into a critical value (equals 1), a flow of jet starts forming to a drop. Therefore, Taylor's cone is formed. A schematic of these steps is presented in Figure 12.6.

$$V_C^2 = 4\frac{H^2}{L^2}\left(\ln\frac{2L}{R} - \frac{3}{2}\right)(0.117\pi\gamma R) \tag{12.1}$$

Many researchers, like Rayleigh, Zeleny, and Taylor, gave insight into the survey of the behavior of liquid jets. Taylor determined that an angle of 49.3° is required to balance the surface tension and the electrostatic force.[14]

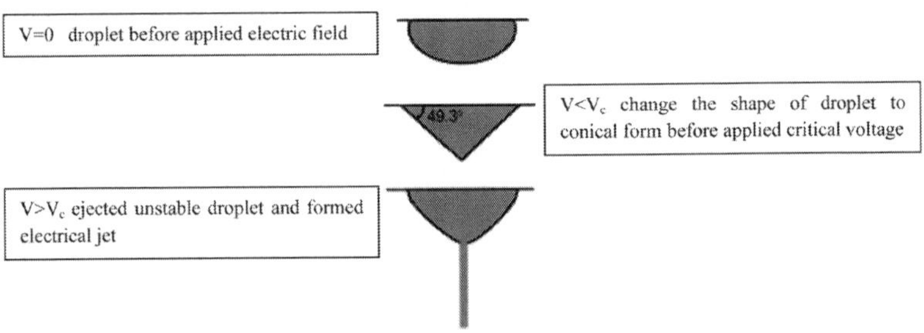

FIGURE 12.6 Changes in the polymer droplet with applied potential.

12.2.2 SECOND STEP: STEADY PART OF JET

As mention before, the jet is initiated from the droplet when the repelling forces of the surface charge overcome the surface tension and viscous forces of the droplet.[7] These repulsive forces between the jet segments will elongate the jet straight in the direction of its axis.[14]

A stable electrospinning jet travels from a polymer solution or melt to a collector. Electricity charges, commonly in the form of ions, tend to act in response to the electrical field that is associated with the potential. The electrical forces which stretch the fiber are resisted by the elongation viscosity of the jet.[12] Figure 12.7 shows stable part of the jet.

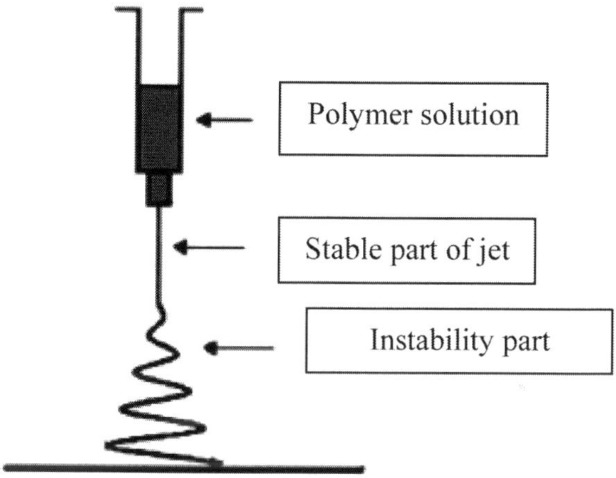

FIGURE 12.7 Instability in electrospinning.

The part of the jet leaving the tip was speeded up as the Coulomb forces, acting along the charges carried with the leading parts, drew out the jet and extended it in its axis toward the collector.[33] During the elongation of the electrified liquid jet, the jet surface area increases dramatically.[6]

Therefore, for a thin, stable jet, it is significant to look for a balance of viscosity charge density, and surface tension. Surface tension tries to lessen specific surface area, by changing jets into spheres. As the viscosity of a solution is increased, bead size increases, and the shape of the beads becomes more spindle shaped than spherical. As net charge density increases the beads become littler and more football shaped as well. Decreasing surface tension makes the beads disappear.[34]

Also, the straight, tapered part of the jet that went forth from the droplet has been mathematically modeled from many points of view. In future, we consider about modeling and simulating process.

12.2.3 THIRD STEP: INSTABILITY PART

After a small distance of stable traveling the jet, it will start unstable behavior and separate into many fibers.[7,12,14] It is necessary to know that the charge on the fiber expands the jet in the radial directions and to extend it in the axial direction. This jet division occurs several more times in rapid sequence and makes many small electrically charged fibers moving toward the collector.[12] In Figure 12.7, instability part of the jet is shown.

Recently, instability in electrospinning has received much attention. Many researchers, like Reneker et al., Yarin et al., and Rutledge et al., have studied jet instability. It is remarked that the jet followed a series of loops in which the loop diameters get larger.[14] Each instability grows at different rates.[34,35]

These instabilities vary and increase with distance, electrical field, and fiber diameter at different rates depending on the fluid parameters and performing conditions. Also, they influence the size and geometry of the deposited fibers.[36] These instabilities generate a looping and spiraling path, and it contributes to the extreme elongation and the acceleration of the electrically charged liquid jet.[6]

12.2.4 DIFFERENT MAIN SECTIONS OF INSTABILITY PART

As mentioned above, when jet spirals toward the collector, higher order instabilities reveal themselves resulting in spinning distance. These instabilities are separated into sections[37]:

1. Bending Instability (Fig. 12.8.a)
2. Rayleigh Instability (Fig. 12.8.b)
3. Whipping Instability (Fig. 12.8.c)

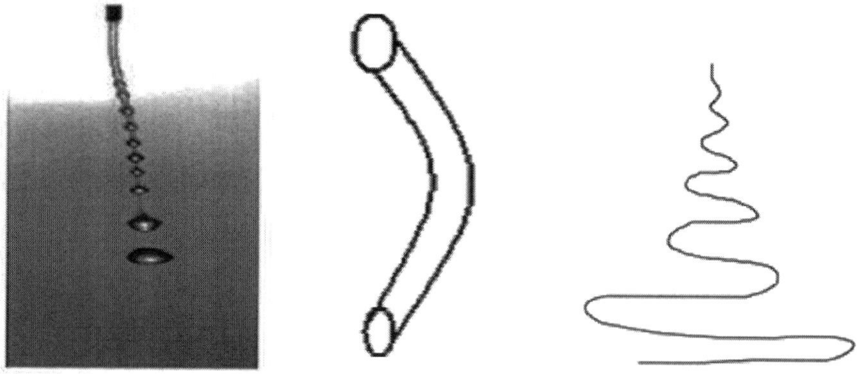

FIGURE 12.8 (a) Rayleigh instability, (b) bending instability, and (c) whipping instability.[38,39]

The polymer jet is influenced by these instabilities. These instabilities arise owing to the charge–charge repulsion between the excess charges present in the jet, which encourages the thinning and elongation of the jet. At high electric forces, the jet is dominated by bending (axisymmetric) and whipping instability (non-axisymmetric), causing the jet to move around and produce waves in the jet. At higher electric fields and at enough charge density in the jet, the axisymmetric (i.e., Rayleigh and bending) instabilities are suppressed and the non-axisymmetric instability increases.[36] Axisymmetric conducting instabilities can be viewed as a direct competition between the surface charges with the surface tension of the fiber, while the fiber is moving.[34] The suits for the noted bending instability are explained and the technique modeled.[40] In the next part, these instabilities are summarized.

12.2.4.1 BENDING INSTABILITY PART

When this jet moves straight toward the collector, then, bending instability develops into a series of loops expanding with time (Fig. 12.9).[33,41–43]

The bent part of the path was drawn out and reduced in diameter. Typically, the electrical bending coil began at a particular distance from the orifice, and the diameter of the turns of the coil grew larger and led toward the collector. The growing bend developed into the first turn of a growing coil. The continuous electrical bending formed a spiral with many turns which expanded in diameter as the jet continued to expand in response to the Coulomb repulsion of the charge. Extension of each section of the electrical force caused by the charge is transported by the jet continued. As the diameter of the jet decreased, the path of the jet again became unstable and a new, smaller diameter electrical bending instability developed. A succession of three or smaller diameter bending instabilities was often caught before the jet solidified. On the other hand, jets got fractal-like shapes, and their length increased enormously as their cross-sectional diameter decreased to a fraction of a micrometer. After several turns were made, a new electrical bending instability formed a smaller coil on a turn of the larger spiral. The turns of the smaller coil transformed into an even smaller coil and so forth until the elongation stopped, usually by solidification of the thin jet. Observations of the conical envelope surrounding the coils suggested that the repulsive forces between the electrical charges on such a jet caused the loops to continue to expand, by cleaning up the multitude of small coils into the larger coils. The variability in the onset and the behavior of the series of bending instabilities can account for a significant part of the variation in diameter that is often noted in electrospun nanofibers.[33]

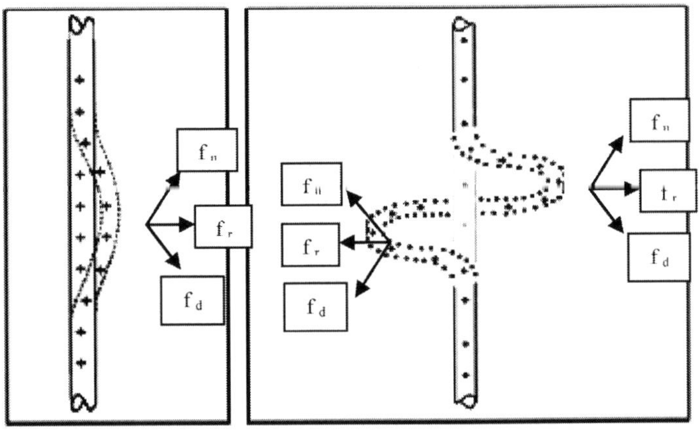

FIGURE 12.9 A part of the electrospun jet with act forces on this part.[33]

12.2.4.2 *RAYLEIGH INSTABILITY PART*

The first instability, also recognized as the Rayleigh instability is axisymmetric and occurs when the strength of electric field is low or when the viscosity of the solution is below the optimal value. Rayleigh instability is suppressed at high electric fields (higher charge densities) or when using higher concentrations of polymer in the solution.[36] It is almost dependent on the surface tension of the material.[34]

12.2.4.3 WHIPPING INSTABILITY PART

Whipping instability plays a key role in electrospinning for producing nanofibers.[9] It is the condition used for the bending instability of this rapidly moving jet.[14] In most operations of interest, the jet experiences a whipping instability, leading to bending and stretching of the jet, noted as loops of increasing size as the instability develops.[8]

When the electrostatic repulsion overcomes to viscoelasticity forces, the jet path was changed to an expanding helix.[7] This instability is a result of small bends in the initial uniformly charged straight fiber formation. As the fiber bends, the surface charges around the circumference of the jet are no longer uniform and a dipole moment is induced standing to the jet. These dipoles begin a localized torque, in response to each other and the applied electric field. This torque bends the jet. The bends expand as the jet moves further downstream, and then become the whipping instability that causes the decreased diameter of the fibers.[34] The whipping jet thins dramatically, while traveling the short distance between the electrodes.[8]

12.2.5 DIFFERENT OTHER SECTIONS OF INSTABILITY PART

As comfortably as the electrical bending instability, other characteristic instabilities were observed[33]:

- Branching
- Formation of physical beads

If the excess charge density on the surface of the jet was high, waves were predicted to take shape along the surface of a cylindrical jet. These waves grew large enough to become unstable and began branching which grew outward from the main jet. The familiar capillary instability that causes a cylindrical jet of liquid to collapse into separated droplets occurred when the surplus electrical charge carried by the jet was reduced. This structure solidified to form beaded nanofibers. It is not unusual for beads or branches to occur on the coils produced by the bending instability, but it is rare for beads and branches to occur at the same time along the same section of the jet. Therefore, the beads occur when the charge per unit area is small and the branches occur when the charge per unit area is great. Empirical observation showed that modification of a single technique parameter sometimes produced predictable changes in the onset of particular instabilities. When an experimental apparatus was producing nanofiber, it was often possible to accommodate a single technique parameter to note changes in some feature of the jet, such as the straight part at which the electrical bending began.[33]

12.2.5.1 BRANCHING DURING ELECTROSPINNING

Forming branches were viewed more often in more concentrated and more viscous solutions, and also in electric fields higher than the minimum area required for producing a single jet. Many branches were observed to get large diameter of the jet, during the electrospinning of polycaprolactone at a concentration of 15% in acetone, in the straight path of the gate or in the first choice of bending instability. Bending and branching began after only a short distance from the tip. Branches that arise from a jet with a large diameter can become long and entangled. Small branches were viewed on fibers electrospun from molten polycaprolactone and on small jets of polyimide electrospun in a vacuum. Interesting branches were noted on jets of polyetherimide in a volatile solvent which tended to form a skin on the jet as it dried. The skin gave way into a ribbon.[33]

12.2.5.2 BEADED FIBERS FORMATION

The capillary instability jet has been cleared since the times of Lord Rayleigh. He recognized the surface energy of a particular volume of fluid in a cylindrical jet is more eminent than the same volume divided into droplets. The excess electrical charges carried by the jet create a strong elongation flow and may either stabilize/destabilize the capillary instability, depending on the wavelength of the bead-forming instability. Stretching the entangled molecules in the strong elongation flow between the growing droplets produced a fluid, beads-on-string structure that solidified, leaving a beaded nanofiber. The beads were often spaced at recurring distances along the fiber and sometimes in repeating patterns of large and small beads. When concentrate the neutralizing counter ions from the corona was small, the electrospun nanofibers were smooth. As concentrate neutralizing ions increased, to a greater extent of the charge on the jet was neutralized by the airborne ions. As the charge density on the surface of the jet decreased, the number of beads per unit length increased. The volume of polymer in the form of nanofibers, about the volume of polymer in beads, decreased. Increasing the viscoelasticity of the electrospinning solution or increasing concentrate dissolved salt in the polyethylene oxide solution stabilizes the jet against forming beads. Beads also formed as the partial pressure of the solvent in the surrounding gas approached saturation. Forming a bead is the effect of a surface tension driven instability which had a cylindrical jet of polymeric fluid to divide into thinner fibers and larger beads by a minuscule extrusion technique.[33]

12.2.6 FOURTH PART: SOLIDIFICATION

As the jet moves toward the collector, it continues to expand by going past through the loops. Jet solidification is based along the traveling distance of the fibers. The distance between the collector and the capillary tip has a direct effect on the jet solidification and fiber diameter. If the nozzle-to-collector distance is long enough or the whipping instability is high, there is more time for fibers to dry before being picked up.[14] The collection part is where the jet is broken off. The polymer fiber that stays after the solvent evaporates may be accumulated on a metal screen. For polymers dissolved in nonvolatile solvents, water, or other suitable liquids can be used to collect the jet, remove the solvent, and develop the polymer fiber. If the jet arrives with a high speed at a stationary collector, the jet coils or folds. Since the jet is charged, fibers lying on the collector repel fibers that come afterward. The charge on the fibers can be changed by ions created in a corona perform and carried to the collection region by air currents. The charge may also be removed by charge migration through the fiber to the conducting substrate, although for dry fibers with low electrical conductivity, this charge migration may be slow.[12]

12.3 CONCLUDING REMARKS

Producing nanofibers by electrospinning is a simple and widely utilized for varied applications. The most significant part of electrospinning is how to control the process. Therefore, we must know about the behavior of every part of process and control instabilities were made in it. Also, the most important tools for controlling process are modeling and simulating. In future, we will review them.

KEYWORDS

- **electrospinning**
- **nanofibers**
- **nanofibers formation**

REFERENCES

1. Gupta, D.; Jassal, M.; Agrawal, A. K. Electrospinning of Poly(Vinyl Alcohol) Based Boger Fluids to Understand the Role of Elasticity on Morphology of Nanofibers. *Ind. Eng. Chem. Res.* **2015,** 7(13), 7351–7356.
2. Carroll, C. P.; et al. Nanofibers from Electrically driven Viscoelastic Jets: Modeling and Experiments. *Korea-Aust. Rheol. J.* **2008,** *20*, 153–164.
3. Fang, J.; Wang, X.; Lin, T. Functional Applications of Electrospun Nanofibers. *Nanofibers—Production, Properties and Functional Applications*; 2011; p. 287–326.
4. Reneker, D. H.; et al., Electrospinning of Nanofibers from Polymer Solutions and Melts. *Adv. Appl. Mech.* **2007,** *41*, 343–346.
5. Karra, S. *Modeling Electrospinning Process and a Numerical Scheme using Lattice Boltzmann Method to Simulate Viscoelastic Fluid Flows*. Texas A&M University, 2007.
6. Šimko, M., J. Erhart, and D. Lukáš, A Mathematical Model of External Electrostatic Field of a Special Collector for Electrospinning of Nanofibers. *J. Electrostat.* **2014,** *72*(2), 161–165.
7. Brooks, H.; Tucker, N. Electrospinning Predictions Using Artificial Neural Networks. *Polymer* **2015,** *58*, 22–29.
8. Fridrikh, S. V.; et al. Controlling The Fiber Diameter During Electrospinning. *Physical Review Letters*; American Physical Society, 2003; pp 144502–144502.
9. Zeng, Y.; et al. Numerical Simulation of Whipping Process in Electrospinning. In WSEAS International Conference. Proceedings. Mathematics and Computers in Science and Engineering, World Scientific and Engineering Academy and Society, 2009.
10. Ciechańska, D. Multifunctional Bacterial Cellulose/Chitosan Composite Materials for Medical Applications. *Fibres Text. East. Europe* **2004,** *12*(4), 69–72.
11. Lu, P.; Ding, B. Applications of Electrospun Fibers. *Rec. Pat. Nanotechnol.* **2008,** *2*(3), 169–182.
12. Reneker, D. H.; Chun, I. Nanometre Diameter Fibres of Polymer, Produced by Electrospinning. *Nanotechnology* **1996,** *7*(3), 216–223.
13. Haghi, A. K. Electrospun Nanofiber Process Control. *Cellulose Chem. Technol.* **2010,** *44*(9), 343–352.
14. Ghochaghi, N. Experimental Development of Advanced Air Filtration Media based on Electrospun Polymer Fibers. *Mechnical and Nuclear Engineering*; Virginia Commonwealth, 2014; pp 1–165.
15. Ziabari, M.; Mottaghitalab, V.; Haghi, A. K. Evaluation of Electrospun Nanofiber Pore Structure Parameters. *Kor. J. Chem. Eng.* **2008,** *25*(4), 923–932.

16. Li, Z.; Wang, C. Effects of Working Parameters on Electrospinning. *One-Dimensional Nanostructures*; Springer, 2013; pp 15–28.
17. Bognitzki, M.; et al. Nanostructured Fibers via Electrospinning. *Adv. Mater.* **2001,** *13*(1), 70–72.
18. De Vrieze, S.; et al., The Effect of Temperature and Humidity on Electrospinning. *J. Mater. Sci.* **2009,** *44*(5), 1357–1362.
19. Sill, T. J.; von Recum, H. A. Electrospinning: Applications in Drug Delivery and Tissue Engineering. *Biomaterials* **2008.** *29*(13), 1989–2006.
20. Angammana, C. J. *A Study of the Effects of Solution and Process Parameters on the Electrospinning Process and Nanofibre Morphology*, University of Waterloo, 2011.
21. Bhardwaj, N.; Kundu, S. C. Electrospinning: A Fascinating Fiber Fabrication Technique. *Biotechnol. Adv.* **2010,** *28*(3), 325–347.
22. Rafiei, S.; et al. New Horizons in Modeling and Simulation of Electrospun Nanofibers: A Detailed Review. *Cellulose Chem. Technol.* **2014,** *48*(5–6), 401–424.
23. Tan, S. H.; et al., Systematic Parameter Study for Ultra-Fine Fiber Fabrication via Electrospinning Process. *Polymer* **2005,** *46*(16), 6128–6134.
24. Fong, H.; Chun, I.; Reneker, D. H. Beaded Nanofibers Formed During Electrospinning. *Polymer* **1999,** *40*(16), 4585–4592.
25. Gupta, P.; Wilkes, G. L. Some Investigations on the Fiber Formation by Utilizing a Side-by-Side Bicomponent Electrospinning Approach. *Polymer* **2003,** *44*(20), 6353–6359.
26. Theron, S. A.; Zussman, E.; Yarin, A. L. Experimental Investigation of The Governing Parameters in The Electrospinning of Polymer Solutions. *Polymer* **2004,** *45*(6), 2017–2030.
27. Sawicka, K.; Gouma, P.; Simon, S. Electrospun Biocomposite Nanofibers for Urea Biosensing. *Sens. Actuat., B: Chem.* **2005,** *108*(1), 585–588.
28. Deitzel, J. M.; et al. The Effect of Processing Variables on The Morphology of Electrospun Nanofibers and Textiles. *Polymer* **2001,** *42*(1), 261–272.
29. Kessick, R.; Fenn, J.; Tepper, G. The Use of AC Potentials in Electrospraying and Electrospinning Processes. *Polymer* **2004,** *45*(9), 2981–2984.
30. Demir, M. M.; et al., Electrospinning of Polyurethane Fibers. *Polymer* **2002,** *43*(11), 3303–3309.
31. Hsu, C. M.; Shivkumar, S. Nano-Sized Beads and Porous Fiber Constructs of Poly (ε-Caprolactone) Produced by Electrospinning. *J. Mater. Sci.* **2004,** *39*(9), 3003–3013.
32. Yarin, A. L.; Koombhongse, S.; Reneker, D. H. Bending Instability in Electrospinning of Nanofibers. *J. Appl. Phys.* **2001,** *89*(5), 3018–3026.
33. Reneker, D. H.; Yarin, A. L. Electrospinning Jets and Polymer Nanofibers. *Polymer* **2008,** *49*(10), 2387–2425.
34. Zhang, S. *Mechanical and Physical Properties of Electrospun Nanofibers.* 2009, pp 1–83.
35. Wu, Y.; et al. *Controlling Stability of the Electrospun Fiber by Magnetic Field. Chaos, Solitons Fractals* **2007,** *32*(1), 5–7.
36. Baji, A.; et al. *Electrospinning of Polymer Nanofibers: Effects on Oriented Morphology, Structures and Tensile Properties. Composites Sci. Technol.* **2010,** *70*(5), 703–718.
37. He, J. H.; Wu, Y.; Zuo, W. W. Critical Length of Straight Jet in Electrospinning. *Polymer* **2005,** *46*(26), 12637–12640.
38. Hohman, M. M.; et al. Electrospinning and Electrically Forced Jets. II. Applications. *Phys. Fluids (1994–Present)* **2001,** *13*(8), 2221–2236.

39. Zhang, S. *Mechanical and Physical Properties of Electrospun Nanofibers.* 2009.
40. Yarin, A. L.; Zussman, E. Electrospinning of Nanofibers from Polymer Solutions, XXI ICTAM.—Warsaw, Poland, 2004, pp 12–15.
41. Sawicka, K. M.; Gouma, P. Electrospun Composite Nanofibers for Functional Applications. *J. Nanopart. Res.* **2006,** *8*(6), 769–781.
42. Yousefzadeh, M.; et al. A Note on The 3D Structural Design of Electrospun Nanofibers. *J. Eng. Fabrics Fibers (JEFF)* **2012,** *7*(2), 17–23.
43. Li, W. J.; et al. Electrospun Nanofibrous Structure: A Novel Scaffold for Tissue Engineering. *J. Biomed. Mater. Res.* **2002,** *60*(4), 613–621.

CHAPTER 13

PRODUCTION OF ELECTRODES FOR MANUAL ARC WELDING WITH USING THE COMPLEX MODIFIERS

D. IL'YASHENKO and S. MAKAROV*

Yurga Institute of Technology, TPU Affiliate, Yurga, Russian Federation

**Corresponding author. E-mail: s.makarov@mail.ru*

CONTENTS

ABSTRACT

This work describes weld and processing characteristics of welding electrodes manufactured with the admixture of complex modifiers (complex nanopowders) and compares them with standard electrodes. It was found that the admixture of complex nanopowders allows increase of mechanical properties of weld metal up to 20% and improvement of chemical composition, particularly, increase of deoxidants content (Si, Mn).

13.1 INTRODUCTION

The beginning of the 21st century was marked by a revolutionary leap in the development of nanotechnologies and nanomaterials. They have already been used by all developed countries worldwide in the most significant human activities (industry, defense, IT, radio electronics, power engineering, transport, biotechnologies, and medicine) in the production of ceramic and composite materials, superconductors, solar batteries, lubricant additives, magnetic pigments, etc.[1]

In the welding industry, nanopowders are used to obtain a fine-grained structure of weld metal, and when crystallized, these additives ensure grain refinement and in the end improve the mechanical properties practically without changing the chemical composition of the alloy.[2]

The goal of our studies is to analyze the influence of complex nanopowders on the characteristics of a welding electrode and arcing process.

13.2 THEORY

Complex nanopowder has been selected as the modifier of electrode compound. This selection is determined by the integrated effect that nanopowders produce on the properties of manufactured electrode, namely:

- Al_2O_3—improves welding process stability, weld formation, and separability of slag crust;
- SiO_2—improves weld metal harness;
- Ni—improves impact strength and moldability;
- TiO_2—provides stable arcing; and
- W—produces solid compounds—carbides that improve hardness and red-hardness of weld metal.

To determine the possibility of using nanopowders in the production of welding electrodes (to enhance the quality characteristics of electrodes, first of all the welding and process features and mechanical properties of weld metal), MP-3 electrodes of Ø4.0 mm were produced. It should be noted that less liquid glass was consumed than with the standard technology (22 kg of glass per 100 kg of dry batch against 24.5 kg of glass per 100 kg of batch in serial production, which is a 12% reduction in liquid glass consumption).[3,4]

A series of experiments was performed to define the processing behavior of the electrodes under study. During these experiments, pins welded on the surface of the plates. During the welding process, the changes in energy (such as amperage and voltage) are expected.

13.3 RESULTS AND DISCUSSION

Steel plates were used for welding: SS type St3, 300 mm long, 50 mm wide, and 4 mm thick. On the surface of the plates a bead of the electrode metal was welded. For welding, electrodes 4 mm in diameter were used, types MP-3. The BA306-U3 rectifier was used as a source of supply. Welding was performed in single-pass under following conditions: amperage 140–160 A, voltage 24–26 V. Mechanical properties and chemical composition of the weld metal are presented in Tables 13.1 and 13.2.

TABLE 13.1 Mechanical Properties of Weld Metal.

MP-3 electrodes of Ø4.0 mm	σв (MPa)	δ (%)	KCU, at 20°C (J/cm²)
Serial	460	25	159
Experimental	492	28	192
GOST 9467-75 requirements	460	18	80

$\sigma_в$—yield stress; δ—elongation; KCU—impact strength.

TABLE 13.2 Chemical Composition of Weld Metal.

MP-3 electrodes of Ø4.0 mm	Mass fraction of elements (%)				
	C	Si	Mn	S	P
Serial	0.07	0.03	0.47	0.025	0.046
Experimental	0.07	0.05	0.61	0.025	0.046
GOST 9467-75 requirements	–	–	–	0.040	0.045

Mechanical properties and chemical analysis of filler metal were measured according to the following methods:

- yield stress and elongation were measured according to GOST 1497-84;
- impact strength of weld seam metal was measured according to GOST 6996-66; and
- chemical analysis of filler metal was tested according to GOST 7122-81.

As Table 13.1 shows, prototype electrodes, compared with series models, provide increase of yield stress by 9%, relative elongation by 11%, and impact strength by 20%. This is due to the admixture of electrode modifiers into electrode compound; they produce integrated effect on weld metal, and technologies of nanopowder admixture into electrode during its production, that is, into soluble glass, using mechanocavitation unit of activation type.

Complex nanopowders promote mechanical properties of filler metal and alloy-transfer efficiency; however, welding materials with nano-structured components can have negative impact on welder's health.[5,6] This fact constrains the widespread use of these materials in welding production and requires additional examination.

13.4 CONCLUSION

Based on the conducted study, it may be concluded that the admixture of complex nanopowders into electrode compound promotes improvement of strength properties and performance of welding connection.

KEYWORDS

- **nanopowder**
- **welding electrode**
- **fusion welding**
- **mechanical properties**
- **alloying element**

REFERENCES

1. Baloyan, B. M.; Kolmakov, A. G.; Alymov, M. I.; Krotov, A. M. *Nanomaterialy. Klassifikatcia, osobennosti svoystv, primenenie and tekhnologii polutchenia* (*Nanomaterials. Classification, Characteristics of Properties, Application and Technology of Production*); Moscow, 2007; 125 p.
2. Makarov, S. V.; Gnedash, E. V.; Ostanin, V. V. Comparative Characteristics of Standard Welding Electrodes and Welding Electrodes with the Addition of Nanopowders. *Life Sci. J.* **2014,** *11*(8s), 414–417.
3. Makarov, S. V.; Sapozhkov, S. B. Use of Complex Nanopowder (Al_2O_3, Si, Ni, Ti, W) in Production of Electrodes for Manual Arc Welding. *World Appl. Sci. J.* **2013,** *22*, 87–90.
4. Makarov, S. V.; Sapozhkov, S. B. Production of Electrodes for Manual Arc Welding Using Nanodisperse Materials. *World Appl. Sci. J.* **2014,** *29*(6), 720–723.
5. Zhang, M.; Jian, L.; Bin, P.; et al. Workplace Exposure to Nanoparticles from Gas Metal Arc Welding Process. *J. Nanopart. Res.* **2013,** *15*(11), 1–14.
6. Guerreiro, C.; Gomes, J. F.; Carvalho, P.; Santos, T. J. G.; Miranda, R. M.; Albuquerque, P. Characterization of Airborne Particles Generated from Metal Active Gas Welding Process. *Inhal. Toxicol.* **2014,** *26*(6), 345–352.

CHAPTER 14

NANOSCALE TECHNOLOGY FOR ENGINEERING PRODUCTS

O. I. SHAVRIN, L. N. MASLOV, and L. L. LUKIN*

Kalashnikov Izhevsk State Technical University, Izhevsk, Russia

Corresponding author. E-mail: lukin@istu.ru

CONTENTS

ABSTRACT

Implementation of nanoscale technology for engineering products is a complex and relevant task. Innovative solution to this task is development of thermal strain processing (TSP) directed to form nanoscale substructure in structural steels. Principles and methods of submicrocrystalline structures forming in steels during wire production stages of sized mill-rolls according to the proposed technology are discussed in this paper. The proposed principle of nanoscale structure forming is based on combination of fast induction heating, plastic flow of complex direction and cooling within the range of scientifically based thermal and speed modes.

14.1 INTRODUCTION

Development of nanoscale technologies in the world practice as innovative activity of scientists has become a new trend.[1,2] Elaboration and development of nanoscale technologies in manufacturing engineering is the most relevant and complex in implementation among various innovations in the sphere of nanoscale technology. Among major goals of nanoscale technology implementation for engineering products can be increased reliability and durability due to strengthening effect of bulk structural metallic materials. Researches in this sphere[3–6] allowed formulation of production methods of nanoscale structural materials.[1,7] Production methods of nanoscale structural materials are divided into four groups in work[1]:

- powder metallurgy (compaction of nanopowders);
- amorphous state crystallization;
- intense plastic flow;
- various methods of overlaying nanostructured coatings.

Similar classification of nanoscale technologies for manufacturing engineering is given in work.[7]

This classification is based on the idea that the process of strain resistance and metal strength are determined by grain sizes or nanopowder particles.

14.2 THEORETICAL OBSERVATION

Authors think that grain sizes, morphology, and texture may vary in accordance with corresponding technological parameters of the nanoscale material manufacturing process. Volume fraction of interface (grain boundaries and triple point joints) increases considerably with grain size decrease, they have significant influence on nanoscale material properties. Structural features of nanocrystalline materials (grain size, considerable interface fraction and their condition, porosity, and other structural flaws) are determined by manufacturing methods and have great influence on their properties. Decreasing grain size increases strength, reveals the low-temperature, and high-speed superplastic effects.

The only criterion for such group division is material grain size and its influence on strength of structural materials. This criterion is physically based, but it does not take into account the influence on plastic flow resistance of metal fine grain texture, that is, subgrains.

During the study of subgrain influence on strength it was found that the empiric dependence of Hall–Petch,[8]

$$\sigma_{T} = \sigma_{0} + k_{y} D^{-1/2},$$

where σ_0 is stress to be applied to overcome lattice friction (Peierls–*Nabarro forces*), k_y is the value of locking and hindering of dislocation degree found for grain sizes in pure metals having no substructure, which is also valid for metals with developed substructure for subgrain sizes ranging from 0.05 μm (50 nm) to 70 μm and subgrain size is used in the formula instead of grain size.

Influence of steel polygonal and cell substructure being formed under high-temperature thermomechanical processing (HTMP) on strength is proved in works[9–13] including pattern alloys retaining austenite state with substructure formed under strain after high-temperature straining and cooling to room temperature.

The research results showed that subgrains contributed much to strengthening due to low-angle subgrain boundaries that become obstacles increasing resistance to dislocation motion inside the crystal. Work[13] proves that beside simple dependence on grain size there is extra effect related to subgrains surrounded by low-angle boundaries that increase flow stresses.

Low-angle subgrain boundaries increase friction inside crystal during dislocation motion without changing grain size.

The available data on substructure role in plastic flow resistance provides further doubt about uniqueness of statements that only grain nanoscale dimensionality with their high-angle boundaries should be taken into account when considering steel strength.

The influence of grains and subgrains on strength characteristics cannot be univocal. It will differ for brittle and ductile steels, for strain resistance characteristics to low and high plastic flows.

Grain and subgrain has special influence on steel resistance to cyclic loading,[14] that is, fatigue strength due to the fact that stress level is lower than the elasticity limit. Subgrains under these conditions contribute to more uniform action of microslip mechanisms in grains, while low-angle subgrain boundaries become more effective obstacles to dislocation motion reducing the ability of critical dislocation piling-up at grain boundaries that cause fatigue crack nucleation.

Substructure strengthening influence on low-plastic flow[15] allows consideration that nanoscale dimensionality of substructure elements is the sign of nanoscale technology and methods of nanoscale substructure forming and can be regarded as a special classification group.

Size of fine structure elements—polygons, cells is determined by techniques and, what is important, by their parameters.

Integrated processes providing various physical actions, for instance, thermal processing and plastic flow, applied in various sequence can become the foundation for such techniques for structural steels. Variations of such techniques to form polygonal substructure can be united into the notion thermal strain processing (TSP),[16] being considered as the fifth method of nanoscale structure forming in structural steels.

One variation of such technique can be implemented on HTMP principles. However, HTMP is applied for production, for example, rolled products according to classical methods[12] and does not provide steel substructure nanoscale dimensionality due to the specific characteristics of the process (high temperatures, fraction strain and duration).

It is required to change process parameters and make them controllable to obtain nanoscale substructure.[17,18]

Control of TSP should be directed to

1. heating temperature minimizing under the condition of process speed, providing homogenization of chemical composition,
2. reducing amount of strain,
3. adjustability of cooling including application of strain soaking.

Implementation of these directions is possible on special equipment to produce several types of metallurgical products (wires, sized steel), workpieces for engineering products or parts.

The advantage of such approach in terms of strengthening is in possibility to form nanoscale structure for parts of different sizes and with different thickness of hardened layer, that is, from surface to full hardening according to the operational load characteristics.

To implement methods of TSP for metallurgical and engineering products a number of methods and equipment were developed:

14.3 THERMAL STRAIN PROCESSING OF METALLURGICAL PRODUCTS

To produce metallurgical products – sized steel 15–26 mm in diameter and wire 2.5–10 mm in diameter, in the result of research and tests pilot-plant equipment was developed and technology was matured.

Developed typical manufacturing method of TSP includes the following operations: initial material surface preparation, induction heating higher than Ac_3 for straining, strain soaking, cooling (quenching) in progressive mode, and tempering.

Hot strain is implemented for wire by drawing, for rolled products by helical drafting in deforming head. Diagrams of pilot-plant equipment for TSP are given in Figures 14.1 and 14.2.

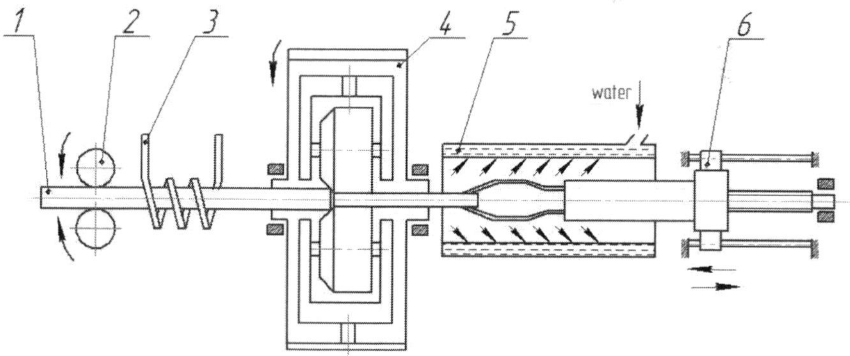

FIGURE 14.1 Design scheme of sized steel TMP machine.

1—Workpiece, 2—charged rollers, 3—induction coil, 4—deforming head, 5—sprayer, and 6—axial movement drive.

14.3.1 MANUFACTURING OF SIZED STEEL

Pilot-plant machine to produce rolled products with TSP shown in Figure 1 consists of series of charging device 2 that feeds workpiece 1, blocks of high-frequency heating 3, helical drafting (HD) deforming head 4, spray-type cooling 5 and pulling mechanism 6. Deforming head consisting of 3 rollers mounted at 120° circumferentially rotates at frequency of 500–700 rpm. Charged rollers 2 and pulling device 6 provide variable speed of rod feed within 5–10 m/min. Heating temperature of workpiece before strain is 950–1020°C, while amount of strain (15–25%) is provided by adjustment of rollers in radial direction. The workpiece being drawn is cooled in sprayer within 2–8 s with water that provides its straightness ($\Delta \leq 0.2$ mm/m) and eliminates further straightening.

Sized strengthened rolled products is manufactured with diameter accuracy of IT11–IT12 and surface roughness $R_a = 2.5–6.3$ according to Russian State Standard GOST 2789-73, curvature does not exceed the allowable values according to Russian State Standard GOST 7417-75.

High efficiency of TSP is proved by results of bench fatigue tests of samples made of steel 38XC (C 0.34–0.42%, Si 1–1.4%, Mn 0.3–0.6%, Ni up to 0.3%, S up to 0.035%, P up to 0.035%, Cr 1.3–1.6%, Cu up to 0.3%). Tests in terms of asymmetric bending under stress values $\sigma_{max} = 1170$ MPa, $\sigma_{min} = 700$ MPa showed that samples after quenching and tempering at 450°C ruptured after (0.27–0.44) × 10^6 cycles of loading, while samples after HTMP (tempering at 270, 320, and 450°C) under basic number of cycles 10^6 did not rupture at any mode of tempering.

High structural strength and operational reliability together with high accuracy make possible application of sized rolled products for manufacturing heavy-duty parts like track bolts, torsion pins of transporting and agricultural machines, large-size springs of car suspensions, railroad cars, buffer springs, axels, etc.

14.3.2 TSP OF WIRE

Pilot production of wire was intended for manufacturing of different purpose springs.

Industrial machine shown in Figure 14.2 consists of drawing bench 1 (model 1/650), two induction coils 11 and 7 connected to loading circuit blocks 10 and 8 of high-frequency units with frequency of 0.066 kHz, die-box 5 with sprayer cooling device 4, wire heating induction coil 2 connected to high-frequency transformer 3, heat insulator 9, ingoing straightening unit 12 and drum 13.

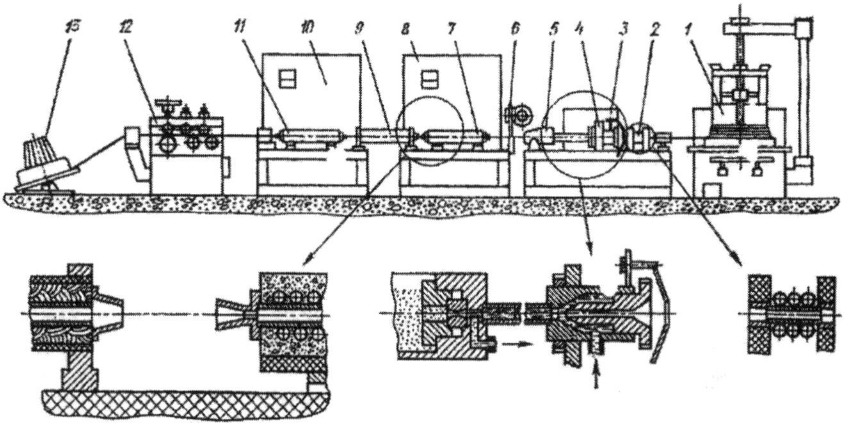

FIGURE 14.2 Design scheme of industrial wire TMP machine.

Intense wire cooling and application of graphite as lubricant provide reliable (without rupture) wire drawing heated up to 920–1020°C. Hardened wire can be supplied in two ways: (1) high-tempered ($T_{tempering}$ = 600–650°C) with ground or polished surface; (2) low-tempered without extra surface treatment. Both wire conditions can be used for spring manufacture under various technology options. Fatigue tests showed not less than tenfold durability increase.

The TSP study gave a number of new results:

1. nanoscale substructure forming was stated, polygonal substructure size of elements corresponds to the criterion of less than 100 nm, according to Figure 14.3;
2. substructural mechanisms of forming steel anisotropy to resist low plastic flows and conditions of obtaining anisotropic properties were established;
3. influence on strain amount providing maximum strengthening effect, part hcating ratc wcrc cstablishcd;
4. range of strain soaking optimal for substructure forming and stability of rupture-free wire drawing were established.

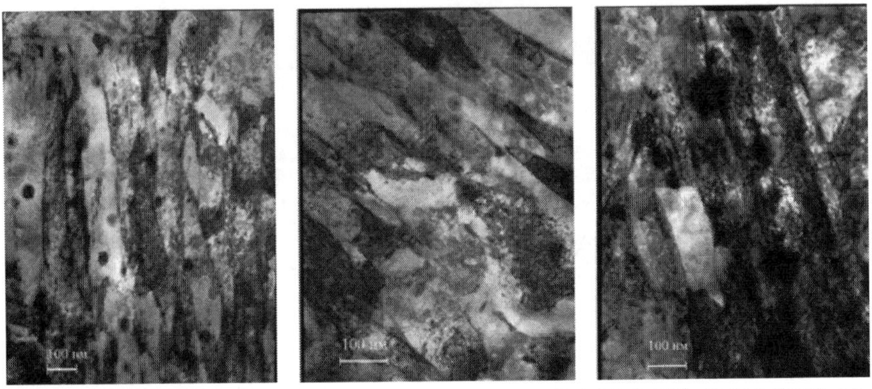

FIGURE 14.3 Wire substructure made of steel 51XΦA (C 0.74–0.55%, Si 015–0.3%, Mn 0.3–0.6%, Ni up to 0.25%, S up to 0.025%, P up to 0.025%, Cr 0.75–1.1%, V 0.15–0.25%, Cu up to 0.2%) after TSP: T_{heating} = 1,000°C, drawing strain amount 20%, $T_{\text{tempering}}$ = 500°C.

The material accumulated during all-round study of TSP allows discussion of setting up manufacturing of strengthened metallurgical products including standard flexible equipment to manufacture strengthened nanostructures products.

14.4 THERMAL STRAIN STRENGTHENING OF MILL ROLLS

Technical and economic efficiency of precision strap directly depends on reliability, durability, and quality of mill rolls. Mill rolls quality affects quality of rolled products: gauge interference, stability of assigned strap profile, surface roughness.

The problem of increasing service durability of multiroll mills is extremely important for rolling industry. Low service durability of rolls requires unexpended roll changes leading to deterioration of technical and economic indices of rolling production.

Rolls of multiroll mills are subjected to high requirements: in terms of high (HRC 60–65) and uniforms hardness of barrel surface at the depth of not less than 3% of the radius, in terms of contact strength ($\sigma_{\text{contact}} \rightarrow$ max), in terms of thermal fatigue strength, in terms of geometrical accuracy (cilindricity) and surface roughness ($R_a \leq 0.32$).

Researches showed that rolls fail both in the result of wear and surface flaws appeared during operation (cuts, dimples, weld-on deposits, delaminations, scabbings, and cracks).

Increasing service durability of rolls can be achieved by increasing surface layer strength together with increasing plasticity. The most effective combination of the material properties of roll surface layer mentioned above is possible due to forming submicrocrystallic (nanoscale) structure in the result of high-temperature TSP. All-round test study of steel grades 9ХФ (C 0.8–0.9%, Si 0.1–0.4%, Mn 0.3–0.6%, Ni up to 0.4%, S up to 0.03%, P up to 0.03%, Cr 0.4–0.7%, V 0.15–0.3%, Cu up to 0.3%), ШХ15(C 0.95–1.05%, Si 0.17–0.37%, Mn 0.2–0.4%, Ni up to 0.3%, S up to 0.02%, P up to 0.027%, Cr 1.3–1.65%, Ti up to 0.01%, Cu up to 0.25%), and 9Х2МФ (C 0.85–0.95%, Si 0.25–0.5%, Mn 0.2–0.7%, Ni up to 0.5%, S up to 0.03%, P up to 0.03%, Cr 1.7–2.1%, Mo 0.2–0.3%, V 0.1–0.2%) widely used for rolls production showed that strengthening effect depends on the scheme of plastic flow, prevailing subgrain direction, degree of martensitic crystal size reduction, content, and volume of deposited fine-dispersed carbides.

Optimal high-temperature TSP mode with complex strain (drafting + tension + torsion) by helical drawing (through deforming rollers mounted radially at 120°) with progressive high-frequency heating and spray cooling (quenching) was worked out.

Industrial comparative tests of rolls were held during strap rolling. Test results showed that consumption of rolls subjected to HTSP for one roll change is 1.8–2.2 times less than for rolls produced according to conventional technique (quenching with high-frequency heating). Radial wear of rolls subjected to TSP under similar conditions is 2 times less than for rolls produced according to conventional technique, although wear type along the barrel is the same as shown in Figure 14.4.

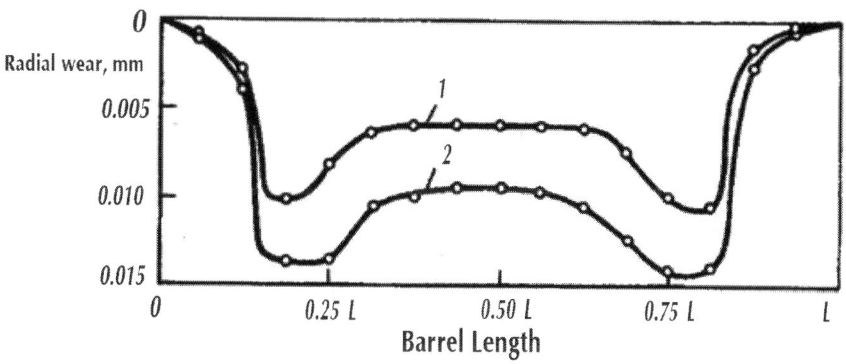

FIGURE 14.4 Typical wear curvatures of effective generating line of mill roll 400 after 5×10^3 revolutions; 1—quenching with high-frequency current heating and 2—HTSP.

There are three processes that can be seen on curvatures describing wear of multiroll mill rolls as shown in Figure 14.5. The first period (classical running-in period) for rolls subjected to TSP is rather weak in terms of wear, initial roughness of effective surface (within $R_a = 0.32$) is almost unchanged. The second period (period of regular wear) represents the operating period of the roll. The third period is characterized by high intensity of wear of strap edge and wear irregularity representing the signal to roll change. The third period comes later for rolls produced in accordance with optimal technological modes and after TSP.

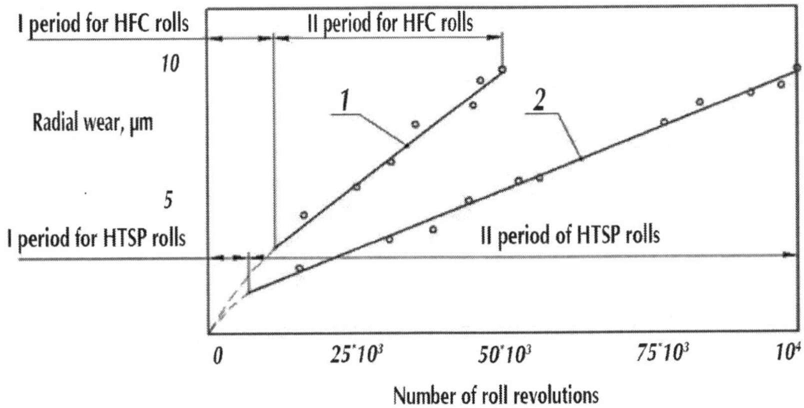

FIGURE 14.5 Radial wear of effective rolls 28 mm in diameter made of steel 9X (C 0.85–0.95%, Si 1.2–1.6%, Mn 0.3–0.6%, Ni up to 0.35%, S up to 0.03%, P up to 0.03%, Cr 0.95–1.25%, Mo up to 0.2%, W up to 0.2%, V up to 0.15%, Ti up to 0.03%, Cu up to 0.3%): 1—quenching HFC, 2—TSP.

Comparative analysis of roll durability of 20-roll mills showed that durability of rolls subjected to TSP is significantly increases (1.7–2.2 times) and, as a consequence, roll consumption reduces: for 20-roll mills 160 and 400 by from 1.25 to 0.82 kg and 0.64 to 0.42 kg per 1 ton of rolled products.

It was found that about one third of rolls produced according to conventional technique fail due to weld-on deposits. Besides, tempering spots of considerable depth appear due to sharp temperature rise during fold rolling on strap tear. The allowable value of mill 400 roll wear is about 30μm per diameter then rolls are subjected to redressing with metal removal rate of 80–90 μm. The metal removal rate grow up to 250–300 μm for rolls quenched by high-frequency current heating and 150–180 μm for rolls strengthened by TSP that contributes the increased wear resistance and improves strengthening efficiency.

Operating experience of 20-roll mill model during several years showed that the main type of roll failure is spalling or surface descaling which is directly related to contact fatigue nucleation. As research proved surface TSP increases contact fatigue strength of roll steels grades 9X (C 0.85–0.95%, Si 1.2–1.6%, Mn 0.3–0.6%, Ni up to 0.35%, S up to 0.03%, P up to 0.03%, Cr 0.95–1.25%, Mo up to 0.2%, W up to 0.2%, V up to 0.15%, Ti up to 0.03%, Cu up to 0.3%) and ШХ15 (C 0.95–1.05%, Si 0.17–0.37%, Mn 0.2–0.4%, Ni up to 0.3%, S up to 0.02%, P up to 0.027%, Cr 1.3–1.65%, Ti up to 0.01%, Cu up to 0.25%) 2.5–10 times that result from many factors during surface layer submicrocrystalline structure forming under TSP.

Testing working rolls of diameter 152 mm on 20-roll mill model made of steel grade 6X6M1ФШ (C up to 0.06%, Cr 6%, Mo 1%, V up to 1% and electroslag remelting) with TSP showed that their durability is 3.5 times higher than durability of rolls made of steel grade 9X2MФШ (C 0.09%, Cr 2%, Mo up to 1%, V up to 1% and electroslag remelting) (batch production technique) and corresponds to quality of rolls manufactured by best foreign companies.

Beside operating performance improvement application of TSP allowed simplification of production technique that includes operations of rolled steel cutting into piece blanks, machining in accordance with roll drawing for surface HTSP, surface HTSP performed on equipment with vibroimpulsive straining, low-temperature tempering, cutting off the process shank, and centering, grinding, and control. The following operations are eliminated from the batch production technique forging, annealing, cleaning, and tempering.

14.5 CONCLUSION

Research of TSP performed on special equipment showed that under certain parameters of production method nanoscale substructure forming in steel, significant increase of steel strength, and operational characteristics of engineering products are provided.

ACKNOWLEDGMENTS

Research is performed at FSBEI HPE Kalashnikov Izhevsk State Technical University with state financial support on behalf of Ministry of Education and Science of the Russian Federation within Federal Targeted Program

"Research and development of priority development fields of Russian science and technology sector for 2014–2020," unique identifier of applied research (project) is RFMEFI57714X0011.

KEYWORDS

- **nanoscale technology**
- **manufacturing engineering**
- **structural steels**
- **thermal strain processing**
- **nanoscale structure**
- **wire**
- **sized mill products**
- **strength**
- **reliability**
- **efficiency**

REFERENCES

1. Lyakishev, N. P.; Alymov, M. I. Nano-scale Structural Materials. *Russian Nanotechnol.* **2006,** *1*(1–2), 71.
2. Alferov, Zh. I. Russia Needs New Nanotechnology. *Russian Nanotechnol.* **2008,** *2*(11–12), 8.
3. Valiev, R. Z.; Alexandrov, J. V.; Zhu, Y. T.; Lowe, T. C. Paradox of Strength and Ductility in Metals Processed by Severe Plastic Deformation. *J. Mater. Res.* **2002,** *17*, 5.
4. Valiev, R. Z. Nanomaterial Advantage. *Nature* **2002,** *419*, 887–889.
5. Wang, Y.; Chen, M.; Zhou, F.; Ma, E. High Tensile Ductility in a Nanostructured Metal. *Nature* **2002,** *419,* 912.
6. Wang, Y. M.; Ma, E. Three Strategies to Achieve Uniform Tensile Deformation in a Nanostructured Metal. *Acta Mater.* **2004,** *52*, 1699–1709.
7. Mashinostroenie. Encyclopedia/Edit. board: K. V. Frolov (chairman), et al.; Moscow, Mashinostroenie. Methods of production of composite materials, plastics, glass and ceramics. Volume III-6. Endorsed by V. S. Bogolyubov. 2006, 576 p., Section 6. Nanotechnology in Manufacturing Engineering, p. 544–555.
8. Ivanova, V. S., Ed. *Role of Dislocations in Metal Strengthening and Rupture*; Nauka: Moscow, 1965; 180 P.
9. Bernstein, M. L. *Steel* **1972,** 2, 157–162.
10. Bernstein, M. L. *Structure of Strained Metals*; Metallurgia: Moscow, 1977; p 432.

11. Bernstein, M. L.; Kaputkina, L. M.; Prokoshkin, S. D.; Dobatkin, S. V. *Izvestia USSR AS. Metals* **1982,** *2*, 94–103.
12. Bernstein, M. L.; Zaymovsky, V. A.; Kaputkina, L. M. *Thermomechanical Steel Processing*; Metallurgia: Moscow, 1983; p 480.
13. Inglish, A. T.; Bakofen, U. A. Influence of Metal Processing on Rupture; M.: Metallurgia, 1976, 309 p.
14. Terentiev, V. F. *Fatigue Strength of Metals and Alloys.* Internet Engineering: Moscow, 2002; p 287.
15. Rahshtadt, A. G. *Spring Steels and Alloys*. Metallurgia: Moscow, 1980.
16. Shavrin, O. I. Nanosized Structure Formation in Machine Part Material. *Vestnik ISTU* **2011,** *1*, 4–6.
17. Shavrin, O. I. Influence of Nanotechnology on Coiled Springs Operational Characteristics. *Nanostructures, Nanomater. Nanotechnol. Nanoindustry* **2014,** 260–272.
18. Shavrin, O. I. *High Strength Springs For The Railway Rolling Stock* [Text], Shavrin, O. I., 2012, 3, 16–18. Vestnik of Institute of Natural Monopoly Problems: Railroad Engineering, 2012, *3*, 71–80.

CHAPTER 15

THE INFLUENCE OF NANOFILLER STRUCTURE ON MELT VISCOSITY OF NANOCOMPOSITES POLYMER/ CARBON NANOTUBES

M. A. MIKITAEV, G. V. KOZLOV*, and A. K. MIKITAEV

Kh. M. Berbekov Kabardino-Balkarian State University, Nal'chik, Russian Federation

Corresponding author. E-mail: kozolov@urfu.ru

CONTENTS

ABSTRACT

As it is known, any real fractal object can be simulated as a bulk or surface fractal. In the first case, the structure fractality is extended to the entire volume of the object, and in the second one—to its surface. For nanocomposites, the matrix of which makes up polymer blend poly(ethylene terephthalate)/poly(butylene terephthalate), the fractal dimensions of both structure and surface of ring-like formations of nanofiller (multiwalled CNT) were calculated. It has been found out that in case of their notion as bulk fractals, that is, matrix polymer penetration in ring-like formations internal regions, the strong enhancement of the nanocomposites melt viscosity is observed and in case of surface fractals this parameter is independent on nanofiller contents.

15.1 INTRODUCTION

Inorganic nanofillers of different types usage for polymer nanocomposites receiving acquires at present wide spreading.[1] However, the indicated nanomaterials melt properties are not studied enough completely. As a rule, when nanofillers application is considered, then the compromise is achieved between improvement of mechanical properties in solid-phase state, enhancement of melt viscosity at processing, nanofillers dispersion problem, and process economic characteristics. Proceeding from this, the relation between nanofiller concentration and geometry and nanocomposites melt properties is an important aspect of polymer nanocomposites study.

It has been shown earlier[2] that nanocomposites polypropylene/carbon nanotubes melt viscosity is not changed practically at nanofiller contents variation within the range of 0.25–3.0 mass %. However, in work [3], the sharp (in 3.5 times) reduction of melt viscosity of nanocomposites on the basis of blend poly(ethylene terephthalate)–poly(butylene terephthalate), filled with multiwalled carbon nanotubes (PET–PBT/MWCT) was found in comparison with the initial matrix polymer blend at MWCT content of 0.45 mass % only. Therefore, the present paper purpose is this effect study with the fractal analysis notions using.

15.2 EXPERIMENTAL

The polymers industrial sorts were used: PET was supplied by firm GE Plastics and PBT of mark S6110SF NC010, supplied by firm DuPont TM

Crastin. As nanofiller MWCT were used with external diameter ranging between 10 and 30 nm and length of 1.5 mcm of firm Sun Nano production.[3]

The blends PET/PBT, having optimal PET content of 20 mass %, were blended with MWCT at nanofiller contents of 0.15–0.45% by weight in melt with usage of twin screw extruder (Bersfort FRG Germany) at temperature of 493 K and screw rotation speed of 150 rpm. After receiving, the extrudate was cooled in water and pelletized. The testing samples were prepared from these pellets using injection molding machine (Windsor, India) at temperature range of 523–558 K.[3]

The uniaxial tension mechanical tests were performed according to ASTM D628 on testing apparate of model AG Rnisd MC of form Shimadzu Autograph at temperature of 296 K and the cross head speed of 50 mm/min that corresponds to the strain rate of 1.67×10^{-2} s^{-1}. Melt flow index (MFI) was measured according to ASTM D1238.[3]

15.3 RESULTS AND DISCUSSION

As it was noted above, for nanocomposites PET–PBT/MWCT very strong increase of viscosity (MFI reduction) was observed in comparison with matrix polymer blend.[3] This effect was fixed at very small MWCT volume content φ_n, which can be determined as follows[1]:

$$\varphi_n = \frac{W_n}{\rho_n}, \quad (15.1)$$

where W_n is nanofiller mass content, ρ_n is its density, which for nanoparticles is determined according to the equation[1]:

$$\rho_n = 188\left(D_{\text{MWCT}}\right)^{1/3}\ (\text{kg/m}^3), \quad (15.2)$$

where D_{MWCT} is MWCT diameter, which is given in nm.

According to the continuum conception,[4] the relation between composites melt viscosity η, matrix polymer melt viscosity η_0 and φ_n can be obtained in the form of two simple relationships:

$$\frac{\eta}{\eta_0} = 1 + \varphi_n, \quad (15.3)$$

$$\frac{\eta}{\eta_0} = \frac{2.5\varphi_n}{1-\varphi_n}. \tag{15.4}$$

Equations (15.3) and (15.4) give the greatest η increasing in comparison with η_0 on 0.8–2.0% that does not correspond to experimentally observed η growth in 3.5 times.

The authors[2] proposed the fractal model of polymer nanocomposites melt viscosity change, the main equation of which is the following one:

$$\eta \sim \eta_0 l^{2-d'_f}, \tag{15.5}$$

where l is flow characteristic linear scale, d'_f is fractal dimension.

Since carbon nanotubes, possessing anisotropy high degree and low transverse stiffness, are formed in polymer matrix ring-like structures with radius R_{CNT},[5,6] then this dimensional parameter was accepted as l, which can be determined with the aid of the equation[7]:

$$\varphi_n = \frac{\pi L_{MWCT} r_{MWCT}^2}{\left(2R_{CNT}\right)^3}, \tag{15.6}$$

where L_{MWCT} and r_{MWCT} are length and radius of carbon nanotubes, respectively.

Since contact polymer matrix-nanofiller is realized through the ring-like structures, then as d'_f the authors[2] have accepted these structures surface dimension d_{surf}, which is determined according to the following technique. First the specific surface S_u was estimated according to the equation[1]:

$$S_u = \frac{3}{\rho_{CNT} R_{CNT}}, \tag{15.7}$$

where ρ_{CNT} is the density of MWCT rang-like structures, which is accepted equal to ρ_n.

Then the value d_{surf} can be determined with the aid of the following formula[1]:

$$S_u = 410\left(R_{CNT}\right)^{d_{surf}-d}, \tag{15.8}$$

where d is dimension of Euclidean space, in which a fractal is considered (it is obvious that in our case, $d = 3$).

In Figure 15.1, the dependence of MFI on MWCT volume contents φ_n is adduced (curve 1) for the considered nanocomposites, where the value MFI is determined as follows[8]:

$$MFI = \frac{2.33}{\eta}, \text{ g/10 min} \tag{15.9}$$

As it follows from the data of Figure 15.1, at the condition $d'_f = d_{surf}$ MFI value is practically constant, which does not correspond to the experimental results.[3]

As it is known,[9] fractals can be divided into two categories: bulk and surface ones. In the first case, fractality property is extended to the object volume, and in the second one—to its surface only. For the bulk fractals their dimension d_f^{CNT}, that is dimension of MWCT ring-like structures, is determined with the aid of the relationship[10]:

$$R_{CNT} \sim \left(\frac{4c_0kT}{3\eta M_0}\right)^{1/d_f^{CNT}} t^{1/d_f^{CNT}}, \tag{15.10}$$

where c_0 is nanoparticles initial concentration, which is accepted equal to φ_n, k is Boltzmann constant, T and t are temperature and duration of processing, respectively, η is melt viscosity, M_0 is mass of separate nanoparticle.

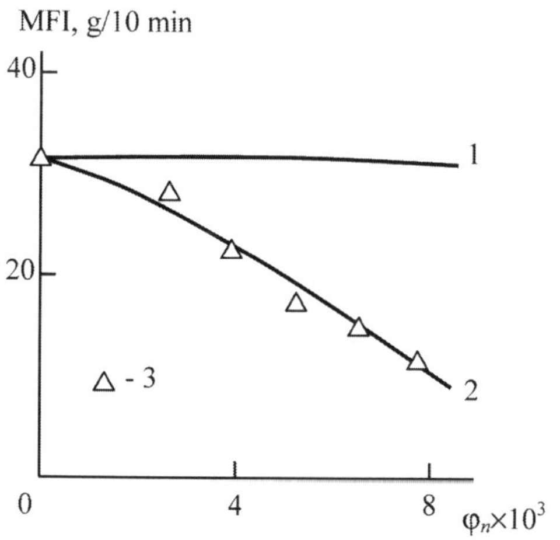

FIGURE 15.1 The dependence of melt flow index (MFI) on nanofiller volume contents φ_n for nanocomposites PET–PBT/MWCT. (1 and 2) The theoretical calculation according to the relationship (15.5) for cases of surface (1) and bulk (2) fractals, (3) experimental data.

Since the parameters k, T, t and M_0 are constant, and $\eta \sim \text{MFI}^{-1}$, then the relationship (15.10) can be simplified as follows:

$$R_{\text{CNT}}^{d_f^{\text{CNT}}} \sim \varphi_n \times \text{MFI}. \tag{15.11}$$

In Figure 15.1, the dependence of MFI(φ_n) for the considered nanocomposites is also adduced, which is calculated according to relationship (15.5) at the condition $d_f' = d_f^{\text{CNT}}$, that is, at simulation of MWCT ring-like structures as bulk fractals. In this case the good correspondence of theory and experiment is obtained.

As it is known,[11] for bulk fractals penetration of polymer matrix macromolecules into their internal regions is typical. The number of macromolecules n, penetrating into these regions, is given by the following relationship[11]:

$$n \sim R_{\text{CNT}}^{\Delta_f}, \tag{15.12}$$

where Δ_f is the fractal dimension of polymer melt macromolecular coil, determined as follows[12]:

$$\Delta_f = (d-1)(1+\nu), \tag{15.13}$$

where ν is Poisson's ratio, estimated according to the mechanical tests results with the aid of the equation[13]:

$$\frac{\sigma_Y}{E} = \frac{1-2\nu}{6(1+\nu)}, \tag{15.14}$$

where σ_Y and E are yield stress and elastic modulus, respectively.

In Figure 15.2, the dependence of MFI(n) for the considered nanocomposites is adduced. As one can see, MFI reduction (melt viscosity growth) at n decreasing is observed. At $n \rightarrow 0$ MFI ≈ 0 and the change of MWCT ring-like structures type from bulk fractal to surface one occurs.

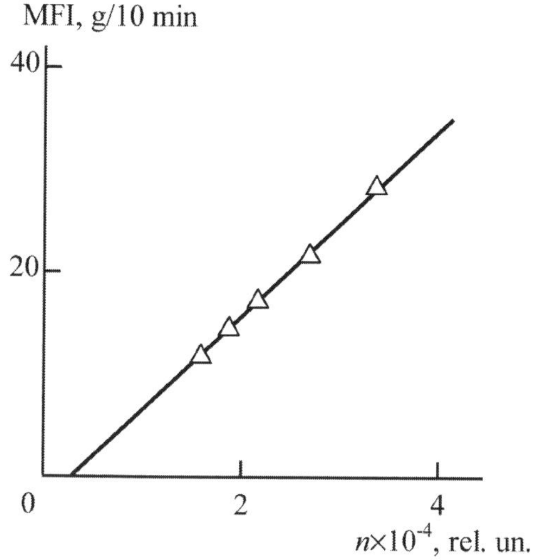

FIGURE 15.2 The dependence of melt flow index (MFI) on polymer matrix macromolecules number n, penetrating in MWCT ring-like structures internal regions, for nanocomposites PET–PBT/MWCT.

15.4 CONCLUSIONS

Thus, the present work results have shown that the dependence of melt viscosity of nanocomposites polymer/carbon nanotubes on nanofiller contents is defined by fractal structures type, which is formed by the nanofiller. In case of surface fractals, very weak dependence of melt viscosity on carbon nanotubes contents is observed, and in case of bulk fractals—its essential (in several times) enhancement. The absence of polymer matrix macromolecules penetration in carbon nanotubes ring-like structures internal regions defines the structures transition from bulk fractals to surface ones.

Work is performed within the complex project on creation of hi-tech production with the participation of the Russian higher educational institution, the Contract of JSC "Tanneta" with the Ministry of Education and Science of the Russian Federation of February 12, 2013 No. 02.G25.31.0008 (Resolution of the Government of the Russian Federation No. 218).

KEYWORDS

- **nanocomposite**
- **blend**
- **carbon nanotubes**
- **melt**
- **viscosity**
- **fractal**

REFERENCES

1. Mikitaev, A. K.; Kozlov, G. V.; Zaikov, G. E. *Polymer Nanocomposites: Variety of Structural Forms and Applications*; Nova Science Publishers, Inc.: New York, 2008; 319 p.
2. Kozlov, G. V.; Zhirikova, Z. M.; Aloev, V. Z.; Zaikov, G. E. *Polym. Res. J.* **2012,** *6*(3), 267–273.
3. Rajakumar, P. R.; Nanthini, R. *Int. J. Adv. Chem.* **2013,** *1*(2), 39–42.
4. Mills, N. J. *J. Appl. Polymer Sci.* **1971,** *15*(11), 2791–2805.
5. Schaefer, D. W.; Justice, R. S. *Macromolecules* **2007,** *40*(24), 8501–8517.
6. Yanovsky, Yu. G.; Kozlov, G. V.; Zhirikova, Z. M.; Aloev, V. Z.; Karnet, Yu. N. *Int. J. Nanomech. Sci. Technol.* **2012,** *3*(2), 99–124.
7. Bridge, B. *J. Mater. Sci. Lett.* **1989,** *8*(2), 102–103.
8. Kalinchev, E. L.; Sakovtseva, M. B. *Properties and Processing of Thermoplasts.* Khimiya: Leningrad, 1983, 288 p.
9. Feder, E. *Fractals*. Plenum Press: New York, 1989; 256 p.
10. Weitz, D. A.; Huang, J. S.; Lin, M. Y.; Sung, J. *Phys. Rev. Lett.* **1984,** *53*(17), 1657–1660.
11. Kozlov, G. V.; Dolbin, I. V.; Zaikov, G. E. *Fractal Physical Chemistry of Polymer Solutions and Melts*; Apple Academic Press: Toronto, New Jersey, 2014; 316 p.
12. Balankin, A. S. *Synergetics of Deformable Body*; Publishers of Ministry Defence SSSR: Moscow, 1991; 404 p.
14. Kozlov, G. V.; Sanditov, D. S. *Anharmonic Effects and Physical-Mechanical Properties of Polymers*; Nauka: Novosibirsk, 1994; 261 p.

CHAPTER 16

SHORT COMMUNICATIONS: UPDATES ON NANOMATERIALS PROPERTIES

A. F. RUBIRA[2], A. J. M. VALENTE[1], A. P. GEROLA[1,2], D. C. SILVA[1], E. C. MUNIZ[2], E. V. KOVERZANOVA[5], E. V. TOMINA[6], G. P. KUZNETSOV[7], I. G. ASSOVSKY[7], I. Y. MITTOVA[6], L. S. ZELENINA[6], N. G. SHILKINA[5], O. BORGES[3,4], O. MUKTAROV[8], S. JESUS[3,4], S. M. LOMAKIN[5], S. V. USACHEV[5], T. MUKTAROVA[8], V. I. KOLESNIKOV-SVINAREV[7] and V. LYASNIKOV[8]

[1]*Department of Chemistry, University of Coimbra, 3004-535 Coimbra, Portugal*

[2]*Grupo de Materiais Polimñricos e Compysitos, GMPC, Chemistry Department, Maringб State University, 87020-900 Maringб, Paranб, Brazil*

[3]*Faculty of Pharmacy, University of Coimbra, Coimbra, Portugal*

[4]*CNC, Center for Neuroscience and Cell Biology, University of Coimbra, 3004-517 Coimbra, Portugal*

[5]*Emanuel Institute of Biochemical Physics of Russian Academy of Sciences, Moscow, Russia*

[6]*Department of Materials Science and Industry of Nanosystems, Voronezh State University, Faculty of Chemistry, Voronezh, Russia*

[7]*Semenov Institute of Chemical Physics RAS, Kosygin Str. 4, Moscow 119991, Russia*

[8]*Yuri Gagarin State Technical University of Saratov, Institute of Electronic Engineering and Mechanical Engineering, Saratov, Russia*

CONTENTS

16.1 pH-RESPONSIVE MODIFIED GUM ARABIC HYDROGELS FOR DELIVERY OF CURCUMIN

Curcumin (CUR) is a polyphenolic compound associated with numerous pharmacological activities, but their application is limited by the low water solubility. Thus, in this work, some inclusion complexes with alpha-cyclodextrin (α-CD) and beta-cyclodextrin (β-CD) in different proportions were prepared to enable the drug solubilization in biological fluids.

The formation of these complexes was confirmed by ^{1}H NMR and thermogravimetric analysis. The stoichiometry of the CUR/α-CD and CUR/β-CD complexes was 1:1 and the association constants were 346 and 4650 mol^{-1} L for α-CD and β-CD, respectively. The major stability of the CUR/β-CD complex was associated with the steric factors, and the CUR solubility increased for the CUR/α-CD (1:4) complex due to the higher solubility of α-CD.

Hydrogels of modified gum arabic containing CUR/α-CD (1:4) were obtained and used for controlled release of CUR in simulated intestinal fluid (SIF) and simulated gastric fluid (SGF). The kinetics of release was pH-responsive and the percentage of CUR released was *c.a.* 97% in SIF and 6.7% in SGF. For the toxicity studies on undifferentiated Caco-2 cells IC_{50} of 63.4 ± 14.4 and 85.2 ± 14.9 μg mL^{-1} for CUR and CUR/α-CD (1:4), respectively, were obtained.

The toxicity of these samples on differentiated Caco-2 was lower than undifferentiated cells due to the presence of mucus layer in the former that acts as a selective barrier. Additionally, the CUR incorporated into hydrogels showed no toxic effect on differentiated and undifferentiated Caco-2, indicating the pharmaceutical potential of three-dimensional matrices of GAm for controlled release of CUR complexed with cyclodextrin.

ACKNOWLEDGMENT

A. P. G. and D. C. S. are grateful for the scholarship provided by CAPES/CSF. Financial support from Fundação para a Ciência e Tecnologia (Portugal) and Coordenação para Aperfeiçoamento de Pessoal de Nível Superior (Brazil), or FCT/CAPES application (Proc. No. 329/13), allowing scientific cooperation between Brazilian and Portuguese researches groups, is gratefully acknowledged.

16.2 METALS COMBUSTION AS METHOD FOR PRODUCTION OF ULTRAPOROUS NANOSTRUCTURAL CERAMICS

This paper is a review of experimental and theoretical investigations of mechanisms and patterns of gas-phase synthesis of nanostructured ultraporous ceramics (aerogels and xerogels) in metal particles combustion carried out in Semenov ICP RAS.

The dependences have been obtained for composition and morphology of the combustion products on the combustion conditions (nature and dispersion of the metal, temperature, pressure, and composition of the gaseous medium).

Some effective methods are presented to influence on the processes of nanostructured aero- and xerogels formation.

The subjects of the study have been: aluminum, magnesium, zinc, titanium, zirconium, and other metals burning in mixtures of oxygen with nitrogen or inert gases (argon, helium), in carbon dioxide, and in air under different pressures.

16.3 POLYAMIDE 6 NANOCOMPOSITES WITH REDUCED WATER PERMEABILITY

From positions of industrial application and due to increasing demand in higher quality control of polymers and polymer composites, the study of moisture absorption effects on the end product properties is of high importance.

Moisture is the basic factor affecting the modulus, strength, and damping properties of polymer materials. Polyamides are highly hygroscopic materials due to presence of amine groups in them. The physical properties of polyamides are dominated by intermolecular hydrogen bond. By hydrogen bonds, water molecules are linked to carbonyl or NH-groups.

Therefore, moisture absorption strongly affects mechanical, thermal, electrical, and other properties of polyamides.

Our study deals with the evaluation of water uptake of the polyamide 6—clay (Cloisite 30B and 93A) nanocomposites as well as correlation of the clay amount and the main physicochemical, thermal, mechanical, and electrical properties of prepared nanocomposites.

16.4 OXIDATION OF GALLIUM ARSENIDE WITH THE MODIFIED NANOSCALE LAYER OF GEL V_2O_5 SURFACE

Nanoscale films, grown as a result of gallium arsenide thermal oxidation, do not satisfy the requirements of semiconductor electronics due to the formation of arsenic in the elemental state during the process. Modification of the semiconductor surface by the oxides of transition metal alters the kinetics and mechanism of the process and, therefore, the composition and properties of formed films.

Vanadium oxide (V) has a pronounced chemical stimulated action because of its nature is capable of exhibiting both transit and catalytic properties in the processes of $A^{III}B^{V}$ semiconductors thermal oxidation. Deposition of V_2O_5 gel on the GaAs surface is a soft method of modifying, because there is no interaction between the layer of chemostimulator and the components of semiconductor substrate before the thermal oxidation.

The purpose of this research is to establish the regularities of thermal oxidation of V_xO_y/GaAs heterostructures synthesized by the deposition of V_2O_5 gel through the aerosol phase on the semiconductor surface.

According to spectroscopic ellipsometry data nanoscale films, formed by oxidation of V_xO_y/GaAs heterostructures, are adequately described by a Cauchy model with a normal dispersion relation of the optical constants.

The effective activation energy of the process of V_xO_y/GaAs heterostructures oxidation calculated using data of laser and spectroscopic ellipsometry is 77 kJ/mol, indicating that the transit mechanism is realized.

As in the oxidation of pure GaAs limiting step is the diffusion of gallium from the substrate to the film. The acceleration of the growth rate of the films during oxidation of V_xO_y/GaAs compared with oxidation of pure GaAs is about 40–50%. According to the AFM data, the maximum height of the relief of the samples after oxidation is increased, and the average surface roughness decreases.

16.5 DEPOSITION OF CARBON NANOTUBES ON THE IMPLANT WITH HYDROXYAPATITE COATING

Carbon nanotubes and nanofibers have several properties that suggest that these materials may be of value in the development of new devices for bone reconstruction. Since they are organic particles, issues related to the metal ions from the implant does not arise. The proportions and the physical form of the carbon fibers mimic the natural crystalline structure of hydroxyapatite bone (i.e., hydroxyapatite crystal size from 50 to 100 nm in length and 1–10 nm in diameter).[1]

Getting nanotubes and fullerenes by chemical vapor deposition is particularly intensively developed in recent years, as it allows us to obtain a large number of identical nanotubes on the surface of the template. A method for fabricating nanostructures, it plays an important role that not only affects the properties of the nanostructure, but also during its life, that is, the period during which the particle is able to exhibit these unique properties.

Hydroxyapatite $Ca_{10}(PO_4)_6(OH)_2$ (HA) has been widely used in dentistry as a component of dental fillings pastes and filling different internal bone cavities, because it has a bioactivity resorbability, stimulate cell division, and activates protein has healing and hemostatic properties.[1] It is known that not stoichiometric HA obtained from animal biological material, unlike synthetic HA have higher bioactive properties because their composition corresponds to the composition of living biological systems.

The results showed that the specific role of producing animal bone hydroxyapatite by thermal treatment depends on the temperature and processing time. Analysis of the results of IR Fourier spectroscopy reveals the identity of the IR spectra with a similar product on the market, "Bio-Oss."

REFERENCES

1. Webster, J.; et al. Selective Bone Cell Adhesion on Formulations Containing Carbon Nanofibers. *Biomaterials* **2003,** *24*, 1877–1887.

16.6 FORMATION OF COMPOSITE COATINGS ON THE BASIS OF BIOLOGICAL HYDROXYAPATITE PLASMA SPRAYING ON TITANIUM SUBSTRATE

In this paper, we experimentally investigate the influence of technological modes of plasma spraying biological hydroxyapatite powder on the formation of nanostructured coatings on titanium-based implant. Used for the deposition of hydroxyapatite powder dispersion of biological origin of 5–20, 30–50, and 50–80 μm. Before spraying, titanium base implant (titanium VT1-00) was subjected to mechanical and chemical cleaning.

Interest in the study plasma sprayed calcium phosphate coatings of hydroxyapatite obtained by chemical synthesis and biological origin is currently part of their wide use as implants to repair damaged tissue.[1]

As a result, plasma spraying of hydroxyapatite powders on titanium substrates formed separate islands molten particles. Then these islands grow in size and formed a continuous coating. Further deposition leads to the formation of a continuous coating with a specific pore structure and nanoscale elements hydroxyapatite. Education nanoscale elements caused by thermophysical and chemical processes are responsible for the cooling rate of the substrate.

Structural and morphological pattern of the hydroxyapatite coating is largely dependent and can vary greatly from technological deposition conditions (arc current, spraying distance, the dispersion of the powder used, the rate of crystallization of the particles), the effectiveness of other types of electrophysical and physicochemical effects.

Thus, it can be assumed that the process on the substrate surface by plasma spraying is largely determined by the excess surface free energy and structuring, and therefore, the appearance of the nanostructured elements is dependent on a set of factors listed above (preparation of the substrate prior to the deposition, sputtering process conditions, etc.

KEYWORDS

- **biological hydroxyapatite**
- **plasma spraying**
- **implant**
- **sputtering process conditions**

REFERENCES

1. Lyasnikov, V. N.; Lyasnikova, A. V.; Pivovarov, A. V.; Antonov, I. N.; Papshev, V. A. Study of Structure of Bioceramic Coatings Obtained by Plasma Spraying of Hydroxyapatites of Synthetic and Biological Origins. *Biomed. Eng.* **2011,** *45*(4), 119–127.

PART V

Applications of Nanostructured Materials and Nanotechnologies

CHAPTER 17

THE USE OF NANOMATERIALS IN GREENHOUSES

E. BASARYGINA*, T. PUTILOVA, and R. PANOVA

Chelyabinsk State Academy of Agricultural Engineering, Faculty of Electrification and Automation of Agricultural Production, Lenin Prospect, 75, Chelyabinsk, Russian Federation

Corresponding author. E-mail: basarygina @urfu.ru

CONTENTS

ABSTRACT

The results obtained from the use of nanomaterials are today able to increase the efficiency of agricultural production and, in particular, growth of the protected ground. Off-season cultivation of plants in greenhouses involves considerable expenditure of energy and resources, and therefore the development of the means of intensification of crop production is an urgent task. In this regard, promising is the use of nanomaterials for high crop yields, combined with environmental friendliness and usefulness of biological products. Application of nanomaterials provides growing vegetables, getting hydroponic forage and ensuring the biological plant protection products on the basis of entomophagous. It is proposed to use silver nanoparticles to create a germicidal coating on the pot designs and supplies; water filtration, nutrient solutions, and fresh air. A method of use of mass entomophage afidimizy midges, characterized in that the cultivation is carried out on plants feed water purified in the filter with the antimicrobial silver nanoparticles. Application of the proposed method increases the yield of cocoons midge 25–30% by providing antimicrobial protection of the root system of the plants that are used in the preparation of larvae and cocoons. Growing plants in water, purified antimicrobial filter with silver nanoparticles will activate all the links in the food chain "plant–aphid–midge larvae": to enhance plant growth and development, stimulate the growth of aphids, and thus increase the yield of cocoons midges. The method may be implemented using a filter designed for purification of liquids. The use of these developments in the greenhouses will increase plant productivity and reduce energy consumption of crop production.

17.1 INTRODUCTION

Results obtained with the use of nanomaterials can already improve agricultural production, including crop-protected soil.[1–4]

Off-season cultivation of plants in greenhouses involves considerable expenditure of energy and resources, and therefore the development of the means of intensification of crop production is an urgent task. In this regard, promising is the use of nanomaterials for high crop yields, combined with environmental friendliness and usefulness of biological products.

17.2 APPLICATION OF NANOMATERIALS IN GREENHOUSE

The paper presents the development, made in the Chelyabinsk State Academy of Agricultural Engineering, in various areas of use in greenhouses silver nanoparticles having antimicrobial activity (Fig. 17.1).

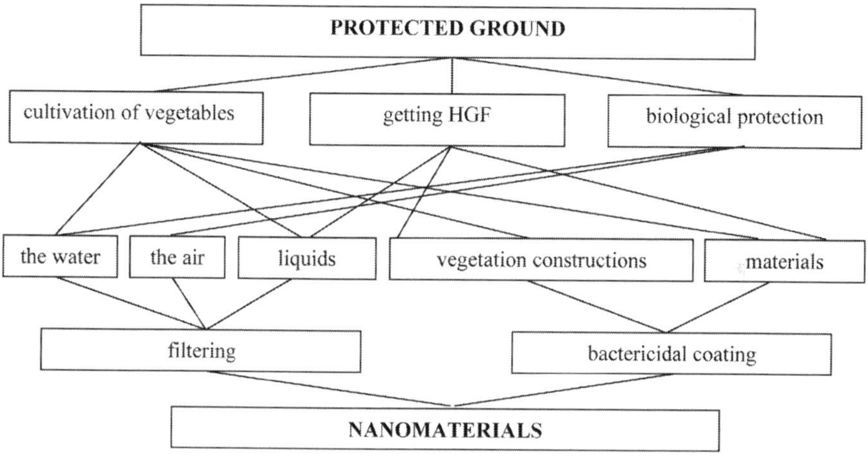

FIGURE 17.1 The use of nanomaterials in greenhouses.

In accordance with scheme (Fig. 17.1) is provided with application nanomaterial of vegetable growing, getting hydroponic green fodder (HGF) and ensuring the biological plant protection products on the basis of entomophagous.

It is proposed to use silver nanoparticles to

- creation of bactericidal coatings on vegetation structures and consumables;
- water filtration, nutrient solutions, and supply air.[5–9]

17.3 HYDROPONIC INSTALLATION

In hydroponic installation, shown as an example in Figure 17.2, apply the coating with silver nanoparticles on consumables vegetation materials.

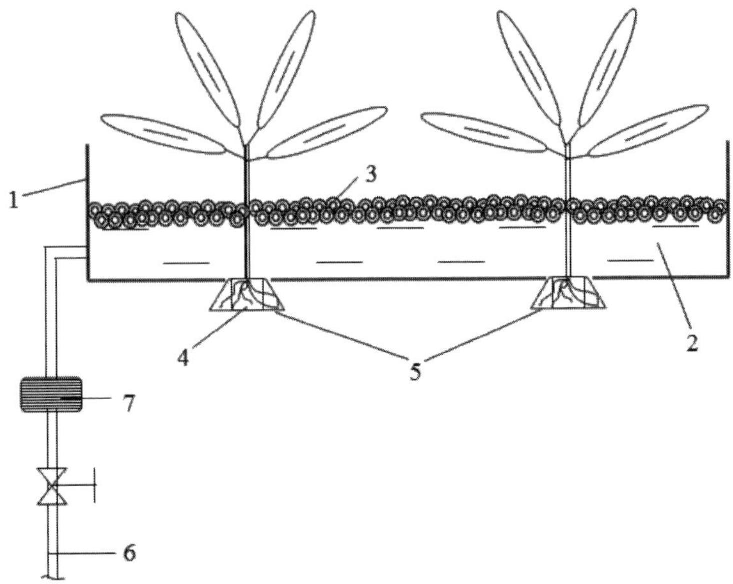

FIGURE 17.2 Hydroponic installation. The installation contains vegetative bath 1 and 2 nutrient solution and placed on the surface 3 of the floating particles is optically opaque material coated antimicrobial layer containing silver nanoparticles. Vegetation 1 bath equipped with four holders of plants, arranged in conical recesses 5 formed in the bottom of a growth bath 1 and line 6 with the filter 7.

Hydroponic system works as follows. The conical recess 5 are placed four holders plants seedlings. The vegetative bath 1 through the conduit 6 to the filter 7 is fed a nutrient solution 2. On the surface of the nutrient solution in a growth bath 2 is applied to one particle 3 of an optically opaque material coated antimicrobial layer containing silver nanoparticles. Buoyancy particles 3 are uniformly distributed on the surface of a nutrient solution of 2, forming a layer of uniform thickness. The layer thickness is chosen, based on the light transmittance layer, so that the solution gets not more than 5 W/m^2. Silver nanoparticles have high antiseptic activity that allows the antimicrobial layer on the surface of the particles 3 as the protection of the root system of plants. After draining, the nutrient solution 2 from the bath 1 particle layer 3 is deposited on the holders 4 plants and is not captured by the root system of plants, which provides the same degree of shading of the nutrient solution 2 and prevents the growth of algae. Filter 7 stops particles 3 and prevents them from entrainment with a stream of a solution of a growth bath 1. When a solution of 2 in 1 bath particle layer 3 again emerges. Thus, in the course of the growing season provided the same degree of shading and nutrient solution is carried out to protect the root system of plants against microbial.[5]

The developed system for growing HGF contains sectional multilevel shelving racks formed 1 (Fig. 17.3), arches 2 and 3 shelves with casters.[6] On the shelves 3 are placed trays 4 germinating seeds, which are grown in hydroponic forage. To protect plants from harmful microorganisms trays 4 are carried out with an antimicrobial coating 5 comprising silver nanoparticles. Shelves 3 pallet 4 mounted on the rotary columns 6 with a double helical support surface 7 having a variable pitch. To irrigate crops has bar with nozzles move along the shelves 3. Supports 1 equipped with the necessary lighting systems.

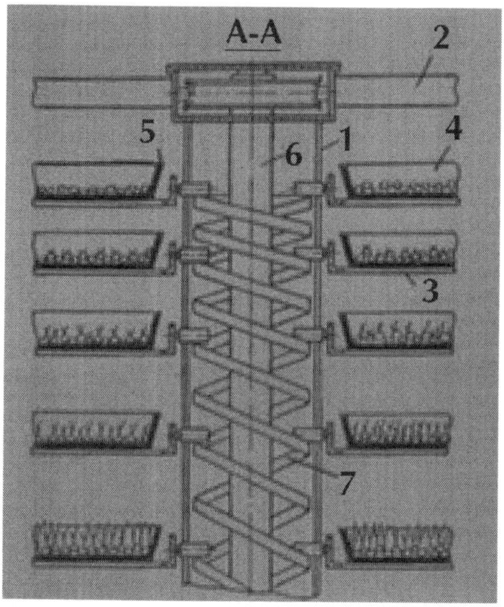

FIGURE 17.3 Installation to grow HGF.

17.4 LIQUID FILTRATION

Proposed filter for liquid[7] is characterized in that the bed of filter material is made of zeolite having an antimicrobial coating of silver nanoparticles. The filter consists of a cylindrical body 1 with the nozzles 2, 3 for supply and discharge of fluid, respectively. Inside the body 1 is fixedly partitions:

upper 4 and lower 5 between the partition 4, 5 is a layer of filter material 6 (Fig. 17.4).

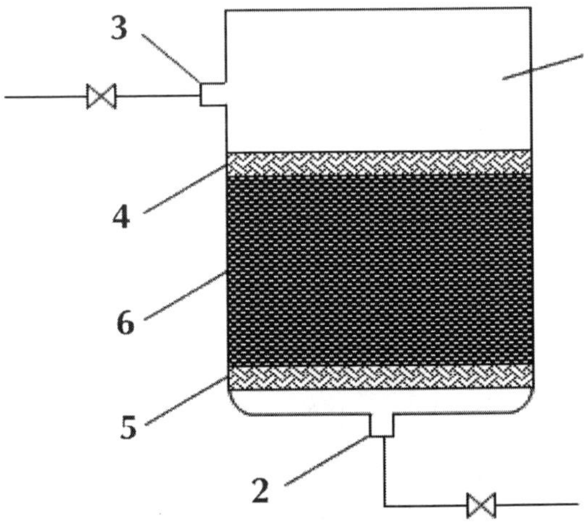

FIGURE 17.4 Filter cleaning fluid.

In the biological plant protection, use of nanomaterials is implemented as follows. Antimicrobial silver nanoparticles filter disposed in the discharge duct is used for air supply in the apparatus for breeding beneficial insects.[8] A method of use of mass entomophage afidimizy midges, characterized in that the cultivation is carried out on plants feed water purified in the filter with the antimicrobial silver nanoparticles.[9] Application of the proposed method increases the yield of cocoons gall midges 25...30% by providing antimicrobial protection of the root system of the plants that are used in the preparation of larvae and cocoons. Growing plants in water, purified antimicrobial filter with silver nanoparticles will activate all the links in the food chain "forage plant–aphid–gall midges larvae": enhance plant growth and development, stimulate the growth of aphids and thus increase the yield of cocoons gall midge. The method may be implemented using a device developed.[7]

17.5 CONCLUSION

The use of these developments in the greenhouses will increase plant productivity and reduce energy consumption of crop production.

KEYWORDS

- **nanomaterials**
- **protected ground**
- **silver nanoparticles**
- **hydroponic installation**
- **filter**
- **vegetation structure**

REFERENCES

1. Blednykh, V. V.; Chernoivanov, V. I.; Kosilov, A. N.; Basarygina, E. M. *Industry Nanosystems and Materials: Prospects for Use in Agriculture*; Chelyabinsk, ChGAU Publ.: Moscow, 2007; 400 p.
2. Orsik, L. S.; Blednykh, V. V.; Basarygina, E. M. *Nanofiltration Water for Agricultural Enterprises*; Chelyabinsk Publ., ChGAU: Moscow, 2008; 300 p.
3. Orsik, L. S.; Basarygina, E. M. The Problems of Agriculture in Light of the Application of Nanotechnology. *Nanotechnics* **2009,** *18*, 3–7.
4. Chetyrkin, Y. B.; Blednykh, V. V.; Ananyan, M. A. The Use of Nanotechnology in Agriculture. *Nanotechnics* **2009,** *18*, 82–85.
5. Blednykh, V. V.; Basarygina, E. M.; Basarygina, T. A. Hydroponic Plant. Patent RF, no. 86072, 2010.
6. Blednykh, V. V.; Basarygina, E. M.; Basarygina, T. A. Installation for the Hydroponic Cultivation of Food. Patent RF, no. 67392, 2008.
7. Blednykh, V. V.; Basarygina, E. M.; Basarygina, T. A. Filter Cleaning Liquids. Patent RF, no. 67467, 2008.
8. Blednykh, V. V.; Basarygina, E. M.; Basarygina, T. A. Set for Growing Insects. Patent RF, no. 67401, 2007.
9. Blednykh, V. V.; Basarygina, E. M.; Basarygina, T. A. Afidimizy midge breeding method. Patent RF, no. 2340176, 2012.

CHAPTER 18

NANOSTRUCTURED FIBERS VIA ELECTROSPINNING (PART I)

G. E. ZAIKOV

Russian Academy of Sciences, Moscow, Russia

CONTENTS

ABSTRACT

In this chapter formation of electrospun nanofibers are discussed in detail in order to attain uniform nanofibers consistently and reproducibly.

Symbols	Definitions
C_i	the morphological transition from bead-only structure
C_f	the morphological transition from bead-free structure
M_w	the polymer molecular weight
M_e	the entanglement molecular weight
ϕ_p	the polymer volume fraction
$[\eta]C$	Berry number
$[\eta]$	the intrinsic viscosity
C	the solution concentration
(C/C_e)	normalized concentration
C_e	the entanglement concentration
ρ	polymer density
rpm	round per minute
MES	magnetic electrospinning
PEO	polyethylene oxide
PVA	polyvinyl alcohol
PMMA	polymethyl methacrylate
PANI	polyaniline
CA	cellulose acetate
PVP	polyvinyl pyrrolidone
PAA	polyacrylic acid
PA6	polyamide-6
PEO	polyethylene oxide
PAA	polyacrylic acid
PU	polyurethane
PCL	polycapro lactone
PS	polystyrene
DMF	dimethylformamide
THF	tetrahydrofuran

18.1 ESSENTIAL PARAMETERS FOR CONTROLLING ELECTROSPINNING PROCESS

In recent decades, electrospinning of polymeric materials has gained much attention mainly because of the cheapest and the simplest of this method.[1–5] In this procedure, as shown in Figure 18.1, the polymer solution receives electrical charges from a high voltage supply; when the repulsive force between the charged ions overcomes the fluid surface tension, an electrified liquid jet could be formed and elongate toward the collector. At end, formatting nanofibers jets are collected on the surface of screen when the solvent is evaporated.[6,7]

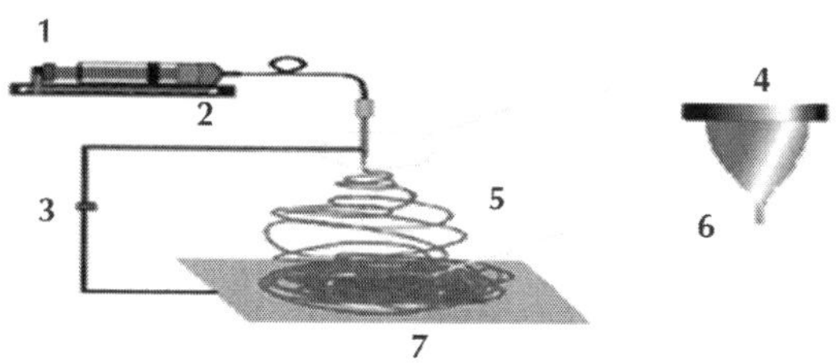

FIGURE 18.1 Sections of electrospinning process: (1) Polymer solution, (2) syringe, (3) high voltage, (4) Taylor cone, (5) whipping instability, (6) nanofibers formation, and (7) collector.

Many parameters affect the features of different parts of electrospun jet like the straight jet part, the instability region and the jet path.[8]

The most significant challenge in this process is to attain uniform nanofibers consistently and reproducibly.[9–12] In addition, the mechanics of this process deserving a specific attention and necessary to predictive tools or way for better understanding and optimization and controlling process.[13]

Also, fiber diameter is an important characteristic for electrospinning, because of its direct influence on the properties of the produced webs.[12,14,15] Depending on several solution parameters, different results can be obtained using the same polymer and electrospinning setup.[16]

Many parameters effect fiber formation. These factors that are studied to have a primary effect on the formation of uniform fibers are the process parameters, environmental parameters and solution parameters.[8,9,17–19] These data are presented in Figure 18.2.

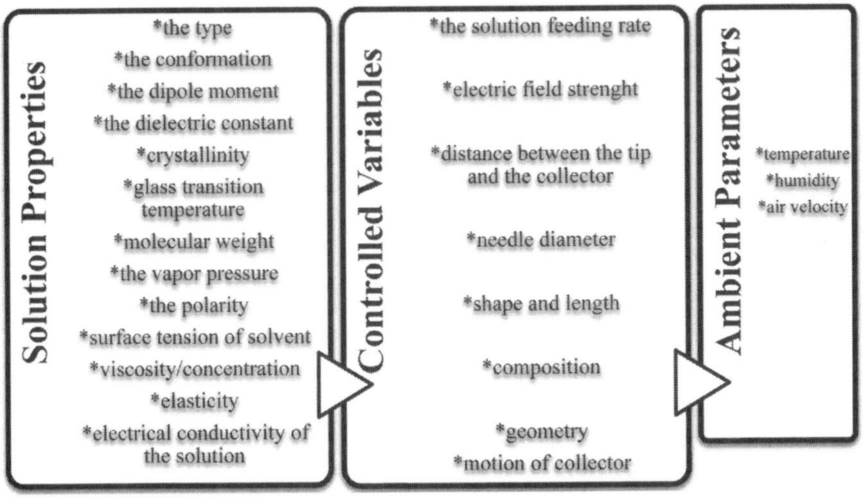

FIGURE 18.2 Parameters affect morphology and size of electrospun nanofibers.

Several general relationships between these parameters and fiber morphology can be drawn; this relationship will differ for each polymer or solvent.[12,16,20,21]

In addition, many researcher studies effect of these parameters on final fibers. A summary of most important parameters with their effects are bring in Tables 18.1 and 18.2.

TABLE 18.1 The Effects of Some Parameters on Electrospinning Nanofibers in Researcher's Studies.

Parameters	Year	Name of Researcher	Effect	Reference
Needle to collector distance	1999	Fong et al.	Inversely proportional to bead formation density Inverse to the electric field strength	[22]
	2003	Gupta & Wilkes	Inversely proportional to bead formation density Inversely proportional to fiber diameter	[23]
	2004	Theron et al.	Exponentially inverse to the volume charge density Inverse to the electric field strength	[24]

TABLE 18.1 *(Continued)*

Parameters	Year	Name of Researcher	Effect	Reference
Flow rate	2004	Theron et al.	Directly proportional to the electric current	[24]
			Inversely related to surface charge density	
			Inversely related to volume charge density	
	2005	Sawicka et al.	Directly proportional to the fiber diameter	[25]
Voltage	2001	Deitzel et al.	Direct effect on bead formation	[26]
	2003	Gupta & Wilkes	Inversely related to fiber diameter	[23]
	2004	Theron et al.	Inversely proportional to surface charge density	[24]
	2004	Kessick et al.	AC potential improved fiber uniformity	[27]
Concentration of polymer	2001	Deitzel et al.	Power law relation to the fiber diameter	[26]
	2002	Demir et al.	Cube of polymer concentration proportional to diameter	[28]
	2003	Gupta & Wilkes	Directly proportional to the fiber diameter	[23]
	2004	Hsu & Shivkumar	Parabolic-upper and lower limit relation to diameter	[29]
Ionic strength	2002	Zong et al.	Directly proportional to charge density	[30]
			Inversely proportional to bead density	
Temperature	2002	Demir et al.	Inversely proportional to viscosity	[28]
			Uniform fibers with less beading	
Solvent	2004	Theron et al.	Effects volume charge density	[24]
			Directly related to the evaporation and solidification rate	
Viscosity	2004	Hsu & Shivkumar	Parabolic relation to diameter and spinning ability	[29]

TABLE 18.2 The Effects of Some Parameters on Electrospinning Nanofibers.

	Parameters	Effect	References
Ambient parameters	Humidity	Increasing ambient humidity the diameter decreases and finally results in beaded fibers in PEO	[12,20]
		The influence of humidity on the formation and the properties of nanofibers are studied using CA and PVP	
		The humidity increases, the average fibre diameter of the CA nanofibers increases, while for PVP the average diameter decreases	
		Average diameter of nanofibers made by electrospinning change significantly through variation of humidity	
		Results in appearing circular pores on the fibers	
	Temperature	Higher solution temperature is produce thinner nanofibers because of the decrease in viscosity	[12,26]
		The influence of temperature on the formation and the properties of nanofibers are studied using CA and PVP	
		Solvent evaporation rate that increases with increasing temperature	
		The viscosity of the polymer solution that decreases with increasing temperature	
		Average diameter of nanofibers made by electrospinning change significantly through variation of temperature	
	Voltage	Is strongly correlated with form bead defects in the fibers	[20,21]
		Can conduct to form beads	
		Decreasing the electrical field decreases the bead density, regardless of concentrate the polymer in the solution	
	Distance between the capillary and the fiber collector	Affects the fiber drying and wants a minimum distance to give the fibers enough time to dry before achieving the fiber collector	[20]
		Increasing the distance, decreases the bead density, regardless of concentrate the polymer in the solution	

TABLE 18.2 *(Continued)*

	Parameters	Effect	References
Controlled variables	Geometry and composition of the collector	affect the surface of fiber, where the porous collector produces more porous fiber and metallic collector smooth fibers	[20]
	Fiber diameter	According to a power law relationship increasing with increasing solution concentration	[20,21,31]
		Larger dielectric constant of solvent, result finer electrospun fibers	
		A solvent with a larger dielectric constant has a higher net charge density in ejected jets	
		Increases with the increase of the concentration or viscosity of the polymer solution and higher conductivity yields smaller fibers in general, except for the technique using the polymers PAA and polyamide-6	
Solution properties	Addition of salts	Adding small amounts of cationic or anionic polyelectrolyte into PEO solution could reduce the average diameter of electrospun PEO fibers and narrow their distributions by increasing the charge density in ejected jets	[31]
	Surface tension	Effect on the morphology of nanofiber and size of electrospun fiber have been found out, however there is no results on this issue yet	[20]
	Solution viscosity	Decreases with increasing temperature and reduces fiber diameter	[20]
	Solution concentration	Has been found to most strongly affect fiber size	[21]
	pH	May be affect on the morphology and diameterof electrospun fibers	[31]
		The charge density in ejected jets affected in PVA solution	

Processing conditions play a major character in the electrospinning technique because the electrospinning technique is governed by the external electric field produced by the applied voltage caused by charges on the jet surface. It can be divided in two sections: processing parameters and type of collector that summarized in Figures 18.3 and 18.4. In addition, the processing conditions and external parameters also have a significant result on the diameter and morphology of the nanofibers. The electric field between the needle tip and the target can be controlled through these varying parameters[8]:

- The applied voltage
- The distance between needle tip and target
- The shape of the collector
- The diameter of the needle.

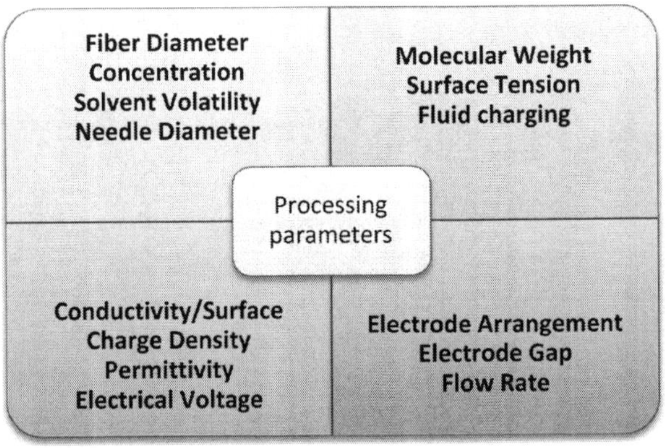

FIGURE 18.3 Important processing parameters.

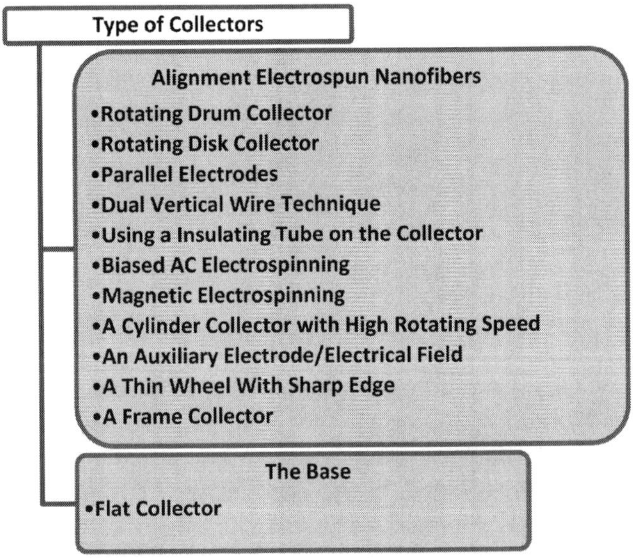

FIGURE 18.4 Type of collectors.

Solution polymer parameters affect forming the nanofibers, causing defects in the fibers in the pattern of beads and junction by their low concentration or viscosity, low conductivity, and by their decrease of molecular weight. Ambient parameters play a role in the properties of the polymeric solution and so affect the morphology of the nanofibers.[20]

The solution must also have these feathers to prevent the gate from collapsing into droplets before the solvent has evaporated.[32]

- A surface tension, low enough
- A charge density, high enough
- A viscosity, high enough.

In addition, major factors that control the diameter of the fibers are[33]

- Concentration of polymer in their solution
- Type of solvent used
- Conductivity of their solution
- Feeding rate of the solution.

In the next part, we discussed about these parameters.

18.1.1 CONCENTRATION

The concentrations of polymer solution play a significant role in the fiber formation during the electrospinning technique. Four critical concentrations from low to high should be remarked in these points[10,21,30,34]:

- As the concentration is low, polymeric micro or nanoparticles will be produced. Now, electrospray occurs instead of electrospinning owing to the low viscosity and high surface tensions of the solution.
- As the concentration is little higher, a mixture of beads and fibers will be receiving.
- When the concentration is suitable, smooth nanofibers can be obtained.
- If the concentration is high, not nanoscale fibers, helix-shaped micro ribbons will be watched.

Normally, increasing concentrates solution, the fiber diameter will increase if the solution concentration is suitable for electrospinning. Also, solution viscosity can be also tuned by setting the solution concentration.[10]

In the electrospinning technique, a minimum solution concentration is needed for fiber formation. It has been found that at low concentration, beads and fibers are obtained. When the solution concentration increases, the shapes of beads changes and finally uniform fibers are finding. There is an optimum solution concentration for the electrospinning technique. (At low concentrations, beads are formed and at higher concentrations continuous fibers are banned.) Researchers have tried to determine a relationship between solution concentration and fiber diameter. They found a power law relationship. Increasing concentrate solution causes increasing fiber diameter with gelatin in electrospinning. Solution surface tension and viscosity also play important roles in determining the range of concentrations from continuous fibers which can be obtained in electrospinning.[17]

Polymer concentration determines the spinnability of a solution, that is to say, whether a fiber form or not. The solution must have a high enough polymer concentration for chain entanglements to occur; however, the solution cannot be either too dilute or too centralized. On the other hand, polymer concentration influences both viscosity and surface tension of the solution technique. If the solution is to dilute, then the polymer fiber will give away up into droplets before reaching the collector (due to the effects of surface tension). If the solution is too concentrated, then fibers cannot be made (due to the high viscosity), because it is difficult to controlling solution flow rate through capillaries. Thus, an optimum range of polymer concentrations is called for.[16]

The molecular properties of polymer play a vital part in controlling fiber initiation and stabilization. Several investigators have attempted to establish optimum ranges for concentration and molecular weight in order to insure stable fiber formation. At any molecular weight, the effect of concentration on the breakdown of the solution jet can be distinguished by two critical concentrations, C_i and C_f. Below C_i, only beads may be developed (due to insufficient chain entanglements in the solution). Above C_i, a combination of beads and fibers is observed. When the concentration is increased above C_f, complete fibers are created. C_i is typically about the entanglement concentration C_e, at which chain entanglements in the solution become significant. Hence, C_i is a transition concentration at which fibers begin to come forth from the beads and C_f is the concentration at which a fibrous structure is stabilized. Shenoy et al. have developed the solution entanglement number $(n_e)_{soln}$, for depicting the transition points for fiber initiation and complete fiber formation[35]:

$$(n_e)_{so\ln} = \frac{\phi_p M_w}{M_e} \tag{18.1}$$

Lyons et al. have reported that fiber diameter varies exponentially with a molecular weight for melt electrospun polypropylene. The fiber diameter, normalized, with deference to the Berry number is plotted. A power law relationship is observed between D and $[\eta]C$ as follows[35]:

$$D(mm) = 18.6\,([\eta]C)^{1.11} \tag{18.2}$$

Various investigators have also correlated the fiber diameter with a normalized concentration defined as[35]

$$C_e = \frac{\rho M_e}{M_w} \tag{18.3}$$

18.1.2 SOLUTION VISCOSITY

One of the most significant parameters that influence the diameter and the morphology of the fiber is the viscosity of the resolution, which is indirectly affected by polymer characteristics such as molecular weight and concentration. When a polymer with a higher molecular weight dissolved in a solvent, the viscosity of the polymer solution is higher than a solution of the same polymer with a lower molecular weight. Similarly, the viscosity of a polymer solution increases with an increased concentration of polymer in that solution. It is obvious that the viscosity of solution and polymer chain entanglements have a direct relationship. It has been proven that continuous and smooth fibers cannot become in low viscosity, whereas high viscosity results in the hard ejection of jets from solution, namely there is a need of suitable viscosity for electrospinning. The viscosity range of different polymer solution at electrospinning is different. It is important that viscosity, polymer concentration, and polymer molecular weight are linked to one another. For solution with low viscosity, surface tension is the dominant factor and just beads or beaded fiber formed. If the solution is of suitable viscosity, continuous fibers can be made. Also, the shape of the beads changes from spherical to elliptical when the viscosity of solution varies from low to high. Higher viscosity also results in larger diameter fibers and smaller deposition areas.[8,10,17,36]

Taken together, these studies suggest that there exist polymer-specific, optimal viscosity values for electrospinning, and this property possesses a remarkable influence on the morphology of the fibers.[17]

18.1.3 MOLECULAR WEIGHT

Molecular weight of polymer also has an important impression on the morphology of electrospun fiber. In principle, molecular weight reflects entangling polymer chains in solutions, namely the solution viscosity. Keep the concentration fixed, lowering the molecular weight of polymer trends to form beads rather than smooth fiber. By increasing the molecular weight, smooth fiber will be obtained. Further by increasing molecular weight, microribbon will be received. Also, the authors establish that as the molecular weight is high, some patterned fibers can also be obtained at low concentration.[10]

It has been remarked that too low a molecular weight solution, foam beads rather than fibers. A high molecular weight solution gives fibers with larger average diameters. Chain entanglement plays an important role in technique electrospinning. It has been discovered that high molecular weights are not always essential for the electrospinning technique if enough intermolecular interactions can provide a substitute for the interchange connectivity got through chain entanglements.[17]

In addition, this parameter plays a vital role in controlling fiber beginning and stabilization. Several investigators have tried to demonstrate the ideal ranges for concentration and molecular weight to insure stable fiber formation. The molecular weight of the polymer has a significant purpose of proving the structure in the electrospun polymer. At a constant concentration, the structure changes from beads, to beaded fibers, to complete fibers and two flat ribbons as the molecular weight is increased.[35]

18.1.4 SURFACE TENSION

Surface tension is an important factor in electrospinning. Different solvents may contribute different surface tensions. With the concentration fixed, reducing the surface tension of the solution, beaded fibers can convert into smooth fibers. The surface tension and solution viscosity can be adjusted by varying the mass ratio of solvents mix and fiber morphologies. Surface tension decides the upper and lower boundaries of the electrospinning window if all other conditions specified.[10]

A lower surface tension of the spinning solution helps electrospinning to occur at a lower electric field.[8,17,37] However, not necessarily a lower surface tension of a solvent will always be more suitable for electrospinning. Also, this parameter determines the upper and lower boundaries of the electrospinning window if all other variables are held constant.[17,37]

Adding a surfactant to the polymer solution also changes the surface tension. If all other variables are held constant, surface tension decides the upper and lower bounds of the electrospinning technique.[8]

It is caused by the attraction between the molecules in a liquid. In the most of liquid, each molecule is attracted equally in all directions by neighboring liquid molecules, resulting in a net force of zero. At the surface of the liquid, the molecules are subjected to a net inward force to balancing only by resisting liquid to compression. The net effecting causes the surface area to reduce it until controlling the possible lowest ratio of surface area to volume. In electrospinning, the charges on the polymer solution must be high enough to overwhelm the surface tension of the solution. As the electrician jet speeds up from the needle to the aim, the polymer jet is stretched. Then, surface tension of the solution may cause the jet to break up into droplets. In addition, if there is a lower concentration of polymer molecules, the surface tension causes beaded fibers formed.[8]

18.1.5 CONDUCTIVITY/SURFACE CHARGE DENSITY

Solution conductivity is mainly decided by the polymer type, solvent sort, and the salt. Usually, natural polymers are polyelectrolyte in nature, subjecting to higher tension under the electric field, resulting in the poor fiber formation. Also, the electrical conductivity of the solution can be tuned by adding the ionic salts. With the aid of ionic salts, nanofibers with small diameter can be produced. Sometimes high solution conductivity can be also accomplished by using organic acid as the solvent. An increase in the solution conductivity favors forming thinner fibers.[10]

Also, solutions with high conductivity will cause a greater charge carrying capacity than solutions with low conductivity. Therefore, the fiber jet of conducive solutions will be subjected to a greater tensile force in the mean of an electric field than will a fiber jet from a solution with a low conductivity.[16]

The minimum voltage for electrospinning to occur can also be cut back if the conductivity of the polymer solution is increased. Also, higher solution conductivity results in greater bending instability and produces a larger deposition area of collecting fibers. It has been reported that the size of the ions in the solution has an important impact on the electrospun fiber diameter besides the charges carried by the jet. Ions with a smaller atomic radius have a higher charge density. Thus, a higher mobility under an external electric field is present.[8]

However, conductivity solution is unstable in the presence of strong electric fields, which results in a dramatic bending instability as well as a broad diameter distribution. Electrospun nanofibers with the smallest fiber diameter can be taken with the highest electrical conductivity and it has been found, there is a drop in the size of the fibers is because of the increased electrical conductivity. The ions increase the charge holding capacity of the jet with it subjecting it to higher tension. Thus, the fiber forming ability of the gelatin is less compared to the synthetic ones.[17] For example, Zong et al. have proved the effect of ions by adding ionic salt on the morphology and diameter of electrospun fibers.[17]

Stanger et al. found that an increase in charge density results in a reduction in the mass deposition rate and initial jet diameter during the electrospinning. In addition, a theory was proposed where they correlated reducing the curvature diameter of the Taylor cone with increasing charge density. Decreasing total electrostatic forces causes a smaller effective area. Similarly, other researchers have described the different behavior of the Taylor cone with compared to the Taylor's observation of ionic liquids.[8]

18.1.6 SOLVENT VOLATILITY

Selecting a suitable solvent or solvent as the carrier of a particular polymer is fundamental for optimizing electrospinning. On the other hand, it is critical in determining the critical minimum solution concentration to allow the transition from electrospraying to electrospinning, by significantly affecting solution spinnability and the morphology of the electrospun fibers.[34]

In addition, choice of solvent is too critical about whether the fibers are forming, as well as influencing fiber porosity. For enough solvent evaporation to happen between the capillary tip and the collector a volatile solvent must be used. As the fiber jet travels through the air toward the collector a phase separation occurs before the solid polymer fibers deposited, a technique that is influenced by the volatility of the solvent.[16]

Megelski et al. studies the properties of polystyrene fibers electrospun from solutions containing various ratios of DMF and THF were examined. Electrospinning solutions from 100% THF (more volatile) explained a high density of pores, which increased the surface area of the fiber depending on the fiber diameter. Solutions electrospun from 100% DMF (less volatile) proved almost a loss of micro texture with forming smooth fibers. Between these two extremes, it was viewed that pore size increased with decreased pore depth (thus decreasing pore density) as the solvent volatility decreased. For volatile solvents, the region close to the fiber surface can be saturated with solvent in the vapor phase, which further limits penetrating nonsolvent. This can hinder skin formation leading to developing a porous surface morphology.[16]

18.1.7 FLUID CHARGING

In electrospinning, generation of charging within the fluid, usually occurs under contract with and flow across an electrode held at high (positive or negative) potential, referred to as induction charging. Depending on the nature of the fluid and polarity of the applied potential, free electrons, ions, or ion pairs may be produced as charge carriers in the fluid; the generation of charge carriers can be sensitive to solution impurities. Forming ions or ion pairs by induction, result in forming an electrical double layer. Without flow, the double layer thickness is found by the ion mobility in the fluid; in the presence of flow, ions may be convected away from the electrode and the double layer continually renewed. Charging of the fluid in electrospinning is typically field-limited, with the breakdown field strength in dry air being between flat plates.[38]

18.1.8 PERMITTIVITY

As with conductivity, the permittivity of a solvent has an important influence on the electrospinning technique and fiber morphology. However, not much discussion has been published around these effects. Theron et al. describe a method for deciding the permittivity of an electrospinning solution by measuring the complex resistance of a small cylindrical volume of the fluid. Bead formation and the diameter of the resultant electrospun fibers can be diminished by using a solution with a higher permittivity. With bending instability and the traversed jet in the path, electrospinning jet increase with higher permittivity, which results in a reduction in the diameter of the fiber and a larger fiber deposition area. Solvents such as DMF can be utilized to increase the permittivity of polymer solutions.[8]

18.1.9 ELECTRICAL VOLTAGE

One of the major parameters which affects the fiber diameter to a remarkable extent is the applied electric potential. In general, a higher applied voltage ejects more fluid in a jet, resulting in a larger fiber diameter.[8,14]

If the applied electric potential is higher, a greater amount of charge will cause the jet to speed up faster, and more solution will be drawn out from the tip of the needle. At a critical voltage, the Taylor cone is no longer seen. The jet imminent directly from the nozzle with increasing applied voltage. The resultant electrical field between the needle and the target increases as well, which contributes to greater stretching of the solution because of the larger columbic force between the surface charges.[8]

An increase in the applied voltage therefore leads to a lessening in the diameter of the electrospun nanofibers. Similarly, Pawlowski et al. have proved the drier fibers can be generated if the voltage is increased because of the faster evaporation of the solvent that results. Zhao et al. have verified that a lower voltage leads to a weaker electrical field, which reduces speeding up the jet and increases the flight time of the electrospinning jet, thus producing thinner fibers. Therefore, they suggested that a voltage close to the minimum critical voltage needed for the onset of electrospinning might be efficient for getting thinner fibers. Higher voltages are related to a greater of bead formation, possibly because of the increased instability of the jet as the Taylor cone recedes into the syringe needle with the increased potential. The shape of the beads transforms from a spindle to a spherical shape with increased voltage, and sometimes the beads will get together to form thicker fibers because of the increased density of the beads on the other hand.[8]

Reneker and Chun have proved there is not much effect of electrical field on the diameter of electrospun PEO nanofibers. Several groups suggested that higher voltages simplified form large diameter fiber. For example, Zhang et al. explored the effect of voltage on morphologies and fiber diameter distribution with PVA or a water solution as a model. Several groups suggested that higher voltages can increase the electrostatic repulsive force on the charged jet, favoring the narrowing of fiber diameter. For example, Yuan et al. analyzed the effect voltage on morphologies and fiber alignment with PSF or DMAC or acetone as a model. Beside those phenomena, some groups also established that higher voltage offers the greatest chance of bead formation. Thus, we can find that voltage does the influence fiber diameter, but the meanings vary with the polymer solution concentration and on the distance between the tip and the collector.[10,30]

18.1.10 FLOW RATE

The flow rate decides solution available for electrospinning. Keeping a stable Taylor cone needs a minimum solution flow rate for a given voltage and electrode gap. On the other hands, at low flow rates, the Taylor cone recedes into the needle, and the jet originates from the liquid surface within the needle. In contrast, if the solution flow rate is greater than the electrospinning rate, it causes solution droplets to come from the needle tip because of lack of time for electrospinning the complete droplet to be an electrician. It has been viewed the diameter of the fiber and the size of the bead both increase with an increased flow rate.[16,17]

Also, at high flow rates significant numbers of bead defects were noticeable, because of the inability of fibers to dry before progressing to the collector. Incomplete fiber drying also leads to forming ribbon like (or flattened) fibers compared to fibers with a circular cross-section.[16]

A summary of researchers work are shown in Figure 18.5.[16,17]

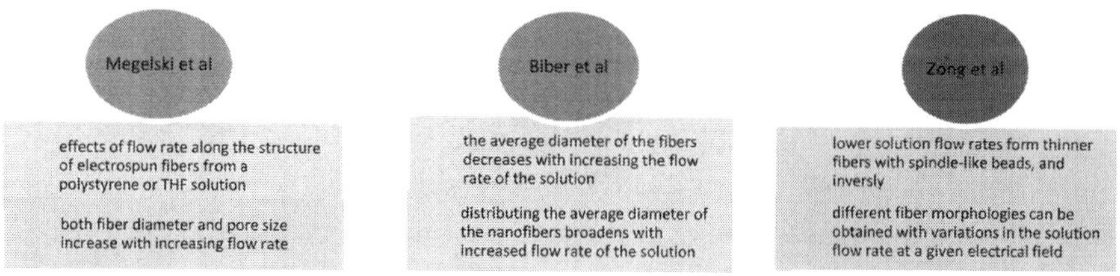

FIGURE 18.5 Summary of researchers work on the flow rate parameter.

18.1.11 NEEDLE DIAMETER

The diameter of the needle has an effect on the electrospinning technique. A smaller needle diameter was found to reduce clogging at the tip of the needle as well as the number of beads in the collected nanofibers because of the lower exposure of the solution to the atmosphere during electrospinning. In addition, using smaller diameter needles means the diameter of the electrospun nanofibers can also be smaller. The jet flying time of the solution between the needle and the collector plate can also be increased if the needle diameter is cut because the surface tension of the droplet is increased and the jet acceleration, decreased. Few studies worked in this field that summarized in Figure 18.6.[8]

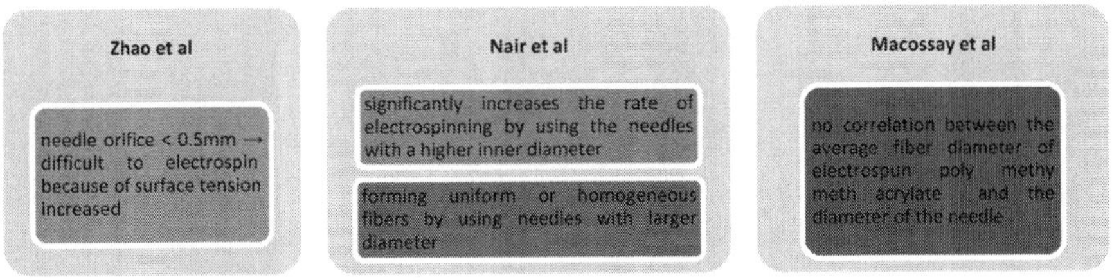

FIGURE 18.6 Summary of researchers study.

18.1.12 DISTANCE BETWEEN THE COLLECTOR AND THE TIP OF THE SYRINGE

Varying the distance between the needle and the target causes a modification in the behavior of the electrospun jet and the morphology of the resultant nanofibers. Shortening the distance between the two electrodes causes an increase in the electrical field strength between the needle and the target and speeds up the electrospinning technique, therefore reducing the time available for evaporation. It has been also reported spreading the diameter of the nanofibers becomes narrower when the space between the two electrodes is increased. Conversely, in other cases, it was considered as the average diameter of the fiber increases with increased distances because of the decreased strength of the electric field.[8]

As a brief summarized, it has been proven the distance between the collector and the tip of the syringe can also affect the fiber diameter and morphology. In brief, if it is too short, the fiber will not have adequate time to solidify before reaching the collector, because dryness from the solvent is an important parameter on electrospun fiber. If the distance is too long, bead fiber can be obtained. It has been reported that flatter fibers can be brought out at closer distances. However, spinning distance is not important effect on fiber morphology.[16,17] Also, many researchers studied this parameter affect on fibers morphology and diameters as shown in Figure 18.7.[16]

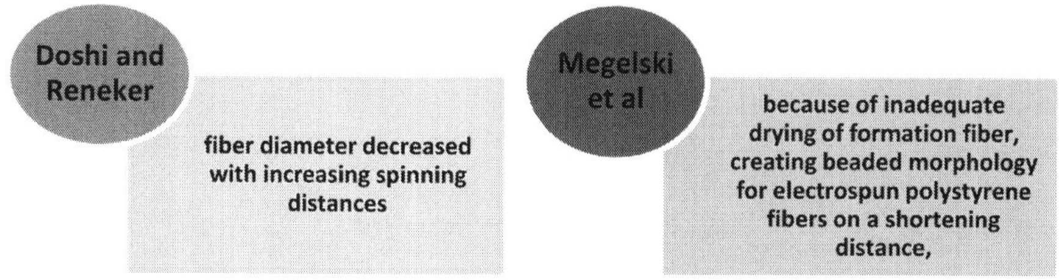

FIGURE 18.7 Summary of researchers results on this parameter.

18.2 CONTROL-ALIGNED FORMATION FIBERS

Aligned fibers have found importance in many engineering applications, such as tissue engineering, sensors, nanocomposites, filters, electronic devices. Some commonly used techniques to align the fibers are discussed in the subsections below.[33] There are various ways to control aligning the depositing fiber. One way is to use different collector like a rotating wheel instead of foil sheet collector.[39]

Recently, it was decided the nature and type of the collector influences significantly the morphological and the physical characteristics of spun fibers. The density of the fibers per unit area of the collector and fiber arrangement is affected by the degree of charge dissipation on fiber deposition. The most commonly used targets are the conductive metal plate that results in collection of randomly oriented fibers in the nonwoven form. The use of metal and conductive collectors helped dissipate the charges and reduced the repulsion among the fibers. Therefore, the fibers collected are smooth and densely compacted. However, the fibers collected on the nonconductive collectors do not fritter away the charges which repel one another. In addition, the fibers can also be collected on specially designed collector to get aligned fibers or arrays of fibers.[33]

In the next parts, we discussed about types of collectors.

18.2.1 TYPES OF COLLECTORS

In the former stages of electrospinning, researchers used a needle and a flat collector plate as electrodes. However, with developing electrospinning technology, many electrode arrangements have been tried as a means of changing the electric field and getting wanted nanofiber morphologies.

Collector electrode arrangements have all been ground out as ways (summarized in Fig. 18.8), to produce aligned fibers.[8]

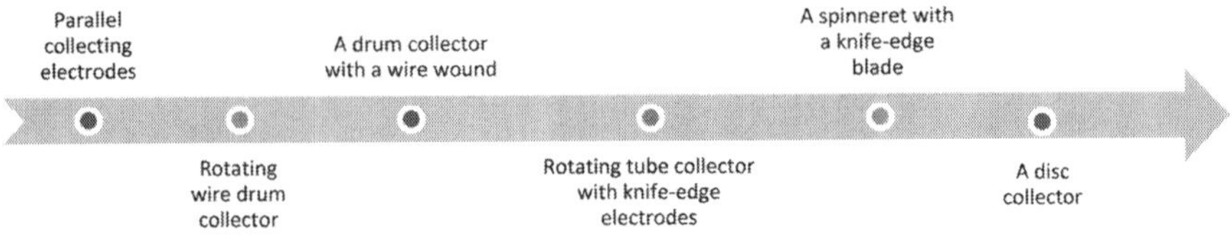

FIGURE 18.8 A summary of electrode arrangements.

One important of the electrospinning technique is the type of collector used. In electrospinning technique, a collector serves as a conductive substrate where the nanofibers are collected. Aluminum foil is used as a collector, but because of difficulty in transferring of collecting fibers and with the need for aligning fibers for various applications, other collectors are also common types of collectors today as summarized in Figure 18.9.[17]

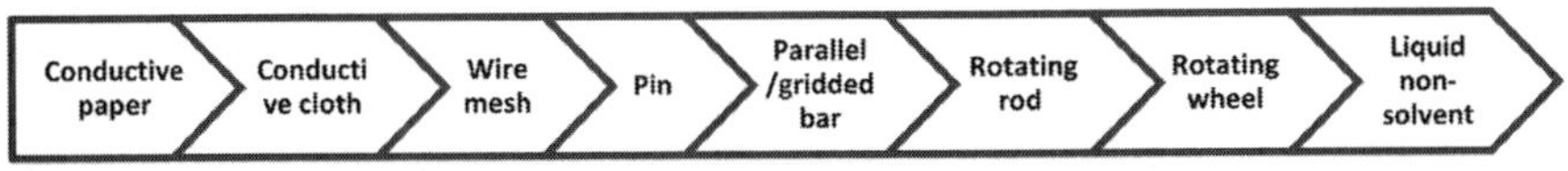

FIGURE 18.9 A summarized on other common types of collectors.

In addition, the fiber alignment is found out by the type of the target or collector and its rotation speed. The created nanofibers are deposited in the collector as a random mass because of the bending instability of the charged jet. Nowadays, several types of collectors, such as a rotating drum or a rotating wheel-like bobbin or metal frame, have been utilized for getting aligned nanofibers.[17] In the next section, we discussed briefly about these types of collectors.

18.2.1.1 FLAT COLLECTOR

During the electrospinning technique, collectors usually acted as the conductive substrate to collect the charged fibers. As usual, a foil is used as a collector.[10]

In addition, this is the most widely utilized methods of fiber collection. The collector can be either a solid metal, foil, or screen. Other materials can also be localized between the capillary and the collector.[39]

However, it is difficult to transfer the collected nanofibers to other substrates for several applications. With the need of fibers transferring, diverse collectors have been developed including[10]:

- Wire mesh
- Pin or grids
- Parallel or gridded bar
- Rotating rods or wheel
- Liquid bath

18.2.1.2 ROTATING DRUM COLLECTOR

This method normally used to collect aligned arrays of fibers. Also, the diameter of the fiber can be controlled and tailored based on the rotational speed of the drum. The cylindrical drum is rotating at high speeds (a few 1000 rpm) and of orienting the fibers circumferentially. Ideally, the linear rate of the rotating drum should match the evaporation rate of the solvent; such the fibers deposited and held up on the surface of the drum. The alignment of the fibers is induced by the rotating drum and the degree of alignment improves with the rotational velocity. At rotational speeds slower than the fiber take-up speed, randomly oriented fibers are obtained along the drum. At higher speeds, a centrifugal force is prepared near the vicinity of the circumference of the rotating drum, which elongates the fibers before being collected on the drum. However, at much higher speed, the take-up velocity breaks the depositing fiber jet and continuous fibers are not taken.[33]

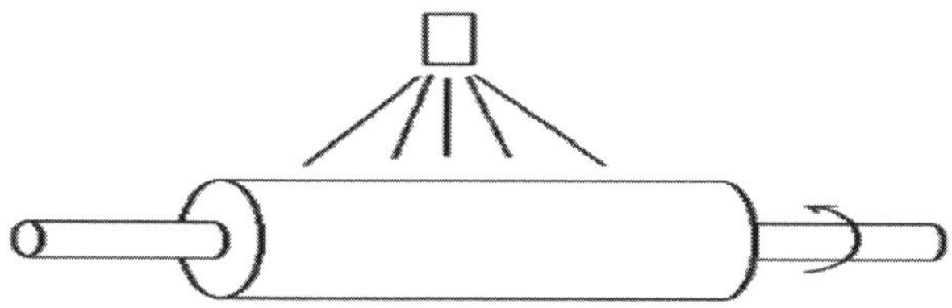

FIGURE 18.10 Rotating collector for electrospun nanofibers.

18.2.1.3 ROTATING DISK COLLECTOR

The rotating disk collector is a variation setup of the rotating drum collector and is practiced to obtain unaxially aligned fibers. The advantage of using a rotating disk collector over a drum collector is that most of the fibers deposited on the sharp-edged disk and are collected as aligned patterned nanofibers. The jet travels in a cone and inverse conical path with the utilization of the rotating disk collector as opposed to a conical path got when using a drum collector. During the first level, the jet follows the usual envelope cone path which is because of the instabilities influencing the jet. At a point above the disk, the diameter of the loop decreases as the conical form of the jet starts to shrink. This results in the inverted cone appearance, with the top of the cone resting on the disk. The electrical field applied concentrated on the tapered edge of the magnetic disc. Therefore the charged polymer jet pulled toward the edge of the wheel, which explains the inverted conical form of the jet at the disk edge. The fibers that attracted to the edge of the disk are wound round the perimeter of the disk owing to the tangential force acting on the fibers produced from rotating the disk. This force further stretches the fibers and reduces their diameter. The quality of fiber alignment got using the disk is more beneficial than the rotating drum. However, only a small quantity of aligned fibers can be got since there is just a small area at the tip of the disk.[33]

18.2.1.4 PARALLEL ELECTRODES

There has already been several groups produced well-aligned nanofibers by using two grounded parallel electrodes, such as aluminum strips with a 1-cm gap used in Yi Xin's group and the metal frame in Dersch's group et al. This apparatus used in this method is uncomplicated. The same as the rotating drum method, it operates by varying the collectors. The drawbacks of both these two methods are taught[39]:

- They can only produce aligned fiber in a small area.
- Fibers fabricated by this method cannot be conveniently transferred to different types of substrates.

The advantage of utilizing this technique lies in the simplicity of the set-up and the ease of collecting single fibers for mechanical testing. Good alignment has been contracting with this technique. The air gap between electrodes creates residual electrostatic repulsion between spun fibers, which helps align fibers. Two nonconductive strips of materials are placed along a straight line and an aluminum foil is placed on each of the strips and connected to the ground. This technique enables fibers to be deposited at the end of the strips so the fibers cling to the strips in an alternate fashion and collected as aligned arrays of fibers. A similar technique by Teo and Ramakrishna used double-edge steel blades on a line to collect aligned arrays of fibers. The fibers were deposited at the gap between the electrodes; however, few fibers were found to deposit along the blades. It was solved by applying a negative voltage between the blades, resulting in the deposition of fibers between the blades.[33]

18.2.1.5 DUAL VERTICAL WIRE TECHNIQUE

The dual vertical wire technique used in Surawut's study is a variance of the parallel electrode technique, comprised two stainless steel wires, used as the secondary target, and the grounded aluminum foil, used as the primary target. The two stainless steel wires are mounted vertically in parallel to each other along a center line between the top of the needle and the grounded aluminum foil. Both the needle and the foil were tilted about 45° from a vertical baseline. They also tried to utilize the secondary electrodes alone, but much smaller amounts of aligned fibers were good. This was because most of the fibers would instead deposit randomly around the first wire electrode. Based on this observation, both the primary and secondary electrodes were essential for making good-aligned fibers for the present set-up. The mechanism for the depositing fibers to extend across the wire electrodes is similar to the parallel electrodes described previously. Both aligned fibers between the parallel vertical wires and a randomly aligned fiber mat on the aluminum foil could be reached.[39]

18.2.1.6 USING AN INSULATING TUBE ON THE COLLECTOR

Yang et al. produced a large area of oriented fibers by putting an insulating tube on the target for a long time. The changed electrical field made the jet bends around the pipe, so the oriented fibers could be received with a suitable tube. Based on different height and diameter of the pipe, there were three kinds of collection of aligned fibers formed[39]:

- Only a round mat within the tube area.
- A round belt outside the tube and a round mat within the tube area.
- Only a ring belt outside the tube area.

Still, since there is only one electrode as the target, the repelling force is not large enough on the tube to keep the jet falling around the tube all the time. In this font, the jet is not coaxial with the tube and results in disordered fibers.[39]

18.2.1.7 BIASED AC ELECTROSPINNING

Biased AC electrospinning is a new method used by Sarkar et al. It employs a combination of DC and AC potentials. The aimed of this study is lessening the inherent instability of the fiber itself, compared to all techniques relying on lessening the fiber instability by using external forces on the fibers during electrospinning. By introducing a DC biased AC potential instead of either a pure AC or DC potential, alternating positively and negatively charged regions in the fiber result in a reduction of electrostatic repulsion and increase in fiber stability, and stability can improve electrospun fiber quality.[39]

18.2.1.8 MAGNETIC ELECTROSPINNING

In this method, the polymer solution is magnetized by adding a few magnetic nanoparticles. The magnetic field stretches the fibers across the gap to make a parallel array as they land on the magnets. When the fibers fall down, the parts of the fibers close to the magnets are attracted to the surface of the magnets, finally the fibers land on the two magnets and suspend over the gap. This method is fundamentally different from all previously discussed methods in preparing aligned fibers, because the driving force is the magnetic field in it, while electrostatic interaction plays the function as driving force in the other methods. In addition, this method has several advantages[39]:

- The magnetic field can be manipulated accurately.
- The resultant nanofibers can be transferred on to any substrate with full retention.
- The area of the aligned fibers is large compared to other techniques.

18.2.1.9 A CYLINDER COLLECTOR WITH HIGH ROTATING SPEED

It has been suggested that by rotating a cylinder collector at a high-speed up to thousands of rpm, electrospun nanofibers could be oriented circumferentially. Researchers from Virginia Commonwealth University have utilized this technique summarized in Figure 18.11.[14]

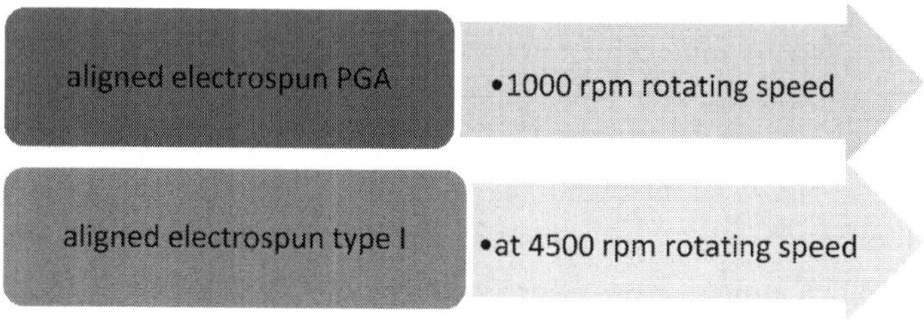

FIGURE 18.11 The speed of this parameter for aligned fibers.

When a linear velocity of rotating cylinder surface matches that of evaporated jet depositions, fibers are taken up on the surface of the cylinder tightly circumferentially, resulting in a fair alignment. Such a speed can be predicted as an alignment speed. If the surface velocity of the cylinder is slower than the alignment speed, randomly deposited fibers will be collected, as it is the fast chaos motions of jets control the final deposition manner. On the other hand, there must be a limit rotating speed above which continuous fibers cannot be collected from the over fast take-up speed will stop the fiber jet. The reasons a perfect alignment is difficult to achieve can be applied to the fact the chaotic motions of polymer jets are not probable to be consistent and are less controllable.[14]

18.2.1.10 A THIN WHEEL WITH SHARP EDGE

A significant advancement in collecting aligned electrospun nanofibers has been recently constructed. The tip-like edge substantially concentrates the electrical field so the as-spun nanofibers are almost all attracted to and can be continuously wound on the bobbin edge of the revolving wheel. It was explained that before getting to the electrically grounded target the nanofibers keep enough residual charges to repel each other. As a result, once a nanofiber is attached to the wheel tip, it will exert a repulsive force on the next fiber attracted to the tip. This repulsion is from one another results in a detachment between the deposited nanofibers. The variation in the separation distances is because of varying repulsive forces related to nanofiber diameters and residual charges.[14]

18.2.1.11 A FRAME COLLECTOR

In order to make an individual nanofiber for experimental characterizations, we recently developed another approach to fiber alignment by simply placing a rectangular frame under the spinning jet. In addition, different frame materials result in different fiber alignments (i.e., aluminum frame favors better fiber alignments than a wooden form). More investigation is undergoing to understand the alignment characteristics in varying the configuration and size of frame rods, the distance between the frame rods, and the inclination angle of a single frame. These will be useful in deciding how many positions would be best suitable making a polygonal multiframe structure.[14]

18.3 CONCLUDING REMARKS

Producing nanofibers by electrospinning is a simple and widely utilized for varied applications. As mentioned above, aligned fibers have found importance in many applications. The most significant part of electrospinning is how to control the process. Therefore, we must know about the behavior of every part of process and control instabilities were made in it. Some commonly used techniques to align the fibers are discussed in this chapter. In the next chapter, we discussed about the behavior formation of nanofiber jet. In addition, the most important tools for better controlling process are modeling and simulating. In near future, we will review them.

KEYWORDS

- **electrospun nanofibers**
- **aligned nanofibers**
- **process parameters**
- **controlling process**

REFERENCES

1. Šimko, M.; Erhart, J.; Lukáš, D. A Mathematical Model of External Electrostatic Field of a Special Collector for Electrospinning of Nanofibers. *J. Electrostat.* **2014,** *72*(2), 161–165.
2. Brooks, H.; Tucker, N. Electrospinning Predictions using Artificial Neural Networks. *Polymer* **2015,** ***58***, 22–29.
3. Fridrikh, S. V.; et al. Controlling The Fiber Diameter during Electrospinning. *Phys. Rev. Lett.* **2003,** 144502–144502.
4. Zeng, Y.; et al. Numerical Simulation of Whipping Process in Electrospinning. in WSEAS International Conference. Proceedings. Mathematics and Computers in Science and Engineering, World Scientific and Engineering Academy and Society, 2009.
5. Ciechańska, D. *Multifunctional Bacterial Cellulose/Chitosan Composite Materials for Medical Applications. FibresText. East. Europe* **2004,** *12*(4), 69–72.
6. Ghochaghi, N. Experimental Development of Advanced Air Filtration Media based on Electrospun Polymer Fibers. In *Mechnical and Nuclear Engineering*; Virginia Commonwealth, 2014; pp 1–165.
7. Ziabari, M.; Mottaghitalab, V.; Haghi, A. K. Evaluation of Electrospun Nanofiber Pore Structure Parameters. *Korean J. Chem. Eng.* **2008,** *25*(4), 923–932.
8. Angammana, C. J. *A Study of the Effects of Solution and Process Parameters on the Electrospinning Process and Nanofibre Morphology*. University of Waterloo, 2011.
9. Lu, P.; Ding, B. Applications of Electrospun Fibers. *Rec. Pat. Nanotechnol.* **2008,** *2*(3), 169–182.
10. Li, Z.; Wang, C. *Effects of Working Parameters on Electrospinning*. In *One-dimensional Nanostructures*; Springer, 2013; pp 15–28.
11. Bognitzki, M.; et al. Nanostructured Fibers via Electrospinning. *Adv. Mater.* **2001,** *13*(1), 70–72.
12. De. V. S.; et al. The Effect of Temperature and Humidity on Electrospinning. *J. Mater. Sci.* **2009,** *44*(5), 1357–1362.
13. Yarin, A. L.; Koombhongse, S.; Reneker, D. H. Bending Instability in Electrospinning of Nanofibers. *J. Appl. Phys.* **2001,** *89*(5), 3018–3026.
14. Huang, Z.-M.; et al. A Review on Polymer Nanofibers by Electrospinning and their Applications in Nanocomposites. *Compos. Sci. Technol.* **2003,** *63*, 2223–2253.
15. Haghi, A. K. Electrospun Nanofiber Process Control. *Cellulose Chem. Technol.* **2010,** *44*(9), 343–352.
16. Sill, T. J.; von Recum, H. A. Electrospinning: Applications in Drug Delivery and Tissue Engineering. *Biomaterials* **2008,** *29*(13), 1989–2006.
17. Bhardwaj, N.; Kundu, S. C. Electrospinning: A Fascinating Fiber Fabrication Technique. *Biotechnol. Adv.* **2010,** *28*(3), 325–347.
18. Rafiei, S.; et al. New Horizons in Modeling and Simulation of Electrospun Nanofibers: A Detailed Review. *Cellulose Chem. Technol.* **2014,** *48*(5–6), 401–424.
19. Tan, S. H.; et al. Systematic Parameter Study for Ultra-fine Fiber Fabrication via Electrospinning Process. *Polymer* **2005,** *46*(16), 6128–6134.
20. Zanin, M. H. A.; Cerize, N. N. P.; de Oliveira, A. M. Production of Nanofibers by Electrospinning Technology: Overview and Application in Cosmetics, In *Nanocosmetics and Nanomedicines*; Springer, 2011; pp 311–332.

21. Deitzel, J. M.; et al. The Effect of Processing Variables on the Morphology of Electrospun Nanofibers and Textiles. *Polymer* **2001,** *42*(1), 261–272.
22. Fong, H.; Chun, I.; Reneker, D. H. Beaded Nanofibers Formed During Electrospinning. *Polymer* **1999,** *40*(16), 4585–4592.
23. Gupta, P.; Wilkes, G. L. Some Investigations on The Fiber Formation by Utilizing a Side-by-Side Bicomponent Electrospinning Approach. *Polymer* **2003,** *44*(20), 6353–6359.
24. Theron, S. A.; Zussman, E.; Yarin, A. L. Experimental Investigation of the Governing Parameters in the Electrospinning of Polymer Solutions. *Polymer* **2004,** *45*(6), 2017–2030.
25. Sawicka, K.; Gouma, P.; Simon, S. Electrospun Biocomposite Nanofibers for Urea Biosensing. *Sens. Actuat. B: Chem.* **2005,** *108*(1), 585–588.
26. Deitzel, J. M.; et al. The Effect of Processing Variables on the Morphology of Electrospun Nanofibers and Textiles. *Polymer* **2001,** *42*(1), 261–272.
27. Kessick, R.; Fenn, J.; Tepper, G. The Use of AC Potentials in Electrospraying and Electrospinning Processes. *Polymer* **2004,** *45*(9), 2981–2984.
28. Demir, M. M.; et al. Electrospinning of Polyurethane Fibers. *Polymer* **2002,** *43*(11), 3303–3309.
29. Hsu, C. M.; Shivkumar, S. Nano-sized Beads and Porous Fiber Constructs of Poly(ε-Caprolactone) Produced by Electrospinning. *J. Mater. Sci.* **2004,** *39*(9), 3003–3013.
30. Zong, X.; et al. Structure and Process Relationship of Electrospun Bioabsorbable Nanofiber Membranes. *Polymer* **2002,** *43*(16), 4403–4412.
31. Keun. S. W.; et al. Effect of Ph on Electrospinning of Poly(Vinyl Alcohol). *Mater. Lett.* **2005,** *59*(12), 1571–1575.
32. Frenot, A.; Chronakis, I. S. Polymer Nanofibers Assembled by Electrospinning. *Curr. Opin. Colloid Interface Sci.* **2003,** *8*(1), 64–75.
33. Baji, A.; et al. Electrospinning of Polymer Nanofibers: Effects on Oriented Morphology, Structures and Tensile Properties. *Compos. Sci. Technol.* **2010,** *70*(5), 703–718.
34. Luo, C. J.; Nangrejo, M.; Edirisinghe, M. A Novel Method of Selecting Solvents for Polymer Electrospinning. *Polymer* **2010,** *51*(7), 1654–1662.
35. Tao, J.; Shivkumar, S. Molecular Weight Dependent Structural Regimes During the Electrospinning of PVA. *Mater. Lett.* **2007,** *61*(11), 2325–2328.
36. Huang, Z. M.; et al. A Review on Polymer Nanofibers by Electrospinning and their Applications in Nanocomposites. *Compos. Sci. Technol.* **2003,** *63*(15); 2223–2253.
37. Reneker, D. H.; Yarin, A. L. Electrospinning Jets and Polymer Nanofibers. *Polymer* **2008,** ***49***(10), 2387–2425.
38. Rutledge, G. C.; Fridrikh, S. V. Formation of Fibers by Electrospinning. *Adv. Drug Deliv. Rev.* **2007,** *59*(14), 1384–1391.
39. Zhang, S. *Mechanical and Physical Properties of Electrospun Nanofibers;* 2009; pp 1–83.

CHAPTER 19

NANOSTRUCTURED FIBERS VIA ELECTROSPINNING (PART II)

G. E. ZAIKOV

Russian Academy of Sciences, Moscow, Russia

CONTENTS

ABSTRACT

In this chapter advanced techniques and new methods for formation of electrospun nanofibers are presented in detail. It is shown the steps should be taken in order to attain uniform nanofibers consistently and reproducibly. New Experimental techniques are presented as well.

Symbols	Definitions
V_c	critical voltage
H	distance between the capillary exit and the ground
L	length of the capillary
R	radius
γ	surface tension of the liquid/ solution

Abbreviations	Symbols
Polyethylene oxide	PEO
Polyvinyl alcohol	PVA
Polymethyl methacrylate	PMMA
Polyaniline	PANI
Cellulose acetate	CA
Polyvinyl pyrrolidone	PVP
Polyacrylic acid	PAA
Polyamide-6	PA6
Polyethylene oxide	PEO
Polyacrylic acid	PAA
Polyurethane	PU
Polycapro lactone	PCL
Polystyrene	PS
Dimethylformamide	DMF
Tetrahydrofuran	THF
Multifunctional polymethylsilsesquioxane	PMSQ

19.1 INTRODUCTION TO NANOTECHNOLOGY IMPORTANCE

Nanotechnology with unique physical, chemical, and biological properties is interested by many scientists in everywhere for novel applications in recent years.[1–3] So, researchers started to analysis these properties.[3,4] When the diameters of polymer fiber materials reduce from micrometers to nanometers, several characteristics are changed compared to other known form of the material in many research fields. These characteristics are[5]:

- A large surface area to volume ratio
- Flexibility in surface functionalities
- High porosity
- Superior mechanical performance (e.g., stiffness and tensile force)

Nanofibers can produce from a spacious range of polymers.[2,6,7]

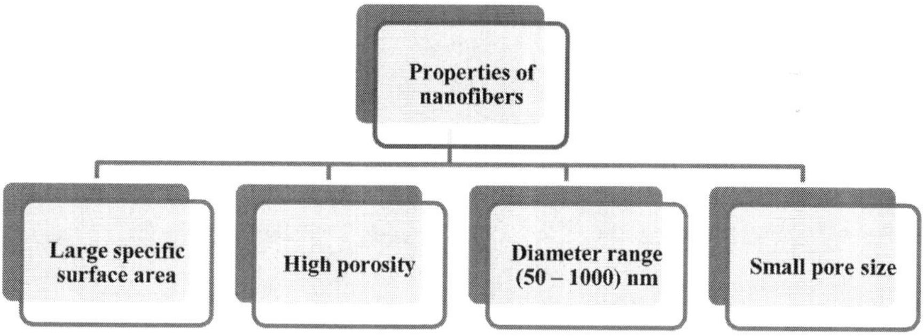

FIGURE 19.1 Properties of nanofibers.

These desirable properties make the polymer nanofibers best candidates for many important applications[1–3,6–23]:

- Environmental engineering and biotechnology
    - Medical science
        - Tissue engineering
            - Wound healing
            - Tissue template
        - Drug delivery
        - Release control
- Composites
- Defense
    - Protective clothing

TABLE 19.1 Utilized Polymer Fibers in Electrospinning Process in Different Applications for Tissue Engineering.[24]

Fiber diameter	Solvent	Polymer	Application
200–350 nm	2,2,2-Trifluo-roethanol Water	Poly(ε-caprolactone) (shell) + poly(ethylene glycol) (core)	**Drug Delivery System**
1–5 μm	Chloroform DMF Water	Poly(ε-caprolactone) and poly(ethylene glycol) (shell) Dextran (core)	
500–700 nm	Chloroform DMF Water	Poly(ε-caprolactone) (shell) and poly(ethylene glycol) (core)	
~4 μm	DCM PBS	Poly(ε-caprolactone-*co*-ethyl ethylene phosphate)	
260–350 nm	DMF	Poly(D,L-lactic-*co*-glycolic acid), PEG-b-PLA, PLA	
1–10 μm	DCM	Poly(D,L-lactic-*co*-glycolic acid)	
690–1350 nm	Chloroform	Poly(L-lactide-*co*-glycolide) and PEG-PLLA	
2–10 nm	Chloroform Methanol	Poly(ε-caprolactone)	**General Tissue Engineering**
500–900 nm	Chloroform DMF	Poly(ε-caprolactone) (core) + zein (shell)	
500 nm	2,2,2-Trifluo-roethanol	Poly(ε-caprolactone) (core) + collagen (shell)	
500–800 nm	DMF THF	Poly(D,L-lactic-*co*-glycolic acid) and PLGA-b-PEG-NH2	
1–4 mm	DMF Acetone	Poly(ethylene glycol-*co*-lactide)	
0.2–8.0 mm	2-Propanol and water Water	Poly(ethylene-*co*-vinyl alcohol)	
180–250 nm	HFP	Collagen	
0.29–9.10 mm	2,2,2-Trifluo-roethanol	Gelatin	
120–610 μm	HFP	Fibrinogen	
130–380 nm	HFP	Poly(glycolic acid) and chitin	

TABLE 19.1 *(Continued)*

Fiber diameter	Solvent	Polymer	Application
0.2–1 nm	Chloroform DMF	Poly(ε-caprolactone)	**Vascular Tissue Engineering**
200–800 nm	Acetone	Poly(L-lactide-*co*-ε-caprolactone)	
5 μm	Chloroform	Poly(propylene carbonate)	
300 nm	1,4-dioxane DCM	Poly(L-lactic acid) and hydroxylapatite	
0.163–8.77 m	HFP	Chitin	

19.2 USUAL METHODS OF PRODUCING NANOFIBERS

Various common techniques can be used for preparing polymer nanofibers such as[7,10,15]

1. Drawing
2. Template synthesis
3. Phase separation
4. Self-assembly
5. Electrospinning

At first these techniques are introduced, and then special ways for producing nanofibers are proceeding.

19.2.1 DRAWING

The drawing technique (Fig. 19.1) is associated with evaporating the solvent from viscous polymer liquids directly, leading to solidification of the fiber. In this method, the nanofiber has an order of microns.[7,10,23] Figure 19.2 shows drawing technique, each fiber is made from a micro droplet of polymer solution using a micropipette.[7,25]

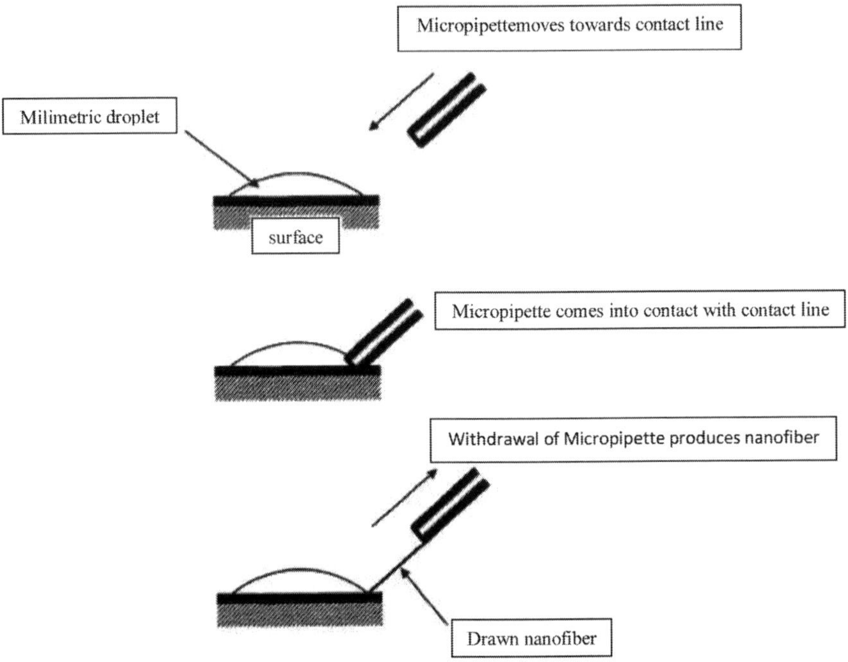

FIGURE 19.2 Schematic of drawing technique (each fiber is drawn from a microdroplet of polymer solution using a micropipette.[7,25]

19.2.2 TEMPLATE SYNTHESIS

The template synthesis uses templates with pores (Fig. 19.3) such as membranes to make solid or hollow form of nanofibers. This technique is similar to the extrusion in manufacturing.[7,25] This method cannot produce one-by-one continuous nanofibers. The most significant advantage of this method is that various materials use for making nanofibers. These materials are as follows[8,15]:

- Conducting polymers
- Metals
- Semiconductors
- Carbons

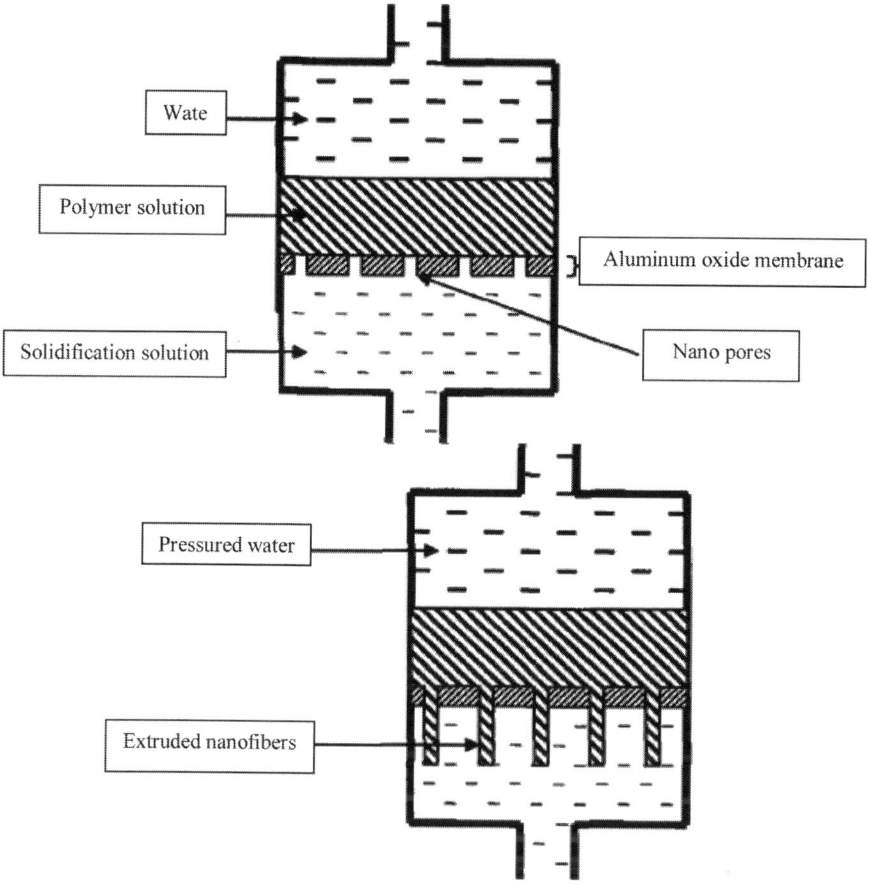

FIGURE 19.3 Schematic of template technique. Polymer extrudes through a nanoporous template by applying pressure.[7,25]

19.2.3 PHASE SEPARATION

The phase separation involves four levels:

1. Prepare a solution of polymer in solvent
2. Do polymer gelatination with low temperature
3. Get rid of solvent by immersion in water
4. Do freezing and freeze-drying

This technique calls for so long time. Figure 19.4 shows forming nanofiber by phase separation.[8,10,15,23,25]

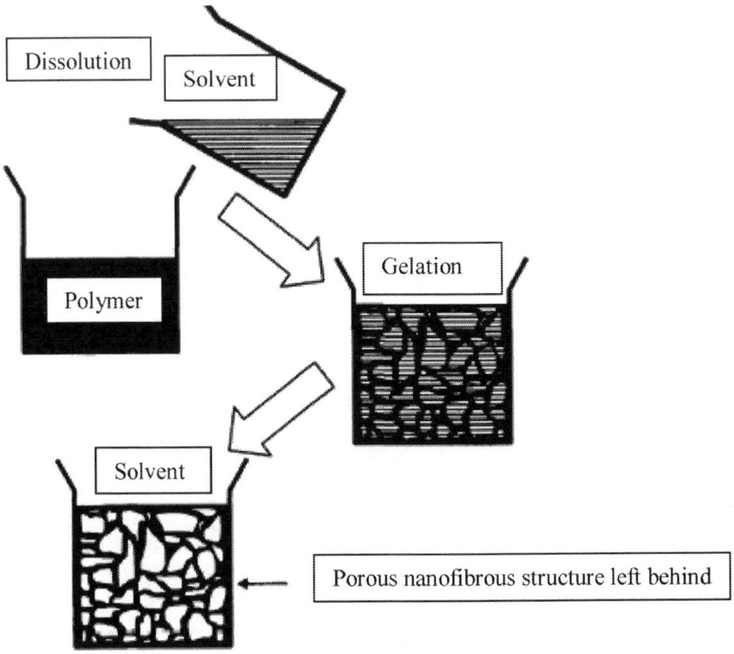

FIGURE 19.4 Formation of nanofiber by phase separation.[10,25]

19.2.4 SELF-ASSEMBLY

The self-assembly (Fig. 19.5) is another technique for producing nanofibers. In this technique, preexisting items make up into favorable patterns. Although this technique similar to the phase separation technique. The best feature of this technique is time-consuming for producing continuous polymer nanofibers. In the self-assembly technique, nanofibers are hung up molecule by molecule to bring out specific structures and functions.[7,8,10,15,23,25]

FIGURE 19.5 (A) Molecular structure and (B) nanostructure of self-assembling peptide amphiphile nanofiber network.[10]

19.2.5 ELECTROSPINNING

Electrospinning (Fig. 19.6) is the most favorite technique for creating more efficient nanofibers.[12,26,27] This technique (Tables 19.2, 19.3, 19.4) is a simple, cheap, and straightforward method to produce nanofibers.[5,6,28–30] It

1. creates continuous fibers with diameters in nanorange;
2. is applicable for abroad range of materials (e.g., synthetic, natural polymers, metals as well as ceramics and composite);
3. and prepares nanofibers with low cost.

An ordinary electrospinning setup contains three main parts[1,10]:

- A high power supply voltage
- A syringe with a needle and a pump
- A collector

Likewise, this technique distinguished by four main sections[31]:

- Taylor Cone
- Steady Jet
- Instability part
- Base part

Nanofibers are formed from polymer solution or melt with a high potential power source. Then this liquid is passed from capillary and collected along the collector[10,32].

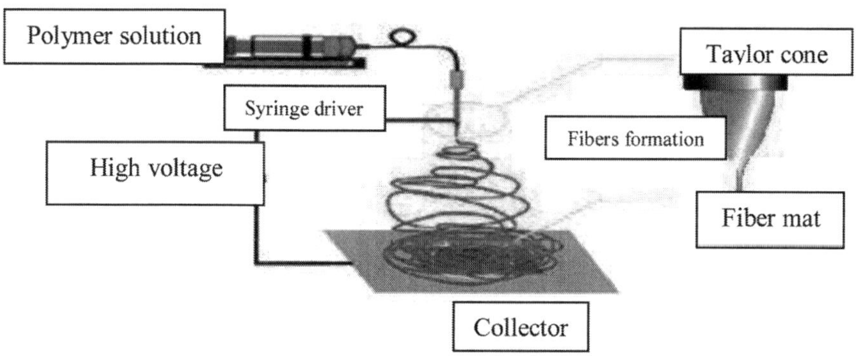

FIGURE 19.6 Standard electrospinning setup.

TABLE 19.2 Compresion of Common Technology for Producing Nanofibers.

Repeatability	Controllability	Simplicity	Scalability	Technology	Technique
No	Yes	Yes	No	Laboratory scale	Drawing
Yes	Yes	Yes	Yes	Laboratory scale	Template synthesis
No	Yes	Yes	No	Laboratory scale	Phase separation
No	No	Yes	No	Laboratory scale	Self-assembly
Yes	Yes	Yes	Yes	Industrial process	Electrospinning

19.3 NEW METHODS OF PRODUCING NANOFIBERS

19.3.1 GELATION TECHNIQUE

Initially, a gel is made using predetermined amounts of polymer and solvent followed by phase separation and gel formation. Finally, nanofiber forms when the gel is frozen and freeze-dried.[7]

19.3.2 BACTERIAL CELLULOSE TECHNIQUE

Cellulose nanofibers produced by bacteria have been long used in diverse applications, including biomedical.[10,27,33] Cellulose synthesis by *Acetobacter* involves polymerization of glucose residues into chains, followed by the extracellular secretion, assembly and crystallization of the chains into hierarchically comprised ribbons (Fig. 19.7).[10,27,33]

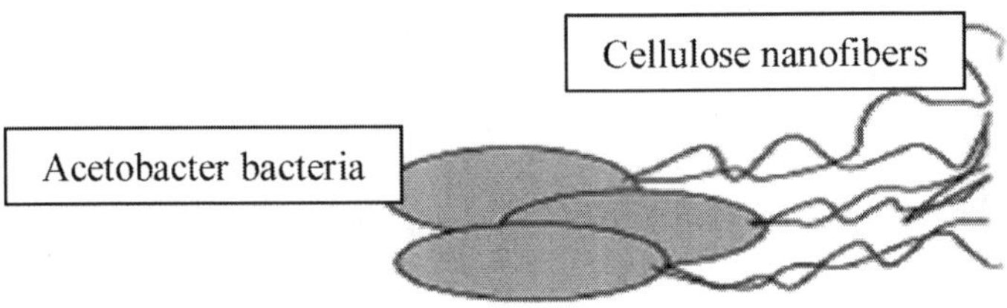

FIGURE 19.7 *Acetobacter* bacteria cells depositing cellulose nanofibers.[10]

Networks of cellulose nanofibers with diameters less than 100 nm are readily made. Fibers with different characteristics may be developed by different strains of bacteria. Copolymers have been created by adding polymers to the growth media of the cellulose producing bacteria.[10,27,33] Bacterial cellulose is mixed by the acetic bacterium *Acetobacter xylinum*. The fibrous structure of bacterial cellulose consists of a three-dimensional network of microfibrils containing glucan chains bound by hydrogen bonds.[27,33]

19.3.3 EXTRACTION TECHNIQUE

Nanofibers can be extracted from natural materials using chemical and mechanical treatments. Cellulose fibrils can be sorted out from plant cell walls. In one example, cellulose nanofibers were extracted from wheat straw and soy hull with diameters ranging from 10 to 120 nm and lengths up to a few thousand nanometers. Invertebrates have also been utilized as a source for extracting nanofibers.[10]

19.3.4 VAPOR-PHASE POLYMERIZATION TECHNIQUE

Polymer nanofibers have also been made from vapor-phase polymerization. Plasma-induced polymerization of vapor phase vinyltrichlorosilane produced organosiloxane fibers with diameters around 25 nm and typical lengths of 400–600 nm and cyanoacrylate fibers with diameters from 100 to 400 nm and lengths of hundreds of microns.[10]

19.3.5 KINETICALLY CONTROLLED SOLUTION SYNTHESIS TECHNIQUE

Nanofibers and nanowires have been created in solution using linear aligned substrates as template agents such as iron-cation absorbed reverse cylindrical micelles and silver micelles. PVA–polymethyl methacrylate nanofibers were produced using silver nanoparticle that was linearly aligned in solution with vigorous magnetic stirring. These nanoparticle chain assemblies acted as a template for further polymerization of nanofibers with diameters from 10 to 30 nm and lengths up to 60 μm.[10]

19.3.6 CONVENTIONAL CHEMICAL OXIDATIVE POLYMERIZATION OF ANILINE TECHNIQUE

Chemical oxidation polymerization of aniline is a traditional method for synthesizing poly aniline and during the former stages of this synthesis technique poly aniline nanofibers are formed. Optimization of polymerization conditions such as temperature, mixing speed and mechanical agitation allows the end stage formation of polyaniline nanofibers with diameters in the range of 30–120 nm.[10]

19.4 HISTORY OF ELECTROSPINNING AND NANOFIBERS

The word "fiber" has its root from "fibra" and the nanoterm comes from the definition that has been discussed generously. When the diameter of polymer fiber reduces to the nanoscale, the nanofibers become important in applications.[34] The electrospinning attracts more attention as ultrafine fibers of varied polymers with lower diameters to nanometers in nanotechnology in the recent years.[4,13,18] Employing electrostatic forces to deform materials in the liquid state goes back many centuries[26,35] but, the origin of electrospinning as fiber spinning technique come back to 100 years ago.[9,36] Many researchers work on electrospinning set up and effective factor on this technique. Patents characterizing an experimental set-up for producing polymer between 1934 and 1944.[35,37] Subjection Formhals work, the focus shifted to developing a better understanding process technique of the electrospinning. Here, we get a summary of electrospinning histories in Table 19.1. Several research groups such as Dr. Darrell Reneker and his research group further interest in electrospinning with a series of papers published starting in early to mid-1990s and continuing today up to engage. This renewed interest spread quickly and many secondary academic groups became interested in the field of the electrospinning.[1,14,23]

TABLE 19.3 History of Electrospinning.

Name of researcher	Year	Subject	Reference
Lord Rayleigh	19th century	Understood the technique of electrospinning	[9]
William Gilbert	1600	Discovered first record of the electrostatic attraction of a liquid	[5]

TABLE 19.3 *(Continued)*

Name of researcher	Year	Subject	Reference
Zeleny	1914	Introduced one of the earliest studies of electrified jetting phenomenon	[4]
Formhals	1934	Invented the experimental setup for the practical production of polymer filaments with an electrostatic force	[3]
Vonnegut and Neubauer	1952	Produce streams of uniform droplets and invented a simple tool for the electrical atomization	[2]
Drozin	1955	Examine the dispersion of series of liquids into aerosols under high electric potentials	[1]
Simons	1966	Patented a tool for producing nonwoven fabrics of ultra thin and weightless	[6]
Taylor	1969	Published his work on the shape of the polymer droplet at the tip of the needle with applying an electric field	[7]
Baumgarten	1971	Made a tool for electrospinning acrylic fibers with a stainless steel capillary tube and a high-voltage DC current.	[15]
		Estimated the jet speed by using energy balance when a critical voltage was applied	
Larrondo and Mandley	1981	Produced polyethylene and polypropylene fibers by melting electrospinning successfully	[8]
Hayati et al.	1987	To study effective factors on jet stability and technique of electrospinning	[28]
Reneker and Chun	1996	Has shown the possibility of electrospinning polymer solutions	[26]

The number of publications(Figs. 19.8 and 19.9) and patents in nanofibers fields and electrospinning have grown significantly in recent years.[1,14,18] Secondary academic groups became interested in the subject area of the electrospinning.[1,14,23]

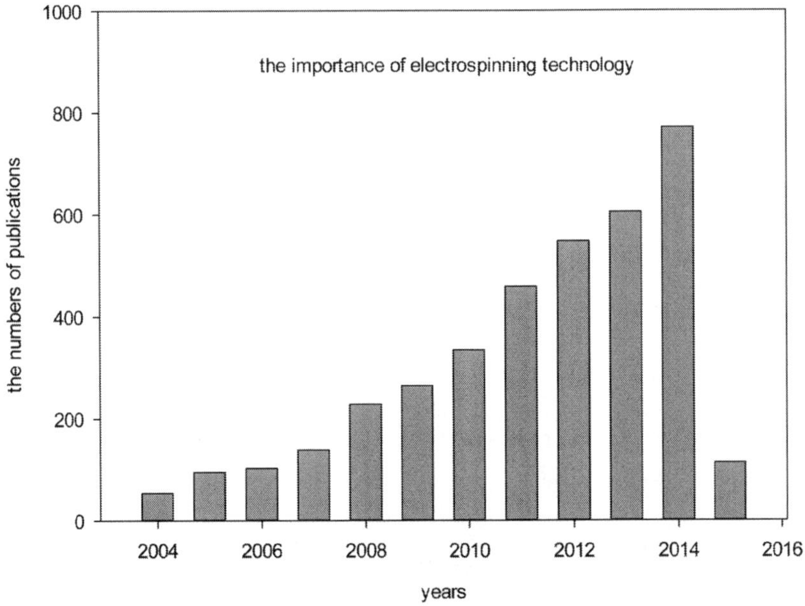

FIGURE 19.8 Numbers of publications about electrospinning.

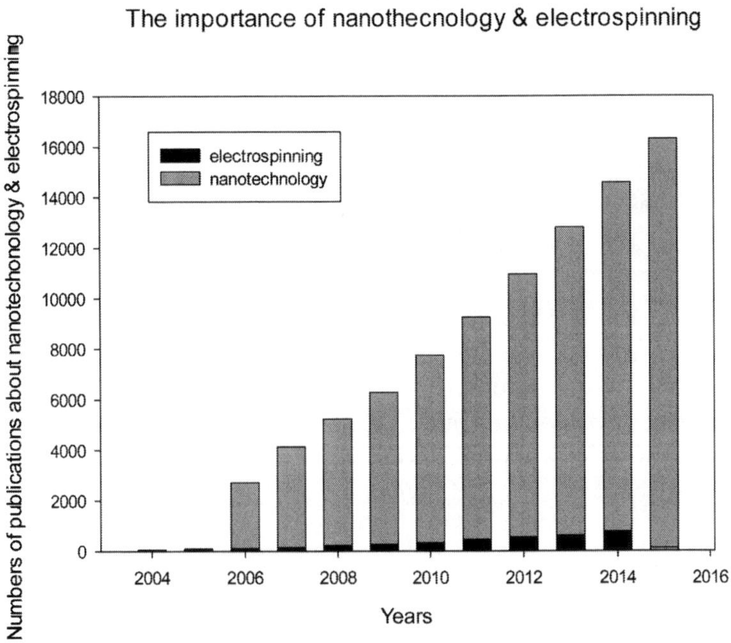

FIGURE 19.9 The importance of nanotechnology and electrospinning.

19.5 THE RULES FOR ELECTROSPINNING OF NANOFIBERS

As it was mentioned before, electrospinning is an efficient and simplest technique for producing of nanofibers with different structures and functionality.[22,34,38–43] The advantages of electrospinning technology are[44]

- Producing a high rate of nanofiber
- Simplest setup and low costs of production

A common electrospinning set-up includes[11,39,43]

- A high voltage power supply
- A syringe
- A needle
- A grounded collector screen

The technique of electrospinning includes several parts[17,22,40]:

- Charging of the fluid
- Formation of the cone-jet (Taylor cone)
- Thinning of the jet with an electric field
- Instability of the jet
- Collection of the jet on target

Here, a simple schematic of the electrospinning technique is shown in particular (figs 19-10, 19-11).

Instead of formal methods of fiber formation (e.g., dry or wet spinning), electrospinning makes nanofibers by electrostatic forces.[17] Ion migrates in the solution or melts with an electric field. When the potential came into a critical value, stream of jet starts formation to throw. This jet moves straight toward the collector then, bending instability develops into a series of loops expanding with time. The solvent evaporates during the jet moves. At last nanofibers are collected on plate.[32,46,47] It is important to observe that it is possible to electrospin all polymers into nanofibers, provided the molecular weight of the polymers is enough large and the solvent can be evaporated quickly enough during the technique.[8,44] The mechanics of this technique deserving a specific attention and necessary to predictive tools or direction for better understanding and optimization and controlling technique. It has been identified that during traveling a solution jet from the tip to collector, the primary jet may show instability during the path. Several videos, graphic, and laser light scattering methods for watching over the three-dimensional path of jets in flight, and for seeing the diameter and rate of parts were developed.[39] On the other hand, as in any liquid, the surface tension reduces the entire surface of the jet thus reduces the free energy of the liquid. If the viscosity is not enough to hold the jet as a continuous shape, what usually occurs is an instability that causes the jet to break up into droplets. This effect is known as Rayleigh instability. Which of these two opposing effects prevails depends on the nature of the fluid, especially its viscosity and surface tension. If the viscosity is enough high with good cohesiveness, the charged jet undergoes a straight jet stage and whipping instability takes place, the amplitude depends on the material and solvent, then dry thin fibers are gathered. Although the setup is straightforward, but controlling of electrospinning is complicated. Some studied done by Taylor on the initial jet formation of electrospinning technique. He gained condition for critical electric potential where surface tension is in equipoise with the electrical force[22,48]:

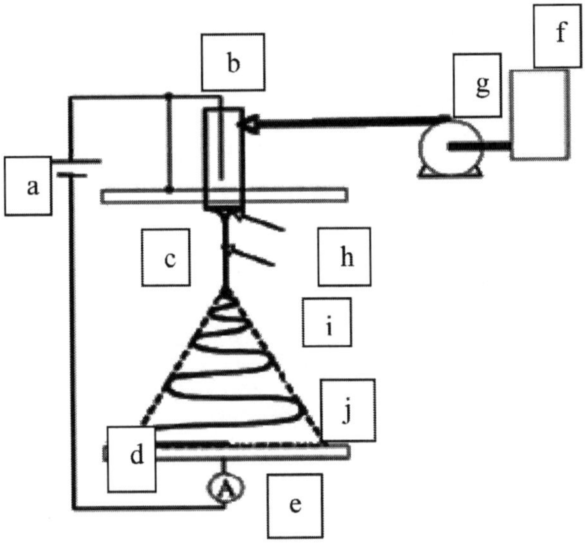

FIGURE 19.10 A simple schematic of electrospinning technique (a) high voltage power supply, (b) charging devices, (c) high potential electrode (e.g., flat plate), (d) collector electrode (e.g., flat plate), (e) current measurement device, (f) fluid reservoir, (g) flow rate control, (h) cone, (i) thinning jet, and (j) instability region.[45]

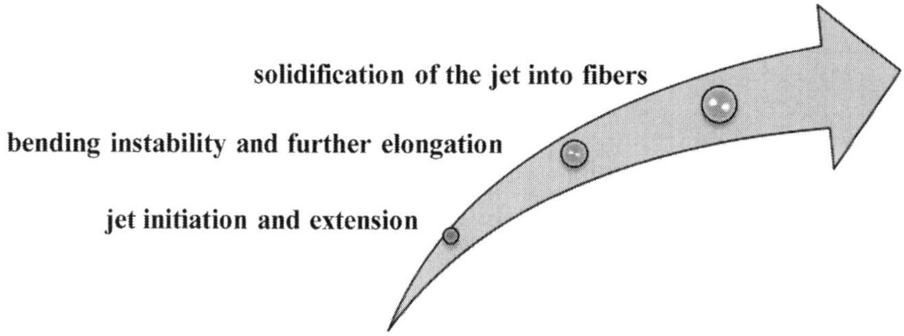

FIGURE 19.11 Basic principle of electrospinning.

$$V_C^2 = 4\frac{H^2}{L^2}\left(\ln\frac{2L}{R} - \frac{3}{2}\right)(0.117\pi\gamma R)$$

Although Taylor cone (fig 19-12) has been viewed in many subjects, the exact shape and the angle of the cone are not fixed and only applicable to slight conducive, monomeric fluids. Researchers studied the initial jet formation through computer simulation and compared with the experimental outcomes. They found that thinning the jet in the initial stage is determined by many features. Viscoelasticity is found to be the key element in the initial jet thinning behavior. Fluid with higher viscoelasticity is thicker. Studies of the whipping motion revealed that in the envelope of the cone it only controls a single jet. The jet undergoes a fast whipping motion and the whipping is so tight that the conventional camera cannot distinguish the splaying with whipping. The bending instability of electrified jet is caused by repulsive forces between the charges carried by the jet.[13,32] The jet remains axisymmetric for some length. Then bending or whipping instability starts. At the onset of this instability, the jet follows a spiral path. As the jet spirals toward the collector, higher order instabilities reveal themselves. This instability makes the jet to loop in spirals with increasing radius.[7,29] The envelope of this closed circuit is a cone. Further, the electric field speeds up the jet. So the jet rate increases. This leads to decreasing in the jet diameter. In addition, the electrostatic repulsion between excess charges in the solution stretches the jet. This stretching also decreases the jet diameter.[7,17]

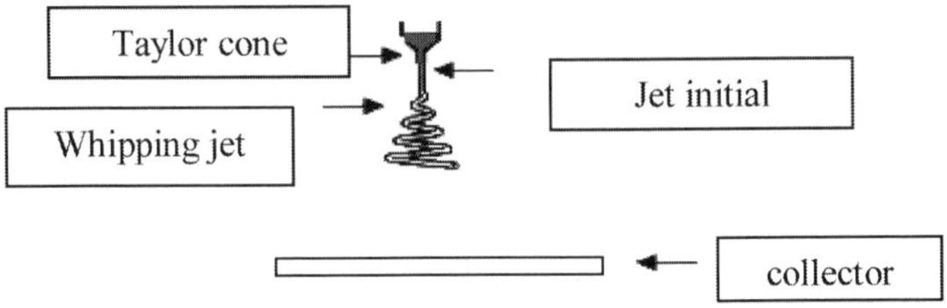

FIGURE 19.12 Whipping instability of jet in electrospinning technique.

19.6 THE USE OF NANOFIBERS IN VARIOUS SCIENCE SEARCHES

The fine electrospun nanofibers make them useful in a wide range of innovative applications (fig 19-13).[22,49] Many materials are used for electrospinning.[8,44]

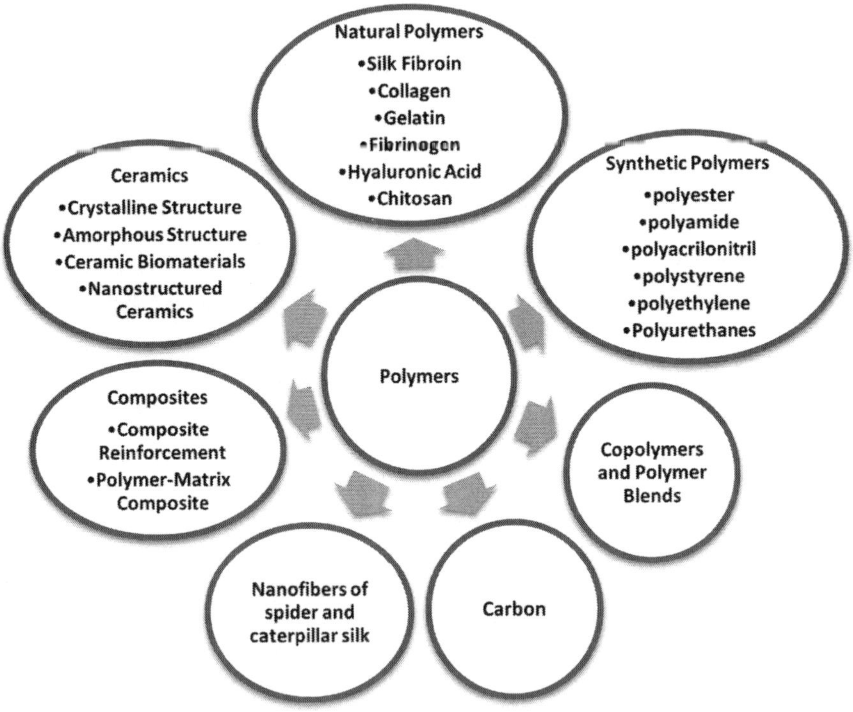

FIGURE 19.13 Varieties of polymers in electrospinning.

TABLE 19.4 Classes of Polymers with Solvents.

Polymer class	Polymer	Solvent
High performance polymers	Polymides	Phenol
	Polyamic acid	*m*-Cresol
	Polyetherimide	Methylene chloride
Liquid crystalline polymers	Polyaramid	Sulfuric acid
	Polygamma-benyzyl-glumate	dimethylformamide
	Polyp-phenylene terephthalamide	Sulfuric acid
Copolymers	Nylon 6-polyimide	Formic acid
Textile fiber polymers	Polyacrylonitrile	Dimethylformamide
	Polyethylene terephthalate	Trifuoroacetic acid and dichloromethane melt in vacuum
	Nylon	Formic acid
	Polyvinyl alcohol	Water
Electrically conducting polymer	Polyaniline	Sulphuric acid
Biopolymers	DNA	Water
	Polyhydroxy butyrate-valerate	Chloroform
	Polycapro lactone	*m*-Cresol, chlorophenol, formic acid

Also new applications have been explored for these fibers continuously(fig 19-14). Main application fields are[41,50,51]

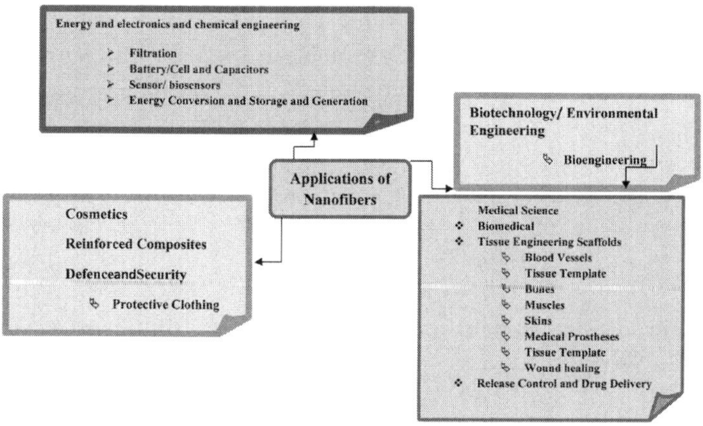

FIGURE 19.14 Potential applications of electrospun fibers.

For selected applications, it is desirable to control not only the fiber diameter, but as well the internal morphology. Porous fibers are of interest for applications such as filtration or prepare nanotubes by fiber templates.[37,40,47,52] Besides, small pore size and high surface area inherent in nanofiber has implications in biomedical applications such as scaffoldings for tissue growth.[37,53] Also, researchers have spun a fiber from a compound naturally present in the blood. This nanofiber can be used as forms of medical applications such as bandages or sutures that ultimately dissolve into the body. This nanofiber minimizes infection rate, blood loss and is also taken up by the body.[44] One of the most significant applications of nanofibers is to be used as reinforcements in composite developments. With these reinforcements, the composite materials can offer superior properties such as high modulus and strength to weight ratios, which cannot be achieved by other engineered monolithic materials alone. Information on the fabrication and structure–property relationship characterization of such nanocomposites is believed to be utilitarian. Such continuous carbon nanofiber composite also has possible applications as filters for[54]:

- Separation of small particles from gas or liquid
- Supports for high temperature catalysts
- Heat management materials in aircraft and semiconductor devices
- Rechargeable batteries
- Super capacitors.

19.6.1 BIOTECHNOLOGY/ENVIRONMENTAL ENGINEERING

Nonwoven electrospun nanofiber meshes are an excellent material for membrane preparation, particularly in biotechnology and environmental engineering applications for this reason[17]:

- High porosity
- Interconnectivity
- Micro scale interstitial space
- A large surface to volume ratio.

Biomacromolecules or cells can be tied to the nanofiber membrane for these applications[17]:

- In protein purification and waste water treatment (affinity membranes)

- Enzymatic catalysis or synthesis (membrane bioreactors)
- Chemical analysis and diagnostics (biosensors).

Electrospun nanofibers can form an effective size exclusion membrane for particulate removal from wastewater.[17] Affinity membranes are a broad class of membranes that selectively captures specific target molecules by immobilizing a specific capturing agent onto the membrane surface. In biotechnology, affinity membranes have applications in protein purification and toxin removal from bioproducts. In the environmental industry, affinity membranes have applications in organic waste removal and heavy metal removal in water treatment. To be used as affinity membranes, electrospun nanofibers must be surface functionalized with ligands. Mostly, the ligand molecules should be covalently attached to the membrane to prevent leaching of the ligands. Also, water pollution is now becoming a critical global issue. One important class of inorganic pollutant of great physiological significance is heavy metals, for example, Hg, Pb, Cu, and Cd. Distributing these metals in the environment is mainly applied to the release of metal containing waste-water from industries. For example, copper smelters may release high quantities of Cd, one of the most mobile and toxic among the trace elements, into nearby waterways. It is impossible to eliminate some classes of environmental contaminants, such as metals, by conventional water purification methods. Affinity membranes will play a critical role in wastewater treatment to remove (or recycle) heavy metal ions in the future. Polymer nanofibers functioned with a ceramic nanomaterial, mention in below, could be suitable materials for fabrication of affinity membranes for water industry applications[17]:

- Hydrated alumina hydroxide
- Alumina hydroxide
- Iron oxides.

The polymer nanofiber membrane acts as a bearer of the reactive nanomaterial that can attract toxic heavy metal ions, such As, Cr, and Pb, by adsorption or chemisorption and electrostatic attraction mechanisms. Again, affinity membranes provide an alternative access for removing organic molecules from wastewater.[17]

19.6.1.1 BIOENGINEERING

In biological viewpoint, almost entirely the human tissues and organs are deposited in nanofibrous forms or structures. Some examples include the following[54]:

- Bone
- Dentin
- Collagen
- Cartilage
- Skin.

All of them are characterized by well-organized fibrous structures realigning in nanometer scale. Current research in electrospun polymer nanofibers has focused one of their major applications on bio engineering. We can easily find their promising potential in various biomedical fields.[54]

19.6.1.1.1 Medical Science

Nanofibers are used in medical applications, which include drug and gene delivery, artificial blood vessels, artificial organs, and medical face masks. For example, carbon fiber hollow nanotubes, smaller than blood cells, have the potential to transport drugs into blood cells.[44]

19.6.1.1.1.1 Biomedical Application

Biomedical field is one of the important application areas among others, using the technique of electrospinning like[11,55]:

- Filtration material
- Protective material
- Electrical applications
- Optical applications
- Sensors
- Nanofiber reinforced composites.

Current medical practice is based almost on treatment regimes. However, it is envisaged that medicine in the future will be based heavily on early detection and prevention before disease expression. With nanotechnology, new treatment will emerge that will significantly reduce medical costs. With recent developments in electrospinning, both synthetic and natural polymers can be produced as nanofibers with diameters ranging from decades to hundreds of nanometers with controlled morphology. The potential of these electrospun nanofibers in human health-care applications is promising, for example[17]:

- In tissue or organ repair and regeneration
- As vectors to deliver drugs and therapeutics
- As biocompatible and biodegradable medical implant devices
- In medical diagnostics and instrumentation
- As protective fabrics against environmental and infectious agents in hospitals and general surroundings
- In cosmetic and dental applications.

Tissue or organ repair and positive feedback are new avenues for potential treatment, avoiding the need for donor tissues and organs in transplantation and reconstructive surgery. In this advance, a scaffold is usually needed that can be fabricated from either natural or synthetic polymers by many techniques including electrospinning and phase separation. An animal model is utilized to study the biocompatibility of the scaffold in a biological system before the scaffold is introduced into patients for tissue-regeneration applications. Nanofibers scaffolds are suited to tissue engineering. These can be made up and shaped to fill anatomical defects. Its architecture can be designed to supply the mechanical properties necessary to support cell growth, growth, differentiation and motility. Also, it can be organized to provide growth factors, drugs, therapeutics, and genes to stimulate tissue regeneration. An inherent property of nanofibers is that they mimic ECM of tissues and organs. The ECM is a complex composite of fibrous proteins such as collagen and fibronectin, glycoproteins, proteoglycans, soluble proteins such as growth factors, and other bioactive molecules that support cell adhesion and growth. One of the aims is to create electrospun polymer nanofiber scaffolds for engineering blood vessels, nerves, skin, and bone. In the pharmaceutical and cosmetic industry, nanofibers are promising tools for controlled these aims[17,44,55]:

1. Delivery of drugs
2. Therapeutics
3. Molecular medicines
4. Body-care supplements.

19.6.1.1.1.2 Tissue Engineering Scaffolds

Successful tissue engineering needs synthetic scaffolds to bear similar chemical compositions, morphological, and surface functional groups to their natural counterparts. Natural scaffolds for tissue growth are three-dimensional networks of nanometer-sized fibers made of several proteins. Nonwoven membranes of electrospun nanofibers are well known for their interconnected, 3D porous structures and large surface areas, which provide a class of ideal materials to mimic the natural ECM needed for tissue engineering. The electrospun nanofibrous support was treated with the cell solution and the nanofiber-cell was cultured in a rotating bioreactor to create the cartilage which controlled compressive strength similar to natural cartilage. The tissue engineered cartilages could be applied in treating cartilage degenerative diseases. The scaffold was applied as biomimic ECM, enzyme, gene, and medicine to revive skin, cartilage, blood vessel, and nerve. The scaffold was helpful in biocompatibility, mechanical property, porosity, degradability in the human physical structure. The electrospun nanofibers showed moderate porosity, excellent mechanical property, and biocompatibility, which could be utilized to repair blood vessels, skin and nervous tissue.[1] Tissue engineering is an emerging interdisciplinary and multidisciplinary research study. It involves the utilization of living cells, manipulated through their extracellular environment or genetically to develop biological substitutes for implantation into the body or to foster remodeling of tissues in some active manners. The purpose of tissue engineering is to renovate, replace, say, or improve the function of a particular tissue or organ. For a functional scaffold, a few basic needs have to be satisfied:

- A scaffold should control a high degree of porosity, with a suitable pore size distribution.
- A large surface area is needed.
- Biodegradability is often needed, with the degradation rate matching the rate of neotissue formation.
- The scaffold must control the needed structural integrity to prevent the pores of the scaffold from collapsing during neotissue formation, with the suitable mechanical properties.
- The scaffold (fig 19-15) should be non-toxic to cells and biocompatible, positively interacting with the cells to promote cell adhesion, growth, migration, and distinguished cell function.

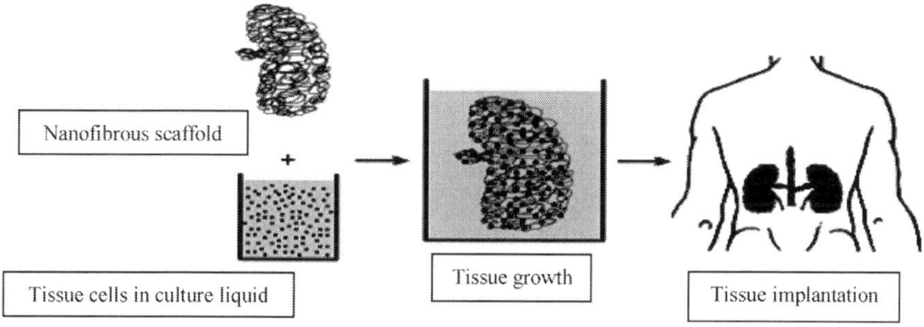

FIGURE 19.15 Principle of tissue engineering.

Among all biomedical materials under evaluation, electrospun nanofibrous scaffolds have presented great performances in cell attachment, increase and penetration.[56] One of the most promising potential applications is tissue scaffolding. The nonwoven electrospun mat has a high surface area and a high porosity. It contains an empty space between the fibers that is approximately the size of cells. The mechanical property, the topographical layout, and the surface chemistry in the nonwoven mat may have a direct effect on cell growth and migration.[31] Ultrafine fibers of biodegradable polymers produced by electrospinning have found potential applications in tissue engineering because of their high surface area to volume ratios and high porosity of the fibers. However, the flexibility of seeding stem cells and human cells on the fibers makes electrospun materials most suited for tissue engineering applications. The fibers produced can be used systematically to design the structures that they perform not only mimic the properties of ECM, but also control high strength and high toughness. For instance, nonwoven fabrics show isotropic properties and support neotissue formation. These mats resemble the ECM matrix and can be applied as a skin-scaffold and wound dressing materials where the materials are needed to be more elastic than stiff. Many natural polymers like collagen, starch, chitin, and chitosan and synthetic biodegradable polymers like PCL, PLA, PLGA have been widely investigated for potential applications in developing tissue scaffolds. These results confirm that electrospinning of natural or synthetic polymers for tissue engineering applications are promising.[9] Tissue engineering is one of the most exciting interdisciplinary and multidisciplinary research fields today, and there has been exponential growth in the number of research publications in this area in recent years. It involves the utilization of living cells, manipulated through their extracellular environment or genetically to develop biological substitutes for implantation into the body or to foster remodeling of tissues in some active manners. The purpose is to repair, replace, maintain, or increase the use of a particular tissue or organ. The core technologies intrinsic to this effort can be organized into three fields[4]:

- Cell technology
- Scaffold frame technology
- Technologies for *in vivo* integration.

The scaffold frame technology focuses on these objectives[4]:

- Designing
- Manufacturing
- Characterizing three-dimensional scaffolds for cell seeding
- In vitro or in vivo culturing

19.6.1.1.1.3 Blood Vessels

Blood vessels vary in sizes, mechanical and biochemical properties, cellar content and ultrastructural organization, depending on their location, and specific role. It is needed that the vascular grafts engineered should have wanted characteristics. Blood vessel replacement, a fine blood vessel (diameter < 6 mm), has stayed a great challenge. Because the electrospun nanofiber mats can give good support during the initial development of vascular smooth muscle cells, smooth film combining with electrospun nanofiber mat could form a good 3D scaffold for blood vessel tissue engineering.[4,56]

19.6.1.1.1.4 Muscles

Collagen nanofibers were first applied to assess the feasibility of culturing smooth muscle cell. The cell growth on the collagen nanofibers was promoted and the cells were easily integrated into the nanofiber network after 7 days of seeding. Smooth muscle cells also adhered and proliferated well on another polymer nanofiber mats blended with collagen, incorporating collagen into nanofibers was observed to improve fiber elasticity and tensile strength, and increase the cell adhesion. The fiber surface wet ability influences cell attachment. The alignment of nanofibers can induce cell orientation and promote skeletal muscle cell morphologenesis and aligned formation.[4]

19.6.1.1.1.5 Medical Prostheses

Polymer nanofibers fabricated by electrospinning have been offered for several soft tissue prosthesis applications, such as blood vessel, vascular, breast, etc. In addition, electrospun biocompatible polymer nanofibers can also be deposited as a slender, porous film onto a hard tissue prosthetic device designed to be implanted into the human body. This coating film with a fibrous structure works as an interface between the prosthetic device and the host tissues. It is anticipated to reduce efficiently the stiffness mismatch at the tissue or Interphase and from here prevents the device failure after the implantation.[54]

19.6.1.1.1.6 Tissue Template

For treating tissues or organs in malfunction in a human body, one of the challenges in the area of tissue engineering or biomaterials is the design of ideal scaffolds or synthetic matrices. They can mimic the structure and biological functions of the natural ECM. Human cells can attach and organize well around fibers with diameters smaller than those of the cellular phones. Nanoscale fibrous scaffolds can provide an ideal template for cells to seed, migrate, and produce. A successful regeneration of biological tissues and organs calls for developing fibrous structures with fiber architectures useful for cell deposition and cell growth. Of particular interest in tissue engineering is creating reproducible and biocompatible three-dimensional scaffolds for cell growth resulting in biometrics composites for various tissue repair and replacement processes. Recently, people have begun to pay attention to making such scaffolds with synthetic polymers or biodegradable polymer nanofibers. It is believed that converting biopolymers into fibers and networks that mimic native structures will eventually improve the usefulness of these materials as large diameter fibers do not mimic the morphological characteristics of the native fibrils.[44,54]

19.6.1.1.1.7 Wound Healing

Wound healing (fig 19-16) is a native technique of regenerating dermal and epidermal tissues. When an individual is wounded, a set of complex biochemical actions take place in a closely orchestrated cascade to repair the harm. These events can be sorted into four groups:

1. Inflammatory

2. Proliferative
3. Remodeling phases
4. Epithelialization.

Ordinarily, the body cannot heal a deep dermal injury. In full thickness burn or deep ulcers, there is no origin of cells remaining for regeneration, except from the wound edges. Dressings for the wound healing role to protect the wound, exude extra body fluids from the wound area, decontaminate the exogenous microorganism, improve the appearance, and sometimes speed up the healing technique. For these functions, a wound dressing material should provide a physical barrier to a wound, but be permeable to moisture and oxygen. For a full thickness dermal injury, when an "artificial dermal layer" adhesion and integration consisting of a 3D tissue scaffold with well cultured dermal fibroblasts will aid there-epithelialization. Nanofiber membrane is a good wound dressing candidate because of its unique properties like[4]

- the porous membrane structure and
- well interconnected pores,

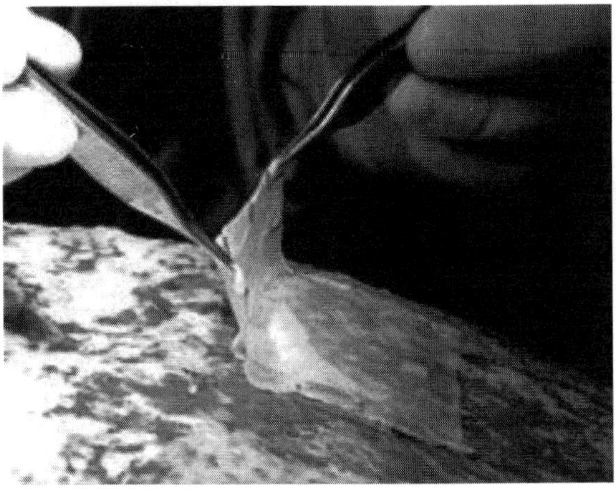

FIGURE 19.16 Nanofiber mats used for medical dressing.

They are important for exuding fluid from the wound. The small pores and high specific surface area not only inhibit the exogenous microorganism invasions, but also assist the control of fluid drainage. In addition, the electrospinning provides a simple path to add drugs into the nanofibers for any possible medical treatment and antibacterial purposes.[4]

For wound healing, an ideal dressing should have certain features:

1. Hemostatic ability
2. Efficiency as bacterial barrier
3. Absorption ability of excess exudates (wound fluid or pus)
4. Suitable water vapor transmission rate
5. Enough gaseous exchange ability
6. Ability to conform to the contour of the wound area
7. Functional adhesion
8. Painless to patient
9. Ease of removal
10. Low cost.

Current efforts using nanofibrous membranes as a medical dressing are still in its early childhood, but electrospun materials meet most of the needs outlined for wound-healing polymer. Because their microfibrous and nanofibrous provide the nonwoven textile with desirable properties.[42] Polymer nanofibers can also be utilized for the treatment of wounds or burns of a human skin, as well as designed for hemostatic devices with some unique characteristics. Fine fibers of biodegradable polymers can spray/spun on to the injured location of the skin to make a fibrous mat dressing. They let wounds heal by encouraging forming a normal skin development and remove form scar tissue, which would occur in a traditional treatment. Nonwoven nanofibrous membrane mats for wound dressing usually have pore sizes ranging from 500 nm to 1 mm, small enough to protect the wound from bacterial penetration by aerosol particle capturing mechanisms. High surface area of 5–100 m^2/g is efficient for fluid absorption and dermal delivery.[44,54] The electrospun nanofibers have been utilized in treating wounds or burns of human skin because of their high porosity which allows gas exchange and a fibrous structure that protects wounds from infection and dehydration. Nonwoven electrospun nanofibrous membranes for wound dressing usually have pore sizes in the range of 500–1000 mm which is low enough to protect the wound from bacterial penetration. High surface area of electrospun nanofibers is efficient for fluid absorption and dermal delivery. Chong invented a composite containing a semipermeable barrier and a scaffold filter layer of skin cells in wound healing by electrospinning.[1] Electrospinning could create scaffold with more homogeneity besides meeting other needs like oxygen permeation and protection of wound from infection and dehydration for use as a wound-dressing materials. Many other synthetic and natural polymers, like carboxyethyl, chitosan or PVA, collagen or chitosan, and silk fibroin, have been electrician to advise them for wound-dressing applications.[11]

19.6.1.1.1.7 Release Control

Controlled release is an effective technique of delivering drugs in medical therapy. It can balance these features:

1. The delivery kinetics
2. Minimize the toxicity
3. Side effects
4. Improve patient convenience

In a controlled release system, the active substance is loaded into a carrier or device first, and then releases at a predictable rate in vivo when governed by an injected or noninjected route. As a potential drug delivery carrier, electrospun nanofibers have showed many advantages. The drug loading is easy to implement by electrospinning technique, and the high applied voltage used in the electrospinning technique had little influence on the drug activity. The high specific surface area and short diffusion passage length give the nanofiber drug system higher overall release rate than the bulk material (e.g., film). The release profile can be finely controlled by modulating of nanofiber morphology, porosity, and composition. Nanofibers for drug release systems mainly come from biodegradable polymers, such as PLA, PCL, PDLA, PLLA, PLGA and hydrophilic polymers such as PVA, PEG, and PEO. Nonbiodegradable polymers, such as PEU, were likewise found out.[4] Nanofiber systems for the release of drugs are needed to fill diverse roles. The mattress should be capable to protect the compound from decomposition and should allow for controlled release in the targeted tissue, over a needed period of time at a constant release rate.[13] Drug release and tissue engineering are closely related regions. Sometimes release of therapeutic causes can increase the efficiency of tissue engineering. Various nanostructured materials are applicable in tissue engineering. Electrospun fiber mats provide the advantage of increased drug release compared to roll-films because of the increased surface area.[11,37]

19.6.1.1.1.8 Drug Delivery and Pharmaceutical Composition

Delivery of drug or pharmaceuticals to patients in the most physiologically acceptable manner has always been an important concern in medicine.

In general, the smaller the dimensions of the drug and the coating material wanted to encapsulate the drug, the better the drug to be assimilated by human being. Drug delivery with polymer nanofibers is based on the rule that the dissolution rate of a particulate drug increases with increasing surface area of both the drug and the similar carrier if needed. As the drug and carrier materials can be mixed for electrospinning of nanofibers, the likely modes of the drug in the resulting nanostructure products are

- Drug as particles attached to the surface of the carrier which is in the form of nanofibers.
- Both drug and carrier are nanofiber form; therefore, the product will be the two kinds of nanofibers interlaced together.
- The blend of drug and carrier materials integrated into one fiber containing both sections.
- The carrier material is electrospun into a tubular frame in which the drug particles are encapsulated.

However, as the drug delivery in the form of nanofibers is still in the early stage exploration, a real delivery mode after production and efficiency has yet to be determined in the future.[53] Drug delivery with electrospun nanofibers is based along the principle that drug releasing rate increases with increasing surface area of both the drug and the similar carrier used. The increased surface area of drug improved the bioavailability of the poor water-soluble drug. Various drugs such as avandia, eprosartan, carvedilol, hydrochloride-thiazide, aspirin, naproxen, nifedipine, indomethacin, and ketoprofen were entrapped into PVP to form pharmaceutical compositions which provided controllable releasing. Not only synthetic polymers but also natural polymers can be applied for modeling drug delivery system.[5] Controlled drug release over a definite period of time is possible with biocompatible delivery matrices of polymers and biodegradable polymers. They mostly used as drug delivery systems to deliver therapeutic agents because they can be well designed for programed distribution in a controlled fashion. Nanofiber mats applied as drug carriers in drug delivery system because of their high functional characteristics. The drug delivery system relies on the rule that the dissolution rate of a particulate drug increases with increasing surface area of both the drug and the similar carrier. Importantly, the large surface area associated with nanospun fabrics allows for quick and efficient solvent evaporation, which provides the incorporated drug limited time to recrystallize which favors forming amorphous dispersions or solid solutions. Depending on the polymer carrier used, the release of pharmaceutical dosage can be designed as rapid, immediate, delayed, or varied dissolution. Many researchers successfully encapsulate drugs within electrospun fibers by mixing the drugs in the polymer solution to be electrospun. Various solutions containing low molecular weight drugs have been electrospun, including lipophilic drugs such as ibuprofen, cefazolin, rifampin, paclitaxel, and Itraconazole and hydrophilic drugs such as mefoxin and tetracycline hydrochloride. However, they have encapsulated proteins in electrospun polymer fibers. Besides the normal electrospinning process, another path to develop drug-loaded polymer nanofibers for controlling drug release is to use coaxial electrospinning and research has successfully encapsulated two kinds of medicinal pure drugs through this process.[42] Electrospinning affords great flexibility in selecting materials for drug delivery applications. Either biodegradable or nondegradable materials can be utilized to control whether drug release occurs by diffusion alone or diffusion and scaffold degradation. Also, because of the flexibility in material selection many drugs can be delivered including:

- Antibiotics
- Anticancer drugs
- Proteins
- DNA

Using the various electrospinning techniques, many different drug loading methods can also be applied:

- Coatings
- Embedded drug
- Encapsulated drug (coaxial and emulsion electrospinning

However, as the drug delivery in the form of nanofibers is still in the early stage exploration, a real delivery mode after production and efficiency has yet to be found in the future.[43]

19.6.2 COSMETICS

The current skin masks applied as topical creams, lotions, or ointments. They may include dusts or liquid sprays and more likely than fibrous materials to migrate into sensitive areas of the body, such as the nose and eyes where the skin mask is being utilized to the face. Electrospun polymer nanofibers have been tried as a cosmetic skin care mask for treating skin healing, skin cleaning, or other therapeutic or medical properties with or without various additives. This nanofibrous skin mask with small interstices and high surface area can make easy far greater utilization and speed up the rate of transfer of the additives to the skin for the fullest potential of the additive. The cosmetic skin mask from the electrospun nanofibers can be applied gently and painlessly as well as directly to the three-dimensional topography of the skin to provide healing or cure treatment to the skin.[53] Electrospun nanofibers have been aimed for use in cosmetic cares such as treating skin healing and skin cleaning with or without various additives in recent years. Despite the growth in the number of electrospun polymer nanofiber publications in the recent years, there is a rare work, including scientific papers and patents, in the cosmetic field about the use of electrospun nanofibers. Developing nanofibers in this field have been focused on skin treatment applications, such as care mask, skin healing and skin cleaning, with active agents (cosmetics) with controlled release from time to time. The cosmetic application included in the biomedicine application, which admits the drug delivery system employing the active agents used in cosmetics, body care supplements. Therefore, the cosmetic and drug delivery are closely interrelated areas. The electrospun nanofibers provide the advantage of the increasing drug release when compared to cast-films because of the increased surface area. Besides, it's agreeable to processing different polymers such as natural, synthetic and blends, according to their solubility or melting point. It is significant that although most of the researches in polymeric nanofiber by electrospinning consider the technique simple, cost-effective and easily scalable from laboratory to commercial production, only a limited number of companies have commercially performed electrospun fibers.[5]

19.6.3 ENERGY AND ELECTRONICS AND CHEMICAL ENGINEERING

The demand for energy use of goods and services has been increasing every year throughout the world. However, it was reported the estimated reserve amounts of petroleum and natural gas in the world are only 41 years and 67 years, respectively. To solve this problem, new, clean, renewable, and sustainable energies have to be ground and used to replace the current nonsustainable energies. Wind generator, solar power generator, hydrogen battery, and polymer battery are among the most popular alternatives to produce new energies. In recent years, electrospun nanofibers have presented their potential in these applications(figs 19-17, 19-18):

- Super capacitors
- Lithium cells
- Fuel cells
- Solar cells
- Transistors

Further, electrospun nanofibers with electrical and electrooptical have also got much interest recently because of their potential applications in creating nanoscale electronic and optoelectronic devices.[5]

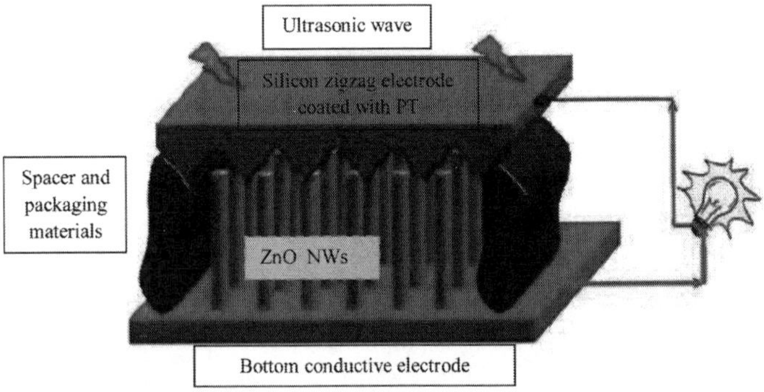

FIGURE 19.17 ZnO Nanofibers in Energy Application.

19.6.3.1 FILTRATION APPLICATION

Filtration is necessary in many engineering fields. It was estimated that future filtration market would be up to US 700 billion US dollars by the year 2020.[1,15,44]

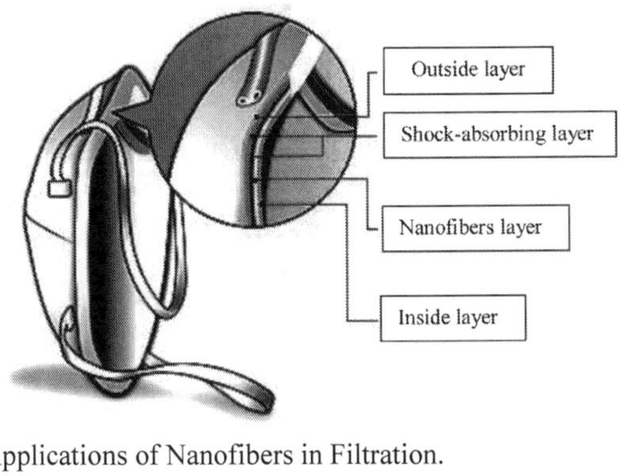

FIGURE 19.18 Applications of Nanofibers in Filtration.

Fibrous materials used for filter media provide advantages of high filtration efficiency and low air resistance. Filtration efficiency, which is closely related with the fiber fineness, is one of the most important concerns for the filter performance. One direct way of developing high efficient and effective filter media are by using nanometer sized fibers in the filter structure.[15,54] With outstanding of polymeric nanofibers properties such as high specific surface area, high porosity, and excellent surface adhesion, they are suited to be made into filtering media for filtering out particles in the sub micron range. Also, this filter system could be used for processing waste water containing active sludge.[1,9] Electrospun fibers are being widely studied for aerosol filtration, air cleaning applications in industry and for particle collection in clean rooms. The advantage of using electrospun fibers in the filtration media is the fiber diameters can be easily controlled and can produce an impact in high efficiency particulate air filtrations.[9] The filtration efficiency is commonly influenced by these parameters[4]:

- The filter physical structure like
  - Fiber fineness
  - Matrix structure
  - Thickness
  - Pore size
  - Fiber surface electronic properties
  - Its surface chemical characteristic
- Surface free energy.

The particle collecting capability is also associated with the size range of particles being collected. Besides the filtration efficiency, other properties such as pressure drop and flux resistance are also important factors to be assessed for a filter media.[4] Filter efficiency increases linearly with the decrease of thickness of filter membrane and applied pressure increase.[42]

19.6.4 REINFORCED COMPOSITES/REINFORCEMENT

Although electrospun fiber reinforced composites (fig 19-19) have significant potential for development of high intensity or high toughness materials and materials with good thermal and electrical con conductivity. Few studies have found out the use of electrospun fibers in composites. Traditional reinforcements in polymer matrices can create stress concentration sites because of their irregular shapes and cracks spread by burning through the fillers or

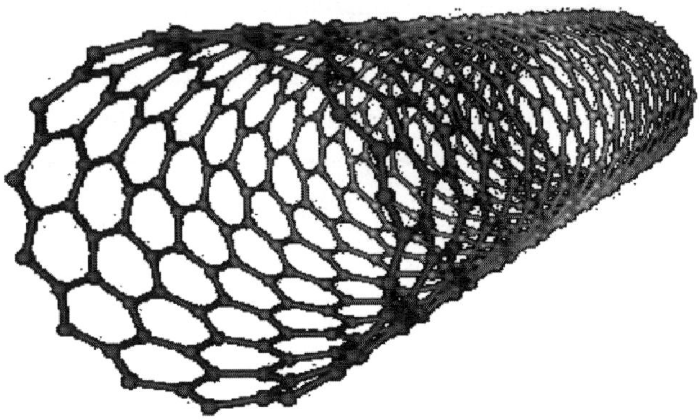

FIGURE 19.19 Carbone nanotubes composite.

traveling up, down and around the particles. However, electrospun fibers have various advantages over traditional fillers. The reinforcing effects of fibers are influenced essentially by fiber size. Smaller size fibers give more efficient support. Fibers with finer diameters have a preferential orientation of polymer chains along the fiber axis. The orientation of macromolecules in the fibers improves with decreasing in diameter, making finer diameter fibers strong. Therefore, the use of nanometer sized fibers can significantly raise the mechanical integrity of polymer matrix compared to micron-sized fibers. However, the high percentage of porosity and irregular pores between fibers can contribute to an interpenetrated structure when spread in the matrix, which also improves the mechanical strength because of the interlocking mechanism. These characteristic features of nanofibers enable the transfer of applied stress to the fiber–matrix in a more serious fashion than most of the commonly used filler materials. Current issues related to the use of electrospun nanofibers as reinforcement materials are the control of dispersion and orientation of the fibers in the polymer matrix. To achieve better reinforcement, electrospun nanofibers, may require to be collected as an aligned yarn instead of a randomly distributed felt so the post-electrospinning stretching process could be applied to further improve the mechanical properties. Further, if crack growth is transverse to the fiber orientation, the crack toughness of the composite can be optimized. So, the interfacial adhesion between fibers and matrix material needs to be controlled such the fibers are deflecting the cracks by fiber–matrix interface debonding and fiber pullout. The interfacial adhesion should not be excessively strong or too weak. Ideal control can only be obtained by careful selective fiber surface treatment. Spreading electrospun mats in the matrix can be improved by cutting down the fibers to shorter fragments. This can be accomplished, if the electrospun fibers are collected as aligned bundles (instead of nonwoven network), which can then be optically or mechanically trimmed to get fiber fragments of several 100 nm in length.[9] Early studies on electrospun nanofibers also included reinforcement of polymers. As electrospun nanofiber mats have a large specific surface area and an irregular pore structure, mechanical interlocking among the nanofibers should occur.[4] One of the most significant applications of traditional fibers, especially engineering fibers such as carbon, glass, and Kevlar fibers, is to be used as reinforcements in composite developments. With these reinforcements, the composite materials can provide superior structural properties such as high modulus and strength to weight ratios, which cannot be attained by other engineered monolithic materials alone. Nanofibers will also eventually find important applications in making nanocomposites. This is because nanofibers can have even better mechanical properties than micro fibers of the same materials, and therefore the superior structural properties of nanocomposites can be anticipated. However, nanofiber reinforced composites may control some extra merits which cannot be shared by traditional (microfiber) composites. For instance, if there is a difference in refractive indices between fiber and ground substance, the resulting composite becomes opaque or nontransparent because of light scattering. This limit, however, can be avoided when the fiber diameters become significantly smaller than the wavelength of visible illumination.[44]

19.6.5 DEFENSE AND SECURITY

Military, firefighter, law enforcement, and medical personal need high-level protection in many environments ranging from combat to urban, agricultural, and industrial, when dealing with chemical and biological threats like[17]:

- Nerve agents
- Mustard gas
- Blood agents, such as cyanides
- Biological toxins such as bacterial spores, viruses, and rickettsiae

Nanostructures with their minuscule size, large surface area, and light weight will improve, by orders of magnitude, our capability to[17]

- Detect chemical and biological warfare agents with sensitivity and selectivity
- Protect through filtration and destructive decomposition of harmful toxins
- Provide site-specific naturally prophylaxis.

Polymer nanofibers are considered as excellent membrane materials owing to their lightweight, high surface area, and breathable (porous) nature. The high sensitivity of nanofibers toward warfare agents makes them excellent candidates as sensing of chemical and biological toxins in concentration levels of parts per billion. Governments across the globe are investing in strengthening the protection levels offered to soldiers in the battlefield. Various methods of varying nanofiber surfaces to improve their capture and decontamination capacity of warfare agents are under investigation. Nanofiber membranes may be employed to replace the activated charcoal in adsorbing toxins from the atmosphere. Active reagents can be planted in the nanofiber membrane by chemical functionalization, post-spinning variation, or through using nanoparticle polymer composites. There are many avenues for future research in nanofibers from the defense perspective. As well as serving protection and decontamination roles, nanofiber membranes will also suffer to provide the durability, wash ability, resistance to intrusion of all liquids, and tear strength needed of battledress fabrics (figs 19-20, 21).[17]

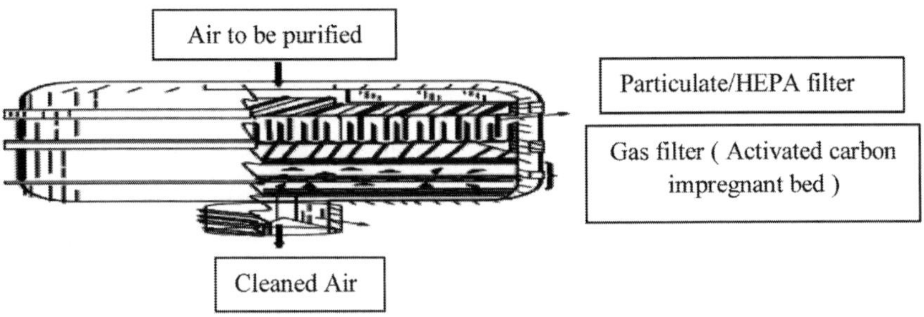

FIGURE 19.20 Cross-sectioning of a facemask canister used for protection from chemical and biological warfare agents.

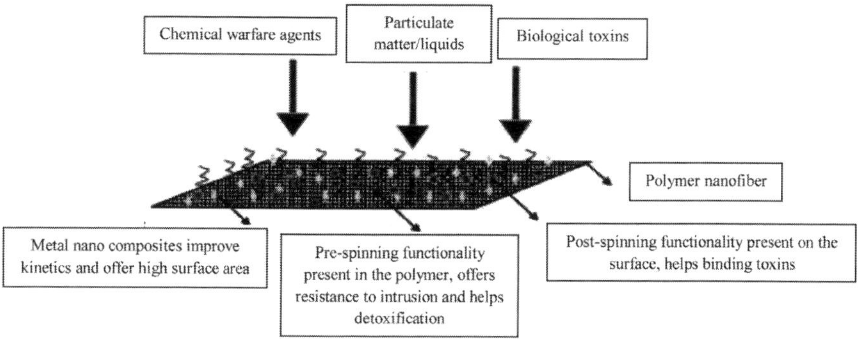

FIGURE 19.21 Incorporating functional groups into a polymer nanofiber mesh.

19.6.5.1 PROTECTIVE CLOTHING

The protective clothing(fig 19-22) in the military is largely expected to help increase the suitability, sustainability, and combat effectiveness of the individual soldier system against extreme climate, ballistics, and NBC warfare. In peace ages, breathing apparatus and protective clothing with the particular role of against chemical warfare agents, such as sarin, soman, tabun, become a special concern for combatants in conflicts and civilian populations in terrorist attacks. Current protective clothing containing charcoal absorbents has its terminal points for water permeability, extra weight-imposed to the article of clothing. A lightweight and breathable fabric, which is permeable to both air and water vapor, insoluble in all solvents and reactive with nerve gases and other deadly chemical agents, is worthy. Because of their large surface area, nanofiber fabrics are neutralizing chemical agents and without impedance of the air and water vapor permeability to the clothing. Electrospinning results in nanofibers lay down in a layer that has high porosity but small pore size, offering good resistance to penetrating chemical harms agents in aerosol form. Preliminary investigations indicate that compared to conventional textiles the electrospun nanofibers present both small impedance to moisture vapor diffusion and efficiency in trapping aerosol particles, as well as show strong promises as ideal protective clothing. Conductive nanofibers are expected to be utilized in fabricating tiny electronic or machines, such as Schottky junctions, sensors, and actuators. Conduct (of electrical, ionic, and photoelectric) membranes also have potential for applications including electrostatic dissipation, corrosion protection, electromagnetic interference shielding, photovoltaic device, etc.[53]

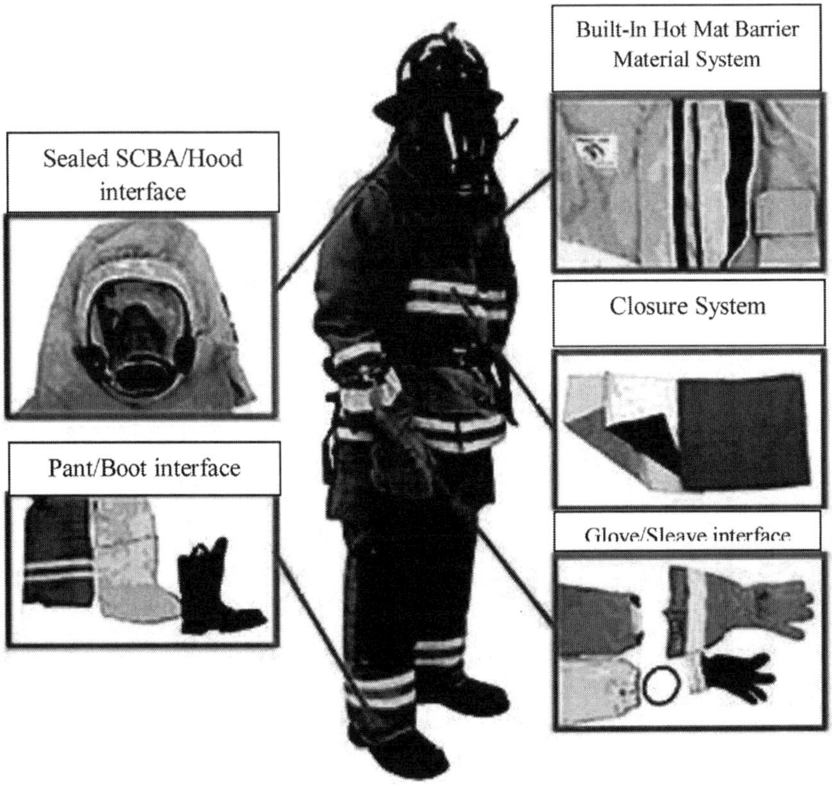

FIGURE 19.22 Protective clothing applications.

Electrospun nanofibers can play an important part in textile applications as protective clothing and other functional fabric materials. The electrospun nanofibrous membranes are capable of neutralizing chemical agents without impedance of the air and water vapor permeability to the clothing because of their high specific surface area and high porosity but small pore size. Preliminary investigations suggest the electrospun nanofibers control both minimal impedance to moisture vapor diffusion and efficiency in trapping aerosol particles compared with conventional textiles. Smith prepared a fabric comprising electrospun PEI nanofibers as lightweight protective clothing which was captured and neutralizing chemical warfare agents. This formed fabric also could be used in protective breathing apparatuses because PEI provides multiple amine sites for the nucleophilic decomposition of mustard gases and fluorophosphates nerve gases. A protective mask was constructed by attaching PC/PS electrospun nanofibrous layer to one side of a moist fabric composed of cellulose and wool. The diameter of the nanofibers in the protective layer was in the range of 100–10,000 nm. A nonwoven fabric composed of a submicrosized fiber, which receives a PC shell and a polyurethane core was made by co-axial electrospinning. The resultant fabric combines the filtration efficiency of the PC and the mechanical effectiveness of polyurethane, which is useful in exposure suits and aviation clothing. A water-resistant and air-permeable laminated fabric was manufactured by utilizing hot-melt polyester as dots onto the surface of the electrospun nylon nonwoven fabric.[5] Ideally, protective clothing should have close to essential properties such as, lightweight, breathable fabric, air and water vapor permeability, insoluble in all solvents and improved toxic chemical resistance. Electrospun nanofiber membranes recognized as potential candidates for protective clothing applications for these causes:

- They're lightweight
- Large surface area
- High porosity (breathable nature)
- Great filtration efficiency
- Resistant to penetration of harmful chemical agents in aerosol form
- Their ability to neutralize the chemical agents without impedance of the air
- Water vapor permeability to the clothing.

Various methods for variation of nanofiber surfaces have been examined to improve protection against toxins. One protection method that has been used includes chemical surface variation and attachment of reactive groups such as axioms, cyclodextrins, and chloramines that bind and detoxify warfare agents.[42]

19.7 CONCLUDING REMARKS

Nanotechnology with unique properties is interested by many scientists in everywhere for novel applications in recent years. Various common techniques can be used for preparing polymer nanofibers. Also, special ways such as gelation and bacterial cellulose were utilized for producing nanofibers. Among these methods, electrospinning has been widely used as a novel technique for generating nanoscale fibers. Therefore, electrospun nanofibers are utilized in a wide range of applications. Also, this work was analyzed the recent advances of this technology in tissue engineering, drug delivery, etc.

KEYWORDS

- **electrospinning**
- **nanofibers**
- **nanotechnology**
- **nanofibers production**
- **nanofibers applications**

REFERENCES

1. Lu, P.; Ding, B. Applications of Electrospun Fibers. *Rec. Pat. Nanotechnol.* **2008,** *2*(3), 169–182.
2. Reneker, D. H.; et al. Electrospinning of Nanofibers from Polymer Solutions and Melts. *Adv. Appl. Mech.* **2007,** *41*, 43–346.
3. Vonch, J.; Yarin, A.; Megaridis, C. M. Electrospinning: A Study in the Formation of Nanofibers. *J. Undergrad. Res.* **2007,** *1*, 1–6.
4. Fang, J.; et al. Applications of Electrospun Nanofibers. *Chin. Sci. Bull.* **2008,** *53*(15), 2265–2286.
5. RAFIEI, S.; et al. Mathematical Modeling in Electrospinning Process of Nanofibers: A Detailed Review. *Cell. Chem. Technol.* **2013,** *47*(5–6), 323–338.
6. Fang, J.; Wang, X.; Lin, T. Functional Applications of Electrospun Nanofibers. *Nanofibers Prod., Properties Funct. Appl.* **2011,** 287–326.
7. Karra, S. *Modeling Electrospinning Process and a Numerical Scheme Using Lattice Boltzmann Method to Simulate Viscoelastic Fluid Flows*. Texas A&M University, 2007.
8. Angammana, C. J. *A Study of the Effects of Solution and Process Parameters on the Electrospinning Process and Nanofibre Morphology*. University of Waterloo, 2011.
9. Baji, A.; et al. Electrospinning of Polymer Nanofibers: Effects on Oriented Morphology, Structures and Tensile Properties. *Compos. Sci. Technol.* **2010,** *70*(5), 703–718.
10. Beachley, V.; Wen, X. Polymer Nanofibrous Structures: Fabrication, Biofunctionalization, and Cell Interactions. *Prog. Polym. Sci.* **2010,** *35*(7), 868–892.
11. Agarwal, S.; Wendorff, J. H.; Greiner, A. Use of Electrospinning Technique for Biomedical Applications. *Polymer* **2008,** *49*(26), 5603–5621.
12. Fridrikh, S. V.; et al. Controlling The Fiber Diameter during Electrospinning. *Phys. Rev. Lett.* **2003,** *90*(14), 144502–144502.
13. Garg, K.; Bowlin, G. L. Electrospinning Jets and Nanofibrous Structures. *Biomicrofluidics* **2011,** *5*(1), 013403-1–013403-19.
14. Haghi, A. K. Electrospun Nanofiber Process Control. *Cell. Chem. Technol.* **2010,** *44*(9), 343–352.
15. Huang, Z. M.; et al. A Review on Polymer Nanofibers by Electrospinning and their Applications in Nanocomposites. *Compos. Sci. Technol.* **2003,** *63*(15), 2223–2253.
16. Kowalewski, T. A.; NSKI, S.; Barral, S. Experiments and Modelling of Electrospinning Process. *Technical Sci.* **2005,** *53*(4), 385–394.

17. Ramakrishna, S.; et al. Electrospun Nanofibers: Solving Global Issues. *Mater. Today* **2006,** *9*(3), 40–50.
18. Reneker, D. H.; Chun, I. Nanometre Diameter Fibres of Polymer, Produced by Electrospinning. *Nanotechnology* **1996,** *7*(3), 216–223.
19. Wang, H. S.; Fu, G. D.; Li, X. S. *Functional Polymeric Nanofibers from Electrospinning. Rec. Pat Nanotechnol.* **2009,** *3*(1), 21–31.
20. Zhang, C.; Ding, X.; Wu, S. *The Microstructure Characterization and the Mechanical Properties of Electrospun Polyacrylonitrile-based Nanofibers;* 2011; pp 177–196.
21. Zhang, S. *Mechanical and Physical Properties of Electrospun Nanofibers.* 2009.
22. Zhou, H. *Electrospun Fibers from Both Solution and Melt: Processing, Structure and Property.* Cornell University, 2007.
23. Zanin, M. H. A.; Cerize, N. N. P.; de, A. M. O. Production of Nanofibers by Electrospinning Technology: Overview and Application in Cosmetics. In *Nanocosmetics and Nanomedicines*; Springer, 2011; pp 311–332.
24. Khan, N. Applications of Electrospun Nanofibers in the Biomedical Field. *Stud. Undergrad. Res. Guelph* **2012,** *5*(2), 63–73.
25. Ramakrishna, S.; Fujihara, K. *An Introduction to Electrospinning and Nanofibers*; 2005; pp 1–383.
26. Zeng, Y.; et al. Numerical Simulation of Whipping Process in Electrospinning. In WSEAS International Conference. Proceedings. Mathematics and Computers in Science and Engineering, World Scientific and Engineering Academy and Society, 2009.
27. Ciechańska, D. Multifunctional Bacterial Cellulose/Chitosan Composite Materials for Medical Applications. *Fibres Text. East. Europe* **2004,** *12*(4), 69–72.
28. Stanger, J. J.; et al. Effect of Charge Density on the Taylor Cone in Electrospinning. *Int. J. Mod. Phys.* B, **2009,** *23*(06).
29. Chronakis, I. S. Novel Nanocomposites and Nanoceramics based on Polymer Nanofibers using Electrospinning Process—A Review. *J. Mater. Process. Technol.* **2005,** *167*(2), 283–293.
30. Wu, Y.; et al. Controlling Stability of the Electrospun Fiber by Magnetic Field. *Chaos, Solitons Fractals* **2007,** *32*(1), 5–7.
31. Zhang, S. *Mechanical and Physical Properties of Electrospun Nanofibers;* 2009; pp 1–83.
32. Li, W. J.; et al. Electrospun Nanofibrous Structure: A Novel Scaffold for Tissue Engineering. *J. Biomed. Mater. Res.* **2002,** *60*(4), 613–621.
33. Brown, E. E.; Laborie, M. P. G. Bioengineering Bacterial Cellulose/Poly(Ethylene Oxide) Nanocomposites. *Biomacromolecules* **2007,** *8*(10), 3074–3081.
34. De. V, S.; et al. The Effect of Temperature and Humidity on Electrospinning. *J. Mater. Sci.* **2009,** *44*(5), 1357–1362.
35. Tao, J.; Shivkumar, S. Molecular Weight Dependent Structural Regimes During The Electrospinning of PVA. *Mater. Lett.* **2007,** *61*(11), 2325–2328.
36. Kowalewski, T. A.; Nski, S. B. Ł. O.; Barral, S. Experiments and Modelling of Electrospinning Process. *Technical Sci.* **2005,** *53*(4).
37. Zong, X.; et al. Structure and Process Relationship of Electrospun Bioabsorbable Nanofiber Membranes. *Polymer* **2002,** *43*(16), 4403–4412.
38. Lyons, J.; Li, C.; Ko, F. Melt-electrospinning. Part I: Processing Parameters and Geometric Properties. *Polymer* **2004,** *45*(22), 7597–7603.
39. Reneker, D. H.; Yarin, A. L. Electrospinning Jets and Polymer Nanofibers. *Polymer* **2008,** *49*(10), 2387–2425.

40. Bognitzki, M.; et al. Nanostructured Fibers via Electrospinning. *Adv. Mater.* **2001,** *13*(1), 70–72.
41. Deitzel, J. M.; et al. The Effect of Processing Variables on the Morphology of Electrospun Nanofibers and Textiles. *Polymer* **2001,** *42*(1), 261–272.
42. Bhardwaj, N.; Kundu, S. C. Electrospinning: A Fascinating Fiber Fabrication Technique. *Biotechnol. Adv.* **2010,** *28*(3), 325–347.
43. Sill, T. J.; von Recum, H. A. Electrospinning: Applications in Drug Delivery and Tissue Engineering. *Biomaterials* **2008,** *29*(13), 1989–2006.
44. Patan, A. K.; et al. Nanofibers-A New Trend in Nano Drug Delivery Systems. *Int. J. Pharma. Res. Anal.* **2013,** *3*, 47–55.
45. Rutledge, G. C.; Fridrikh, S. V. Formation of Fibers by Electrospinning. *Adv. Drug Delivery Rev.* **2007,** *59*(14), 1384–1391.
46. Sawicka, K. M.; Gouma, P. Electrospun Composite Nanofibers for Functional Applications. *J. Nanoparticle Res.* **2006,** *8*(6), 769–781.
47. Yousefzadeh, M.; et al. A Note on The 3D Structural Design of Electrospun Nanofibers. *J. Eng. Fabrics Fibers (JEFF)* **2012,** *7*(2), 17–23.
48. Yarin, A. L.; Koombhongse, S.; Reneker, D. H. Bending Instability in Electrospinning of Nanofibers. *J. Appl. Phys.* **2001,** *89*(5), 3018–3026.
49. Keun. S, W.; et al. Effect of PH on Electrospinning of Poly (Vinyl Alcohol). *Mater. Lett.* **2005,** *59*(12), 1571–1575.
50. Feng, J. J. *The Stretching of An Electrified Non-Newtonian Jet: A Model for Electrospinning.* Physics of Fluids (1994-present), **2002,** *14*(11), 3912–3926.
51. Maleki, M.; Latifi, M.; Amani, M. T. Optimizing Electrospinning Parameters for Finest Diameter of Nano Fibers. *World Acad. Sci., Eng. Technol.* **2010,** *40*, 389–392.
52. Thompson, C. J. *An Analysis of Variable Effects on A Theoretical Model of the Electrospin Process for Making Nanofibers*. University of Akron, 2006.
53. Luo, C. J.; Nangrejo, M.; Edirisinghe, M. A Novel Method of Selecting Solvents for Polymer Electrospinning. *Polymer* **2010,** *51*(7), 1654–1662.
54. Huang, Z.-M.; et al. A review on Polymer Nanofibers by Electrospinning and their applications in Nanocomposites. *Compos. Sci. Technol.* **2003,** *63*, 2223–2253.
55. Frenot, A.; Chronakis, I. S. Polymer Nanofibers Assembled by Electrospinning. *Curr. Opin. Colloid Interface Sci.* **2003,** *8*(1), 64–75.
56. Fang, J.; Wang, X.; Lin, T. Functional Applications of Electrospun Nanofibers. Nanofibers-Production, Properties and Functional Applications; 2009; pp. 287–326.

CHAPTER 20

SHORT COMMUNICATIONS: PROGRESS ON APPLICATION OF NANOTECHNOLOGY

M. R. MOSKALENKO

Ural Federal University, Institute of Humanities and Arts, Department of History of Science and Technology, Ekaterinburg, Russia

CONTENTS

20.1 SOME QUESTIONS OF FORECASTING THE DEVELOPMENT OF NANOTECHNOLOGY

This short communication dedicates to the scientific and technological forecasting in nanotechnology trend. Investigations in scientific and technological forecasts made applicable to the development of nanotechnology.[1-4]

Nowadays, the intermittent development of the sphere of scientific researches, designing and engineering on an atomic scale, which is called nanotechnology, can be observed. Of course, tempestuous development of nanotechnology gives life to the most incredible and bold forecasts of its usage in everyday life and industry, which are stirred up by sensation-seeking journalists. That is why it is necessary to know the general specific of scientific-technical forecasts in order to estimate all the perspectives of the development of nanotechnology.

Usually these standard methods are used in scientific-technical forecasts: analysis of trends and tendencies of development, extrapolation of existing tendencies, elicitation of the processes' cyclicality, studying of principles and regularities of the process or phenomenon, and making of scenarios. Sometimes science fiction is related to a specific type of forecasting where the method of intuitional foreseeing and artistic imagination of artists are used. Several issues should be considered when we speak about scientific-technical forecasting in general and especially when we speak about forecasting of nanotechnology:

1. How do social conditions of any specific society contribute to the development of high-tech economy and integration of know-how? For example, well-developed system of education and the cult of scientific knowledge in the late USSR contributed to the creation of many original development projects and inventions. However, economic system wasn't able to realize them; no required mechanisms and institutions were made to provide the integration. Also, the phenomenon of "resource curse" can also takes place—when the excessive amount of natural resources and corresponding export profits turn all other branches of economy (including high-tech ones) to be secondary and subordinated to the import.
2. It can be quite difficult to value new breakthrough technologies and inventions. For example, airplane appeared later then dirigibles did, despite the fact that before the First World War, dirigibles had been predicted to be the main striking air force, the attitude toward first

machine-guns was skeptical among soldiers, first cars and vacuum cleaners often caused nothing but skeptical smile, etc.

3. There could be some engineering designs that seem to be quite realistic and perspective, but then appear being hardly possible to built in practice: nuclear aircrafts in 1950s; enormous cargo and passenger seaplanes (some development projects could weight up to 1000 t); hovercraft tanks; etc.
4. Tempestuous development of any technology can crucially change industry and everyday life within a short period of time (automotive industry in 1900s, personal computers in 1980–1990s, cellular network in 1990-early 2000s, etc.).
5. The development of technology and technic is often nonlinear: tempestuous, intermittent development can slow down and become more fluent (classic example—the history of tanks and aviation).
6. Tempestuous development of any sphere of science and technology always goes with excessively optimistic forecasts, like how this thing will change the face of humanity; then comes a period of more deliberated valuation and normally forecasts become more restrained.

These peculiarities of scientific-technical forecasting should be considered while forecasting and planning the development of nanotechnology in various branches of industry.

KEYWORDS

- **scientific and technological forecasting**
- **nanotechnology**

REFERENCES

1. Canton, J. *The Strategic Impact of Nanotechnology on the Future of Business and Economics*. http://www.globalfuturist.com/dr-james-canton/insights-and-future-forecasts/stratigic-impact-of-nanotechnology-on-business-and-economics.html.
2. *Nanotechnology: A Technology Forecast*. http://www.tfi.com/pressroom/pr/nano.html.
3. Toffler, A. *Future Shock*; Random House: New York, 1970; 505 p.
4. Zigunenko, S. N. *100 Great Records of Military Technique*. M: Veche, 2012, 432 p.

INDEX

N

Y

Z